DNA Barcoding in Marine Perspectives

Subrata Trivedi · Abid Ali Ansari
Sankar K. Ghosh · Hasibur Rehman
Editors

DNA Barcoding in Marine Perspectives

Assessment and Conservation of Biodiversity

Springer

Editors
Subrata Trivedi
Department of Biology, Faculty of Science
University of Tabuk
Tabuk
Saudi Arabia

Sankar K. Ghosh
Department of Biotechnology
Assam University
Assam
India

Abid Ali Ansari
Department of Biology, Faculty of Science
University of Tabuk
Tabuk
Saudi Arabia

Hasibur Rehman
Department of Biology, Faculty of Science
University of Tabuk
Tabuk
Saudi Arabia

ISBN 978-3-319-82442-0 ISBN 978-3-319-41840-7 (eBook)
DOI 10.1007/978-3-319-41840-7

Printed on acid-free paper

This Springer imprint is published by Springer Nature
The registered company is Springer International Publishing AG Switzerland

"We would like to dedicate this book to our parents and mentors"

Dr. Subrata Trivedi and Dr. Hasibur Rehman

Preface

When I hear of the destruction of a species, I feel just as if all the works of some great writer have perished.

—Theodore Roosevelt

We know that when we protect our oceans we're protecting our future.

—Bill Clinton

The oceans cover more than 70 % of our planet's surface area, and the massive marine and coastal environments are blessed with diverse marine life. To meet the demands of increasing population we are becoming more reliant on the marine bioresources. For example only the marine fish and invertebrates provide more than 2.6 billion people with about 20 % per capita protein consumption. To make the marine environment sustainable, the proper assessment and conservation of marine biodiversity is of prime importance. In the last decade, the molecular technique of DNA barcoding has become an effective tool in the assessment and conservation of biodiversity. The marine ecosystem is threatened by several activities such as overfishing, introduction of invasive alien species, depleting mangrove and sea grass cover, illegal trading of endangered marine species and their body parts, etc. DNA barcoding plays a very significant role in all these aspects along its primary role in the proper and prompt identification of species. In this book we discuss DNA barcoding from the marine perspective.

The present book offers insights into different aspects of DNA barcoding in relation to the marine habitat. The chapters cover diverse marine life including marine plants such as phytoplanktons, marine algae, seagrasses, and also marine animals as marine invertebrates including the primitive nemartines, horse shoe crabs, fishes, etc. Since marine fishery has a very significant role, a special emphasis has been given to DNA barcoding of marine fishes including Antarctic fishes. The chapters also include aspects such as bioinformatics, seafood safety assessment and authentication. Many of the chapters are based on the research projects and case history studies conducted at specific sites and also around the globe. The chapters not only describe the promise of DNA barcoding but also some of its pitfalls. The contribution made by authors from nine different countries has enriched this book.

The editors and the contributing authors think that this book will provide important and interesting insights to DNA barcoding in the diverse and massive marine ecosystem. Till date, only a few books are available on DNA barcoding and we hope this book will fill the lacuna. This is the first book related to DNA barcoding exclusively on marine organisms.

Tabuk, KSA	Subrata Trivedi
Tabuk, KSA	Abid Ali Ansari
Silchar, India	Sankar K. Ghosh
Tabuk, KSA	Hasibur Rehman

Contents

Part I
Overview, Significance and Bioinformatics

DNA Barcoding in the Marine Habitat: An Overview

Subrata Trivedi, Hasibur Rehman, Shalini Saggu, Chellasamy Panneerselvam, Zahid Khorshid Abbas, Iqbal Ahmad, Abid A. Ansari and Sankar K. Ghosh

Abstract Major part of our planet includes the marine habitat which faces severe threat due to overexploitation of its bio resources. Assessment of biodiversity in the massive and diverse marine ecosystem is a challenging task. In this introductory chapter, we give a brief description of the marine habitat and types of marine organisms, followed by the concept of DNA barcoding. We also describe the applications and different initiatives of DNA barcoding in the marine ecosystem. A brief account of DNA barcoding in marine fungi, different groups of animals and plants is also elucidated. This chapter gives a bird's eye view on the DNA barcoding in the marine perspective.

Keywords DNA barcoding · Marine · COI · Biodiversity · Conservation

1 An Introduction to the Marine Environment

The marine water covers about 72 % of our earth surface and the oceans contain about 97 % of the water that is present in the earth. The five world oceans in order of size are as follows: Pacific Ocean, Atlantic Ocean, Indian Ocean, Arctic Ocean and Southern Ocean. The largest ocean is the Pacific Ocean which covers about one-third of the total area of our planet. The deepest trench-Mariana Trench

S. Trivedi (✉) · H. Rehman · S. Saggu · C. Panneerselvam ·
Z.K. Abbas · A.A. Ansari
Faculty of Sciences, Department of Biology,
University of Tabuk, Tabuk 71491, Saudi Arabia
e-mail: strivedi@ut.edu.sa; trivedi.subrata@gmail.com

I. Ahmad
Center for Environmental and Marine Studies,
University of Aveiro, 3810-193 Aveiro, Portugal

S.K. Ghosh
Department of Biotechnology, Assam University,
Silchar 788011, Assam, India

S. Trivedi et al. (eds.), *DNA Barcoding in Marine Perspectives*,
DOI 10.1007/978-3-319-41840-7_1

(10,800 m) is also located in the Pacific Ocean. The Atlantic Ocean covers about one-sixth area of the earth is the second largest ocean but has the longest coastline. In the year 2000, the International Hydrological Organization delineated a new ocean called Southern Ocean surrounding the Antarctica and extending up to 60° S. The ocean currents which are regular movement of water from one part of ocean to another have profound influence on the distribution and abundance of marine organisms and particularly the phytoplanktons and zooplanktons. For example, the confluence of cold Oyashio Current with the warm Curoshio Current off the Japan coast results in huge number of planktons which in turn makes it an ideal fishing ground. The alternate rise and fall of ocean surface water or the tides also play a vital role in the coastal regions like the mangrove ecosystems. Besides the oceans, the marine ecosystem also contains seas, bays, lagoons, salt marshes, estuaries, coral reefs, kelp forests, etc.

The dynamic and massive marine environment is facing severe anthropogenic threat which is severely affecting the marine biotic community which is in turn affecting the health of our planet as a whole. Some of the threats include over-fishing, pollution, global warming, rising sea level, El Niño, depleting mangroves and sea grass cover, coral bleaching and also introduction of invasive alien species.

Figure 1 shows the photograph of beautiful and biodiversity rich Red Sea.

Fig. 1 Red Sea at Haql, Saudi Arabia. (*Photograph* by Dr. SubrataTrivedi)

2 Zonations of the Oceans

The open oceans or the pelagic ecosystem is divided into several zones as follows:

2.1 *Epipelagic Zone*

This zone extends from the sea surface to a depth of about 200 m. This is the photic zone where photosynthesis can take place.

2.2 *Mesopelagic Zone*

This zone is found below the epipelagic zone (200 m) and extends up to a depth of about 1000 m.

2.3 *Bathypelagic Zone*

This zone is found below the mesopelagic zone (1000 m) and extends up to a depth of about 4000 m.

2.4 *Abyssalpelagic Zone*

This zone is found below the bathypelagic zone (4000 m) and extends up to a depth of about 6000 m.

2.5 *Hadalpelagic Zone*

This zone is found below the depth of 6000 m and includes deep sea trenches.

2.6 *Littoral Zone*

Littoral zone is the part of the sea that extends from the high tide water mark through the intertidal region and up to as far as the edge of continental shelf. The littoral zone can be further subdivided into three sections.

2.6.1 Supralittoral Zone

This region is just above the spring high tide line which is also called supratidal zone.

2.6.2 Eulittoral Zone

This is the intertidal region which is also called foreshore.

2.6.3 Sublittoral Zone

This zone is located below the eulittoral zone which is permanently covered with sea water.

2.7 Benthic Zones

The bottom regions of the seas are called benthic zones.

Figure 2 depicts the different zonations of the oceans.

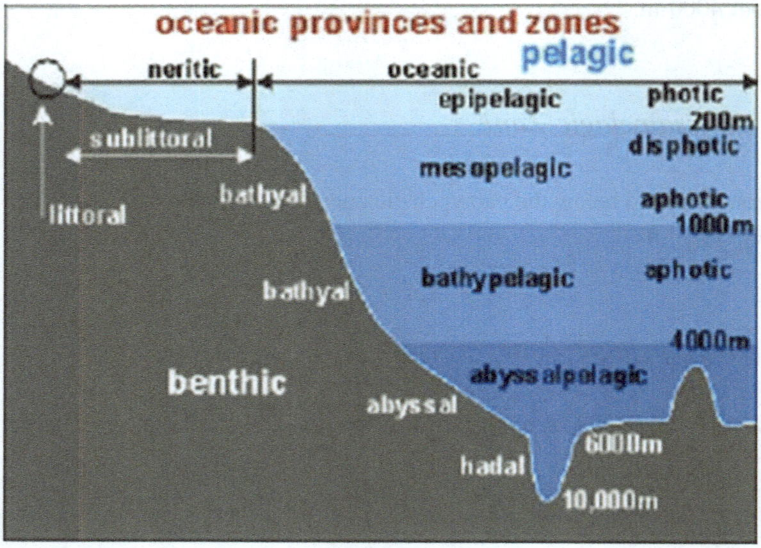

Fig. 2 Different zonations of the oceanic region

3 Types of Marine Organisms

Marine organisms can be broadly divided into the following three categories depending on their habitat and locomotion.

3.1 *Necton*

They include animals that can move actively in the water like fish, turtles, etc. which can propel through water either by swimming or by other means.

3.2 *Plankton*

They include phytoplanktons and zooplanktons that cannot propel through water. It is to be noted that some of the larval stages of nectons can act as plankton since they act as drifting community.

3.3 *Benthos*

These are organisms that dwell at the bottom of the sea floor. They may be free living or attached to the sea floor. They cannot propel through water like the nekton. Examples are echinoderms and molluscs.

4 The Concept of DNA Barcoding

Proper assessment and conservation of marine bio resources is urgently needed in order to make this fragile ecosystem sustainable. The first step towards this effort is the proper and prompt identification of marine species. Identification of species in this huge and dynamic marine ecosystem is a challenging task. In the past, most of the taxonomic identification of species has been based on classical morphological methods. Taxonomic controversies relating to the identification of marine species has been a long standing problem. With the advancement of modern molecular techniques a comparatively new concept of DNA barcoding has been put forward. In this method a short standardized region of a gene or a 'barcode gene' is used for identification of species just like the barcode which is used in the supermarkets. *Cytochrome c oxidase* subunit 1 (COI) gene in the mitochondrial genome was suggested as barcode region for animals (Hebert et al. 2003). The faster mutation

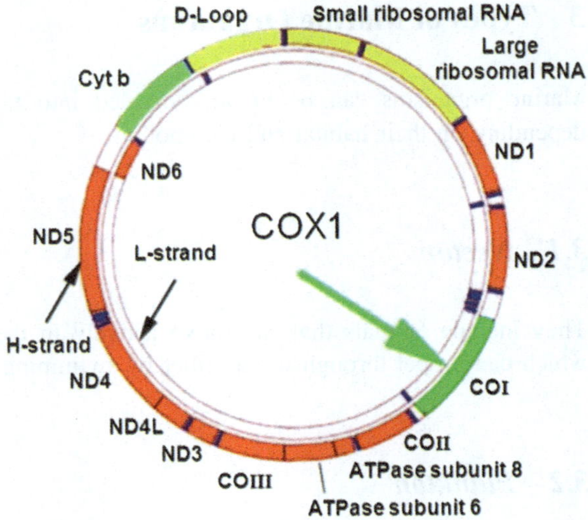

Fig. 3 The mitochondrial genome showing the position of COI gene

rate and sequence conservation in conspecifics in the COI gene makes it ideal for species delimitation. The lack of introns, higher copy number and maternal inheritance makes mitochondrial genome good candidate for barcoding purposes. Figure 3 depicts the circular mitochondrial genome with the location of the COI gene. Besides the *COI* gene, several other molecular markers like *ITS, rbcL, matK, trnH-psbA,* etc. are also used for barcoding purposes.

5 The Procedure of DNA Barcoding

The overall process of DNA barcoding includes the following steps.

5.1 Sampling and Vouchering

This is a very important step of DNA barcoding. Biological samples are collected from the habitat. Museum samples may also be used for barcoding purposes. One very important aspect is the proper vouchering of the samples. The collected samples should be accompanied with thorough documentation. The required information includes name of collector, date of collection, location (latitude and longitude), elevation/ depth, notes on habitat, voucher number, name of identifier, photograph, etc. There are several voucher requirements and data needs for different groups and types of samples.

5.2 *Tissue Extraction*

Usually a small amount of suitable tissue is used for the next step of barcoding.

5.3 *Extraction of DNA*

DNA is extracted by suitable method (e.g. Phenol- Chloroform-Isoamyl alcohol method). The extracted DNA is then quantified by spectrophotometric method and visualized by agarose gel electrophoresis.

5.4 *PCR Amplification of Barcode Region*

Self-designed primer sets or published primer sets are used for PCR amplification of the barcode region. The PCR amplified products are then checked by agarose gel electrophoresis.

5.5 *Purification of PCR Amplicons and DNA Sequencing*

The PCR amplified products of expected size are extracted from the gel and purified.

This DNA is then subjected to DNA sequencing.

5.6 *Submission of Sequencing to Global GenBank or BOLD*

The DNA sequences are submitted to global database like NCBI which provide accession numbers after successful incorporation.

5.7 *Bioinformatics Analysis of the Sequences and Interpretation*

The sequence analysis is done by several bioinformatics tools for species identification, molecular phylogeny study, discovery of new species, etc. Bioinformatics tools include BLAST, CLUSTAL, etc.

6 The Global DNA Barcoding Initiative

To coordinate DNA barcoding globally, The Consortium for the Barcode of Life (CBOL) was established. The Barcode of Life Data Systems (http://www.barcodinglife.org) was launched for the global management of barcode data. For the efficient functioning of International Barcode of Life Project (iBOL), three global nodes were created—Central Nodes, Regional Nodes and National Nodes. The Central Nodes include Canada, China, European Union and United States. The Regional Node includes Australia, Brazil, India, Korea, Mexico, New Zealand, Norway, Russia and South Africa. The National Nodes include countries like Argentina, Colombia, Costa Rica, Kenya, Madagascar, Panama and Papua New Guinea. The iBOL has Working Groups and Core Facilities where institutions as well as individuals can participate once they get affiliated to iBOL Nodes. Some important barcoding initiatives include The Quarintine Barcode of Life (QBOL), The Moorea Biocode Project, Mosquito Barcoding Initiative (MBI), The Bee Barcode of Life Initiative (Bee-BOL), Fish Barcode of Life (FISH-BOL), All Birds Barcoding Initiative (ABBI) etc.

7 DNA Barcoding in the Marine Ecosystem

The Marine Barcode of Life or MarBOL is the international organization devoted to the barcoding of diverse marine organisms. MarBOL is a project affiliated to Census of Marine Life (CoML). CoML undertakes different Ocean Realm Field Projects like Arctic Ocean Diversity (ArcOD), Biogeography of Chemosynthetic Ecosystems (ChEss), Census of Antarctic Marine Life (CAML), Census of Diversity of Abyssal Marine Life (CeDAMar), Census of Marine Zooplankton (CMarZ), Continental Margin Ecosystems on a Worldwide Scale (CoMargE), Global Census of Coral Reef Ecosystems (CREEFS), Global Census of Marine Life on Seamounts (CenSeam), Gulf of Maine Area Program (GOMA), International Census of Marine Microbes (ICOMM), Natural Geography in Shore Areas

Table 1 Marine barcoding initiatives and their websites

No.	Name	Abbreviation	Website
1	Marine Barcode of Life	MarBOL	http://www.marinebarcoding.org
2	Census for Marine Zooplanktons	CMarZ	http://www.cmarz.org
3	Census of the Diversity of Abyssal Marine Life	CeDAMar	http://www.cedamar.org
4	Fish Barcode of Life	FISH-BOL	http://www.fishbol.org
5	Shark Barcode of Life	SharkBOL	http://www.sharkbol.org
6	Sponge Barcoding Project	SBP	http://www.spongebarcoding.org

(NaGISA), Pacific Ocean Shelf Tracking (POST), Tagging of Pacific Pelagics (TOPP) etc. The websites of some initiatives involved in marine DNA barcoding is shown in Table 1.

8 Applications of DNA Barcoding

DNA barcoding has several important applications in identification, assessment and conservation of marine bio resources. Some positive attributes of DNA barcoding are as follows.

8.1 Discovery of New Species

One of the most important applications of DNA barcoding is the discovery of new species. DNA barcoding of reptiles in the Madagascar Island revealed the presence of several new species (Nagy et al. 2012). There are numerous reports of new species revealed through DNA barcoding (Järnegren et al. 2007; Cárdenas et al. 2009; De Wit et al. 2009).

8.2 Safety Assessment and Authentication of Sea Food

DNA barcoding technique can be effectively used for safety assessment and authentication of seafood. Identification of seafood based on morphological methods is very difficult or not possible for processed, cooked, smoked products or for products which contain body parts rather than the whole sample. DNA barcoding can be used for authentication and assessment of such seafood. In fact, many seafood safety assessment agencies routinely use DNA barcoding for this purpose. A barcoding analysis comprising 254 seafood samples from Canada showed that 41 % of the samples were mislabeled (Hanner et al. 2011). Analysis of tuna sushi form several restaurants in USA also highlighted similar problems (Lowenstein et al. 2009).

8.3 Identification and Assessment of Cryptic Species

The cryptic species are relatively abundant in the marine ecosystem. DNA barcoding can be very effective in identification and assessment of cryptic species (Pauls et al. 2010). A study showed the relevance of DNA barcoding in assessment of cryptic species among sea birds (Efe et al. 2009).

8.4 Rapid Detection of Invasive Alien Species

Invasive Alien Species (IAS) poses severe threat to the biodiversity of marine ecosystems. They not only overpower the native species but also cause severe economic consequences. The Great Barrier Reef is facing severe problems due to invasive species which are introduced through fouling, release of ballast waters or by other activities. One such example is Caribbean tubeworm—*Hydroides sanctaecrucis*. DNA barcoding can be used as an effective tool for prompt identification and assessment of invasive alien species (Molnar et al. 2008).

8.5 Identification of Endangered and Threatened Marine Species

Barcoding serves as a handy tool for identification of endangered and threatened marine species and can therefore help in preventing illegal trade of these endangered species, their body parts or larval stages. Shark fins confiscated from illegal fishers in Australia were identified by DNA barcoding (Holmes et al. 2009). Similarly, illegally traded endangered sea turtles were assessed through DNA barcoding of turtle meat, carcasses and eggs (Vargas et al. 2009).

8.6 Linking of Life Cycle Stages of Marine Organisms

Linking of different life cycle stages to the adults of marine organisms is often difficult because the enormous marine ecosystem is the habitat of numerous species and the taxonomists often struggle to link the larval stages to an adult species. DNA barcoding can establish accurate link between the different life cycle stages to the adult.

In study conducted in the Carrabian Sea, DNA barcoding in conjugation with morphology was not only used to link the fish larvae with the adult species but also resulted in the discovery of new species (Baldwin and Johnson 2014).

8.7 Securing Intellectual Property Rights for Marine Bioresources

DNA barcoding can be used in securing Intellectual Property Rights (IPRs) for bioresourses (Chinnappareddy et al. 2012).

8.8 Global Management Plan for Conservation Strategies

DNA barcoding results can be used for framing global management plans for conservation strategies. There is rapid decline of mangrove cover globally which is posing great threat to the stability and sustenance of our coastal regions. It is reported that DNA barcoding ensures the phylogenetic information needed for proper global management plan on mangroves (Daru et al. 2013).

8.9 Elucidating the Feeding Niche

Barcoding of species contained in the gut or molecular gut content analysis can reveal the species that are present in the diet which in turn helps in better understanding of feeding habits of marine organisms.

9 DNA Barcoding of Marine Fungi

Marine fungi play a very significant ecological role as decomposer, as pathogen and also form symbiotic relationship with other marine organisms. An estimate shows about 1500 marine fungal species excluding those species associated with lichens exist in the marine ecosystem.

DNA barcoding can help in rapid global assessment of marine fungal species (Velmuragan et al. 2013; Andreakis et al. 2015).

10 DNA Barcoding of Marine Animals

Marine animals are extremely diverse ranging from the smallest to the largest in our planet. 34 among 35 animal phyla have marine representatives and 14 phyla are represented by exclusively marine animals (Briggs 1994; Gray 1997). Here we will discuss DNA barcoding in reference to some important marine animal groups.

10.1 Marine Zooplanktons

Zooplanktons have immense ecological significance and have representatives from 15 animal phyla. The identification, spatial distribution and assessment of zooplanktons are very important aspects of marine biodiversity studies. A global initiative for barcoding marine zooplanktons is led by Census for Marine Zooplanktons

(CMarZ). The five main DNA barcoding centers of CMarZ are located at University of Connecticut (USA), Ocean Research Institute (Japan), National Institute of Oceanography (India), Bremerhaven (Germany) and Qingdao (China).

Copepods (belonging to arthropoda) are among the most diverse zooplanktons. Several barcoding studies have been conducted on marine copepods which the revelation of many new species and providing greater insights to cryptic species diversity and geographical distribution of varied copepods (Blanco-Bercial et al. 2014; Laakmann et al. 2012; Bucklin and Frost 2009; Blanco-Bercial et al. 2011; Kozol et al. 2012; Böttger-Schnack and Machida 2011). Studies also have been conducted on amphipod zooplanktons, also belonging to arthropoda (Costa et al. 2009). Four molecular markers namely COI, 12S, nuclear ITS, and 28S were used for identification of *Neocalanus* copepods (Machida and Tsuda 2010).

Holo zooplanktons from Arctic Ocean and metazoan zooplanktons from Sargasso Sea were barcoded (Bucklin et al. 2010a, b). DNA barcoding was effective in identification of different Chaetognath species which were difficult to identify morphologically (Jennings et al. 2010b).

10.2 Marine Invertebrates

DNA barcoding has been performed on different groups of marine invertebrates which have provided better insights on the biodiversity and distribution of marine invertebrates.

10.2.1 Porifera

Sponges are very important marine animals that have pharmaceutical applications. Different types of canal systems (asconoid, syconoid and leuconoid) are unique characteristics of sponges. One of the successful global barcoding projects is the Sponge Barcoding Project (SBP). A unique example of corroborating DNA barcoding data to morphological characters can be found in the database available at http://www.spongebarcoding.org. Since this website is linked to World Porifera Database, it provides valuable updates on different species.

In the Antarctic Ocean, sponges are very important and also provide shelter to other benthic community. Recently several Antarctic sponges were barcoded which provided insights to diverse sponge communities in this cold region (Vargas et al. 2015). Researchers have emphasized the study of holotypes in sponge taxonomy (Erpenbeck et al. 2015). Recent barcoding of lithistid demosponges have provided new insights to their evolution and the authors have proposed reallocation of groups (Schuster et al. 2015). A new lithistid species belonging to genus *Isabella* was described (Carvalho et al. 2015). The complete mitochondrial genome of *Aplysina cauliformis* was analyzed to reveal some regions that can be studied in conjugation

Table 2 Publications on DNA barcoding of sponges

Sl. No.	Title of publication	Reference
1	Deceptive desmas: molecular phylogenetics suggests a new classification and uncovers convergent evolution of lithistid demosponges	Schuster et al. (2015)
2	Nothing in (sponge) biology makes sense—except when based on holotypes	Erpenbeck et al. (2015)
3	Diversity in a cold hot-spot: DNA-barcoding reveals patterns of evolution among antarctic demosponges (Class Demospongiae, Phylum Porifera)	Vergas et al. (2015)
4	Lithistid sponges of the upper bathyal of Madeira, Selvagens and Canary Islands, with description of a new species of Isabella	Carvalho et al. (2015)
5	Molecular phylogeny of Abyssocladia (Cladorhizidae: Poecilosclerida) and Phelloderma (Phellodermidae: Poecilosclerida) suggests a diversification of chelae microscleres in cladorhizid sponges	Vergas et al. (2013)
6	Phylogeny of Tetillidae (Porifera, Demospongiae, Spirophorida) based on three molecular markers	Szitenberg et al. (2013)
7	The complete mitochondrial genome of the verongid sponge *Aplysina cauliformis*: implications for DNA barcoding in demosponges	Sperling et al. (2012)
8	Barcoding sponges: an overview based on comprehensive sampling	Vergas et al. (2012)
9	ALG11—A new variable DNA marker for sponge phylogeny: Comparison of phylogenetic performances with the 18S rDNA and the COI gene	Belinky et al. (2012)
10	Congruence between nuclear and mitochondrial genes in Demospongiae: A new hypothesis for relationships within the G4 clade (Porifera: Demospongiae)	Morrow et al. (2012)
11	Mitochondrial DNA of *Clathrina clathrus* (Calcarea, Calcinea): six linear chromosomes, fragmented rRNAs, tRNA editing, and a novel genetic code	Lavrov et al. (2012)
12	The phylogeny of halichondrid demosponges: past and present re-visited with DNA-barcoding data	Erpenbeck et al. (2012)
13	DNA barcoding and traditional taxonomy unified through Integrative taxonomy: a view that challenges the debate questioning both methodologies	Pires and Marinoni (2010)
14	COI barcoding reveals new clades and radiation patterns of Indo-Pacific sponges of the family Irciniidae (Demospongiae: Dictyoceratida)	Pöppe et al. (2010)
15	Morphological description and DNA barcodes of shallow-water *Tetractinellida* (Porifera: Demospongiae) from Bocas del Toro, Panama, with description of a new species	Cárdenas et al. (2009)
16	The sponge barcoding project: aiding in the identification and description of poriferan taxa	Wörheide et al. (2007)
17	Identification of genes suitable for DNA barcoding of morphologically indistinguishable Korean Halichondriidae sponges	Park et al. (2007)

to COI sequences in relation to DNA barcoding of this verongid sponge (Sperling et al. 2012). Table 2 depicts some important publications on DNA barcoding of sponges.

10.2.2 Cnidaria

They contain marine animals like jelly fish, sea anemone and the corals. Their radially symmetrical body contains gastrovascular cavity. DNA barcoding has highlighted the presence of cryptic species among coastal and deep sea cnidarians (Moura et al. 2007).

Constant monitoring of jellyfish abundance and distribution is very important for the detection of non-native species which have adverse effect on the local community. There is exponential increase of jelly fish due to global warming which is causing great economic damage globally. A study was conducted on marine jellyfishes of South Korea in which DNA barcoding in conjugation with microarray analysis was used to develop DNA chip for jellyfish identification (Lee et al. 2011).

The corals are rapidly declining globally and the coral reefs are among most endangered ecosystems in our planet. DNA barcoding has huge potential in assessing the biodiversity and can play a very vital role in conservation of these fragile coral reefs. Coral reefs of the tropics shelter about 25 to 33 % of all marine species which makes the coral reefs among the most diverse marine ecosystems. Since the traditional morphological methods for assessing the biodiversity of the reef areas is highly time consuming, DNA barcoding holds great promise and recent barcoding studies have shown that the reef biodiversity is largely underestimated (Plaisance et al. 2011). A barcoding study on the model coral species *Stylophora pistillata* has revealed the distinct clades and significant variation in this coral "lab-rat" (Keshavmurthy et al. 2013).

10.2.3 Annelida

They include segmented marine worms like polychaetes. Four new species of annelida belonging to the genus *Grania* were revealed through a study conducted in the Great Barrier Reef (De Wit et al. 2009). Several studies have proved the effectiveness of DNA barcoding in the identification of different marine polychaete species (Zhou et al. 2010; Maturana et al. 2011; Lobo et al. 2015).

10.2.4 Platyhelmenthes

Few barcoding studies have been conducted on flatworms including the marine flatworms. Non availability of universal primers for the diverse flatworms and the other problems related to barcoding of flatworms has been highlighted (Vanhove et al. 2013).

10.2.5 Nemathelmenthes

Nematodes play a very important role as indicator of marine pollution. 18S gene was used to barcode marine nematodes with a very high success rate (Bhadury et al. 2006). Derycke et al. (2008) used COI, ITS and D2D3 sequences to study molecular phylogeny of nematodes.

10.2.6 Mollusca

About 23 % of all marine organisms belong to Mollusca thereby making it the largest marine phylum. This makes marine malacology a vast and diverse field. The phylum Mollusca is divided into seven classes namely Gastropoda, Bivalvia, Aplacophora, Monoplacophora, Polyplacophora, Scaphopoda and Cephalopoda. Researchers have stressed the need for DNA barcoding for diversity rich molluscs for new species discovery as well as for conservation (Puillandre et al. 2009). DNA barcoding of marine molluscs from Canada revealed several overlooked species (Layton et al. 2014). Several barcoding studies have been conducted on marine gastropods revealing important insights to their identification, diversity and geographical distribution (Kemppainen et al. 2009; Maas et al. 2013; Krug et al. 2007; Jennings et al. 2010a).

Edible marine oysters are regarded as a sea food delicacy. Identification of oysters are generally based on morphological traits like shell morphology. Research has shown that shell morphology changes with the habitat thereby raising doubt about its role in the identification of species (Tack et al. 1992). This confusion can be solved by the molecular technique of DNA barcoding. Barcoding of marine bivalves have provided clarity in identification and better understanding on their phylogenetic relationships (Kappner and Bieler 2006; Järnegren et al. 2007; Mikkelsen et al. 2007; Trivedi et al. 2015).

10.2.7 Arthropoda

The sub phylum Crustacea includes very diverse marine species and the class Pycnogonida contains exclusively marine sea spiders. Another exclusively marine class Merostomata includes the living fossil—Horse shoe crabs. COI sequences were used to decipher the phylogeny of Horse shoe crabs (Kamaruzzaman et al. 2011). A recent large scale barcoding analysis of crustaceans residing in the North Sea has confirmed the efficacy of DNA barcoding in modern biodiversity census (Raupach et al. 2015). Numerous zooplanktons also are a very important part of marine arthropod community. Barcoding was performed to reveal molecular phylogeny of deep sea copepods (Laakmann et al. 2012). There are several publications depicting the significance of DNA barcoding studies of marine copepods (Bucklin and Frost 2009; Machida and Tsuda 2010; Castellani et al. 2012; Blanco-Bercial et al. 2014) and amphipods (Costa et al. 2009).

10.2.8 Echinodermata

Echinoderms are exclusively marine radially symmetrical animals which are divided into five living classes namely Asteroidea, Ophiuroidea, Echinoidea, Crinoidea, and Holothuroidea. A DNA barcoding study including all these five classes showed very high success rate in identification of echinoderm species (Ward et al. 2008). DNA barcoding was also performed for commercial echinoderm species (Uthicke et al. 2010).

10.3 Lower Chordates

Lower chordates include the urochordates, cephalochordates and hemichordates which are marine dwelling. Ascidians, belonging to urochordata is regarded as model animals to study several complicated bioprocesses like immune system and gene regulation, heavy metal bioaccumulation, heart and embryonic development etc. (Holland and Gibson-Brown 2003; Trivedi et al. 2003; Satoh et al. 2003; Stolfi and Christiaen 2012; Tolkin and Christiaen 2012; Razy-Krajka et al. 2014). New ascidian species was discovered in Japan by using DNA barcoding (Hirose and Hirose 2009). *Ciona savignyi* was first time reported in the Southern Hemisphere through a DNA barcoding study (Smith et al. 2012). DNA barcoding was performed using COI gene for amphioxus (cephalochordate) of West Pacific Ocean (Li et al. 2010).

10.4 Vertebrates

10.4.1 Fishes

Marine and coastal fisheries contribute significantly to the food security, employment and global economy. Proper identification, assessment and conservation of fisheries resources are important to maintain sustainability. Overfishing, pollution and anthropogenic stress is imparting negative impacts on marine fishery resources.

An ambitious and large scale campaign to barcode all fishes was launched (Ward et al. 2009). Two main global barcoding initiatives on fishes are FISH-BOL and SharkBOL.

Several barcoding studies on fishes have been conducted at different oceans and seas like Atlantic Ocean (Weigt et al. 2012), Arctic Ocean (Mecklenburg et al. 2011), Antarctic Ocean (Dettai et al. 2011), Pacific Ocean (Zemlak et al. 2009), Indian Ocean (Dhaneesh et al. 2015), Mediterranean Sea and Cantabric Sea (Ardura et al. 2013), Red Sea (Trivedi et al. 2014), China Sea (Zhang et al. 2013) etc. Besides this, DNA barcoding of marine fishes have been done at different regions and countries like Europe (Kochzius et al. 2010), North America (Steinke et al. 2009b), Australia

(Ward et al. 2005), South Africa (Zemlak et al. 2009), China (Zhang 2011), India (Lakra et al. 2010), Japan (Zhang and Hanner 2011), Argentina (Mabragaña et al. 2011).

Study on fishes belonging to genus *Sebastes* collected from China Sea proved the efficacy of DNA barcoding (Zhang et al. 2013).

10.4.2 Amphibians

Very few barcoding has been conducted on amphibians. Recently an initiative called 'cold code' has been launched with the aim to barcode cold blooded vertebrates including amphibians and reptiles. It is reported that *16S rRNA*gene can be used as universal barcode for amphibians (Vences et al. 2005).

10.4.3 Reptiles

Reptiles are cold blooded vertebrates and extant reptiles belong to four orders namely Crocodilia, Squamata, Testudines and Sphenodontia. A study on marine testudines showed that DNA barcoding is an efficient tool for identification, phylogenetic assessment and conservation of threatened turtles (Naro-Maciel et al. 2009). Another study showed the efficacy COI sequences to identify sea turtles from Brazil (Vargas et al. 2009). Large scale barcoding involving 468 specimens belonging to Squamata and Testudines of Madagaskar—a biodiversity hot spot, revealed several new species (Nagy et al. 2012). Recently, a review on the DNA barcoding in different groups of reptilia has been published (Trivedi et al. 2016a).

10.4.4 Sea Birds

Global barcoding initiative on birds including the seabirds is led by All Birds Barcoding Initiative (ABBI) which was launched in 2005. A large scale DNA barcoding of birds was conducted involving 387 specimens belonging to 147 avian species in Netherlands (Aliabadian et al. 2013). In many places around the globe the mangroves provide shelter for the sea birds. With the rapid depletion of mangrove cover, there is severe habitat loss for the sea birds. More effort is needed for barcoding and conservation of the seabirds as well as mangroves.

10.4.5 Mammals

For the mammals the barcoding initiative is led by Mammalia Barcode of Life (http://www.mammaliabol.org). A barcoding study along the Atlantic coast of France showed that marine mammal biodiversity can be monitored through this technique (Alfonsi et al. 2013).

11 DNA Barcoding of Marine Plants

In contrast to the animals there is no consensus on the barcode gene in plants. Several molecular markers have been used for barcoding different groups of plants. Some researchers have put forward the 'super barcode' concept for efficient plant identification. CBOL's Plant Working Group is devoted to working on standard barcoding regions in plants. DNA barcoding as a tool to assess biodiversity in plants has been reviewed (Trivedi et al. 2016c).

11.1 Seagrasses

Sea grasses are flowering plants that are very significant for their high primary productivity and contain several secondary compounds. They can play a vital role in mitigating global climate change. They are an important component of the 'blue carbon' and are one of the most important marine sinks of organic carbon. Our research on sea grasses of Indian Ocean has shown their high carbon storage potential (Ghosh et al. 2015). It is reported that the sea grass cover is decreasing at a rapid pace of 110 km^2 per year globally (Waycott et al. 2009). The identification of sea grass species is a challenging task because the flower which is one of the most important morphological part for identification is often missing. DNA barcoding on sea grasses was performed using different molecular markers likeITS, trnK, rbcL, 5. 8S rDNA and matK (Waycott et al. 2002; Les et al. 2002; Uchimura et al. 2008; Lucas et al. 2012).

11.2 Marine Algae

The two main groups of marine algae are microalgae (blue green algae, dinoflagellates, bacillariophyta, etc.) and macroalgae (seaweeds including green, brown and red algae). Microalgae are great source of numerous bioactive compounds and seaweeds serve as nutritious food and also source of gums.

Using COI and UPA domain V of the 23S rRNAas marker Clarkston and Saunders (2010) discovered a new species *Euthoratim burtonii*. This study also showed that COI showed better results for species delimitation of Kallymeniaceae —a red alga family. The efficacy of COI, UPA and ITS was tested on a separate study on intertidal red macro algae (Xiaobo et al. 2013). DNA barcoding was effective in identification of red algae belong to the genus Gracilaria (Kim et al. 2010). 16S rRNA and 23S rRNA was used to barcode microalga of Indian Ocean (Ahmad et al. 2013). DNA barcoding is a reliable and quick method for identification of Harmful Algal Bloom (HAB) species making it a useful tool for monitoring of marine and coastal pollution.

11.3 Marine Phytoplanktons

Phytoplanktons play a very significant role in the coastal and marine ecosystems. They occupy the base of marine food pyramid and transfer the atmospheric carbon dioxide to the ocean. Study on the distribution and abundance of marine phytoplanktons is an important aspect of ecology. The rbcL was used to barcode phytoplanktons lesser than 10 μm size (Bhattacharjee et al. 2013).

11.4 Mangroves

The mangroves are halophytic plants found at the coastal regions of the tropical and subtropical regions. They have immense ecological and economic significance. One of the most important roles of the mangroves is that they protect the fragile coastal regions from severe wave action, storm surges and cyclones. The strong root system of mangroves presents coastal soil erosion. One of the largest mangrove blocks in the world is the Sundarbans—spanning the coastal regions of West Bengal in India and Bangladesh. It is UNESCO World Heritage Site (http://whc.unesco.org/en/list). The salinity, nutrient and dissolved oxygen status is changing drastically with changing time thereby directly affecting the abundance and zonation of different mangrove species in this anthropogenically stressed region (Mitra et al. 2015; Amin et al. 2015; Trivedi et al. 2015b). DNA barcoding can provide useful information not only in assessing the mangrove biodiversity but also the mangrove associated flora and fauna.

The importance of DNA Barcoding for conservation of mangroves has been highlighted (Daru et al. 2013).

12 Conclusion

It is reported that huge number of marine species are genetically not anchored to traditional taxonomy based on morphology. The authors have stressed the need for barcoding and metabarcoding approach to bridge this existing gap (Laray and Knowlton 2015). The development of improved faster techniques like high-throughput sequencing will expedite and refine the process of DNA barcoding. Genomic studies along with DNA barcoding assay can be helpful is assessing global marine biodiversity. The role of DNA barcoding in ecology and especially eco-informatics has been highlighted (Joly et al. 2013). The development of newer and improved bioinformatics tools will open new chapters in the field modern taxonomy. DNA Barcoding can play a very significant role in marine biodiversity assessment and also in conservation of marine bio resources (Bucklin et al. 2011; Trivedi et al. 2016b).

Acknowledgments The authors gratefully acknowledge the Deanship of Scientific Research, University of Tabuk, Saudi Arabia, for the support provided through the projects numbered S-1434-0106, S-1435-0112 and S-1436-0252 to Dr. Subrata Trivedi (Principal Investigator).

References

Ahmad I, Fatma Z, Yazdani SS (2013) DNA barcode and lipid analysis of new marine algae potential for biofuel. Algal Res 2:10–15. http://dx.doi.org/10.1016/j.algal.2012.10.003

Alfonsi E, Méheust E, Fuchs S (2013) The use of DNA barcoding to monitor the marine mammal biodiversity along the French Atlantic coast. In: Nagy ZT, Backeljau T, De Meyer M, Jordaens K (eds) DNA barcoding: a practical tool for fundamental and applied biodiversity research. ZooKeys 365:5–24. http://dx.doi.org/10.3897/zookeys.365.5873

Aliabadian M, Beentjes KK, Roselaar CS (2013) DNA barcoding of Dutch birds. In: Nagy ZT, Backeljau T, De Meyer M, Jordaens K (eds) DNA barcoding: a practical tool for fundamental and applied biodiversity research. ZooKeys 365:25–48. http://dx.doi.org/10.3897/zookeys.365.6287

Amin G, Biswas S, Zaman S, Pramanick P, Trivedi S, Mitra A (2015) Prediction of dissolved oxygen in Indian sundarbans: vision 2050. Int Adv Res J Sci Eng Technol 2(12):31–33

Andreakis N, Høj L, Kearns P, Hall MR, Ericson G, Cobb RE et al (2015) Diversity of marine-derived fungal cultures exposed by DNA barcodes: the algorithm matters. PLoS ONE 10(8):e0136130. doi:10.1371/journal.pone.0136130

Ardura A, Planes S, Garcia-Vazquez E (2013) Applications of DNA barcoding to fish landings: authentication and diversity assessment. In: Nagy ZT, Backeljau T, De Meyer M, Jordaens K (eds) DNA barcoding: a practical tool for fundamental and applied biodiversity research. ZooKeys 365:49–65. http://dx.doi.org/10.3897/zookeys.365.6409

Baldwin CC, Johnson GD (2014) Connectivity across the Caribbean sea: DNA barcoding and morphology unite an enigmatic fish larva from the Florida straits with a new species of sea bass from deep reefs off Curaçao. PLoS ONE 9(5):e97661. doi:10.1371/journal.pone.0097661

Belinky F, Amir Szitenberg A, Goldfarb I, Feldstein T, Wörheide G, Ilan M, Huchon D (2012) ALG11—a new variable DNA marker for sponge phylogeny: comparison of phylogenetic performances with the 18S rDNA and the COI gene. Mol Phylogenet Evol 63:702–713

Bhadury P, Austen MC, Bilton DT, Lambshead PJD, Rogers AD, Smerdon GR (2006) Development and evaluation of a DNA-barcoding approach for the rapid identification of nematodes. Mar Ecol Prog Ser 320:1–9

Bhattacharjee D, Samanta B, Danda AA (2013) Temporal succession of phytoplankton assemblages in a tidal creek system of the Sundarbans Mangroves: an integrated approach. Int J Biodiv. http://dxdoi.org/10.1155/2013/824543

Blanco-Bercial L, Álvarez-Marqués F, Bucklin A (2011) Comparative phylogeography and connectivity of sibling species of the marine copepod Clausocalanus (Calanoida). J Exp Mar Biol Ecol 404:108–115

Blanco-Bercial L, Cornils A, Copley N, Bucklin A (2014) DNA barcoding of marine Copepods: assessment of analytical approaches to species identification. PLoS Curr 6. http://dx.doi.org/10.1371/currents.tol.cdf8b74881f87e3b01d56b43791626d2

Böttger-Schnack R, Machida R (2011) Comparison of morphological and molecular traits for species identification and taxonomic grouping of oncaeid copepods. Hydrobiologia 666:111–125

Briggs JC (1994) Species-diversity—land and sea compared. Syst Biol 43:130–135

Bucklin A, Frost BW (2009) Morphological and molecular phylogenetic analysis of evolutionary lineages within Clausocalanus (Copepoda: Calanoida). J Crust Biol 29:111–120

Bucklin A, Hopcroft RR, Kosobokova KN, Nigro LM, Ortman BD et al (2010a) DNA barcoding of Arctic Ocean holozooplankton for species identification and recognition. Deep-Sea Res II 57:40–48

Bucklin A, Ortman BD, Jennings RM, Nigro LM, Sweetman CJ et al (2010b) A "Rosetta Stone" for metazoan zooplankton: DNA barcode analysis of species diversity of the Sargasso Sea (Northwest Atlantic Ocean). Deep-Sea Res II 57:2234–2247

Bucklin A, Steinke D, Blanco-Bercial L (2011) DNA barcoding of marine metazoa. Annu Rev Mar Sci 3:471–508. http://dx.doi.org/10.1146/annurev-marine-120308-080950

Cárdenas P, Menegola C, Rapp HT, Diaz MC (2009) Morphological description and DNA barcodes of shallow-water *Tetractinellida* (Porifera: Demospongiae) from Bocas del Toro, Panama, with description of a new species. Zootaxa 2276:1–39

Carvalho FC, Pomponi SA, Xavier JR (2015) Lithistid sponges of the upper bathyal of Madeira, Selvagens and Canary Islands, with description of a new species of Isabella. J Mar Biol Assoc UK 95:1287–1296. doi:10.1017/S0025315414001179

Castellani C, Lindley AJ, Wootton M, Lee CM, Kirby RR (2012) Morphological and genetic variation in the North Atlantic copepod Centropages typicus. J Mar Biol Assoc UK 92:99–106

Chinnappareddy LRD, Khandagale K, Reddy SHS, Chennareddy CKA, Singh TH (2012) SSR-based DNA barcodes as a tool for identification of eggplant genotypes. Int J Vegetable Sci 18(3):260–271. doi:10.1080/19315260.2011.633976

Clarkston BE, Saunders GW (2010) A comparison of two DNA barcode markers for species discrimination in the red algal family Kallymeniaceae (Gigartinales, Florideophyceae), with a description of *Euthoratim burtonii* sp. Botany 88:119–131. http://dx.doi.org/10.1139/B09-101

Costa FO, Henzler CM, Lunt DH, Whiteley NM, Rock J (2009) Probing marine Gammarus (Amphipoda) taxonomy with DNA barcodes. Syst Biodivers 7(4):365–379. http://dx.doi.org/10.1017/S1477200009990120

Daru BH, Yessoufou K, Mankga LT, Davies TJ (2013) A global trend towards the loss of evolutionarily unique species in mangrove ecosystems. PLoS ONE 8(6):e66686. http://dx.doi.org/10.1371/journal.pone.0066686

De Wit P, Rota E, Erséus C (2009) Grania (Annelida: Clitellata: Enchytraeidae) of the Great Barrier Reef, Australia, including four new species and a re-description of Graniatrichaeta Jamieson, 1977. Zootaxa 2165:16–38

Derycke S, Fonseca G, Vierstraete A, Vanfleteren J, Vincx M, Moens T (2008) Disentangling taxonomy within the Rhabditis (Pellioditis) marina (Nematoda, Rhabditidae) species complex using molecular and morphological tools. Zool J Linn Soc 152:1–15

Dettai A, Lautredou AC, Bonillo C (2011) Theactinopterygian diversity of the CEAMARC cruises: barcoding and molecular taxonomy as a multi-level tool for new findings. Deep-Sea Res II 58(1):250–263

Dhaneesh KV, Kumar TTA, Kumar AB (2015) Barcoding, phylogeography and species boundaries in crownfishes of the Indian Ocean. DNA Barcodes 3:5–16

Efe MA, Tavares ES, Baker AJ, Bonatto SL (2009) Multigene phylogeny and DNA barcoding indicate that the Sandwich tern complex (Thalasseussandvicensis, Laridae, Sternini) comprises two species. Mol Phylogenet Evol 52:263–267

Erpenbeck D, Hall K, Alvarez B, Büttner G, Sacher K, Schätzle S, Schuster A, Vargas S, Hooper JNA, Wörheide G (2012) The phylogeny of halichondrid demosponges: past and present re-visited with DNA-barcoding data. Org Divers Evol 12:57–70. doi:10.1007/s13127-011-0068-9

Erpenbeck D, Ekins M, Enghuber N, Hooper JNA, Lehnert H, Poliseno A, Schuster A, Setiawan E, De Voogd NJ, Wörheide G, Soest RWM (2015) Nothing in (sponge) biology makes sense—except when based on holotypes. J Mar Biol Assoc UK. doi:10.1017/S0025315415000521

Ghosh R, Trivedi S, Pramanick P, Zaman S, Mitra A (2015) Seagrass: a store house of carbon. J Energy Environ Carbon Credits 5(2):23–29

Gray JS (1997) Marine biodiversity: patterns, threats and conservation needs. Biodiv Conserv 6:153–175

Hanner R, Becker S, Ivanova VN (2011) FISH-BOL and seafood identification: geographically dispersed case studies reveal systemic market substitution across Canada. Mitochondrial DNA 22(S1):106–122. http://dx.doi.org/10.3109/19401736.2011.588217

Hebert PD, Cywinska A, Ball SL (2003) Biological identifications through DNA barcodes. Proc Biol Sci R Soc 270(1512), 313–321. http://dx.doi.org/10.1098/rspb.2002.2218

Hirose M, Hirose E (2009) Barcoding in photosymbiotic species of Diplosoma (Ascidiacea: Didemnidae), with the description of a new species from the Southern Ryukyus Japan. Zool Sci 26(8):564–568. http://dx.doi.org/10.2108/zsj.26.564

Holland LZ, Gibson-Brown JJ (2003) The *Ciona intestinalis* genome: when the constraints are off. BioEssays 25(6):529–532

Holmes BH, Steinke D, Ward RD (2009) Identification of shark and ray fins using DNA barcoding. Fisher Res 95:280–288

Järnegren J, Schander C, Sneli J-A, Rønningen V, Young CM (2007) Four genes, morphology and ecology: distinguishing a newspecies of Acesta (Mollusca; Bivalvia) from the Gulf of Mexico. Mar Biol 152:43–55. http://dx.doi.org/10.1007/s00227-007-0651-y

Jennings RM, Bucklin A, Ossenbrugger H, Hopcroft RR (2010a) Species diversity of planktonic gastropods (Pteropoda and Heteropoda) fromsix ocean regions based on DNA barcode analysis. Deep-Sea Res II 57:2199–2210. http://dx.doi.org/10.1016/j.dsr2.2010.09.022

Jennings RM, Bucklin A, Pierrot-Bults A (2010b) Barcoding of arrow worms (Phylum Chaetognatha) from three oceans: genetic diversity and evolution within an enigmatic phylum. PLoS ONE 5(4):e9949. http://dx.doi.org/10.1371/journal.pone.0009949

Joly S, Davies TJ, Archambault A, Bruaeau A, Derry A, Kembel SW, Peres–Neto P, Vamosi J, Wheeler TA (2013) Ecology in the age of DNA barcoding: the resource, the promise and the challenges ahead. Mol Ecol Resour. doi:10.1111/1755-0998.12173

Kamaruzzaman BY, John BA, Zaleha K, Jalal KCA (2011) Molecular phylogeny of horseshoe crab. Asian J Biotechnol 3:302–309

Kappner I, Bieler R (2006) Phylogeny of venus clams (Bivalvia: Venerinae) as inferred from nuclear and mitochondrial gene sequences. Mol Phylogenet Evol 40(2):317–331

Kemppainen P, Panova M, Hollander J, Johannesson K (2009) Complete lack of mitochondrial divergence between two species of NE Atlantic marine intertidal gastropods. J Evol Biol 22 (10):2000–2011. http://dx.doi.org/10.1111/j.1420-9101.2009.01810.x

Keshavmurthy S, Yang S-Y, Alamaru A, Chuang Y-Y, Pichon M, Obura D et al (2013) DNA barcoding reveals the coral "laboratory-rat", *Stylophora pistillata* encompasses multiple identities. Sci Rep 3:1520. doi:10.1038/srep01520

Kim MS, Yang MY, Cho GY (2010) Applying DNA barcoding to Korean Gracilariaceae (Rhodophyta) Cryptogamie. Algologie 3(4):387–401

Kochzius M, Seidel C, Antoniou A (2010) Identifying fishes through DNA barcodes and microarrays. PLoS ONE 5(9):e12620. http://dx.doi.org/10.1371/journal.pone.0012620

Kozol R, Blanco-Bercial L, Bucklin A (2012) Multi-Gene analysis reveals a lack of genetic divergence between *Calanusagulhensis* and C. *sinicus* (Copepoda; Calanoida). PLoS ONE 7: e45710

Krug PJ, Ellingson RA, Burton R, Valdés A (2007) A new poecilogonous species of sea slug (Opisthobranchia: Sacoglossa) from California: comparison with the planktotrophiccongener-modesta (Lovén 1844). J Mollus Stud 73(1):29–38. doi:10.1093/mollus/eyl025

Laakmann S, Auel H, Kochzius M (2012) Evolution in the deep sea: biological traits, ecology and phylogenetics of pelagic copepods. Mol Phylogenet Evol 65:535–546

Lakra WS, Verma WS, Goswami M (2010) DNA barcoding Indian marine fishes. Mol Ecol Resour 11:60–71

Lavrov DV, Pett W, Voigt O, Wörheide G, Forget L, Franz Lang B, Kayaly E, (2012) Mitochondrial DNA of *Clathrina clathrus* (Calcarea, Calcinea): six linear chromosomes, fragmented rRNAs, tRNA Editing, and a novel genetic code. Mol Biol Evol 30(4):865–880. doi:10.1093/molbev/mss274

Layton KKS, Martel AL, Hebert PDN (2014) Patterns of DNA barcode variation in canadian marine molluscs. PLoS ONE 9(4):e95003. http://dx.doi.org/10.1371/journal.pone.0095003

Lee G, Park SY, Hwang J, Lee Y-H, Hwang SY, Lee S, Lee T-K (2011) Development of DNA chip for jellyfish verification from South Korea. BioChip J 5(4):375–382. doi:10.1007/s13206-011-5412-9

Leray M, Knowlton N (2015) DNA barcoding and metabarcoding of standardized samples reveal patterns of marine benthic diversity. PNAS 112(7):2076–2081

Les DH, Moody ML, Jacobs SWL, Bayer RJ (2002) Systematics of seagrasses (Zosteraceae) in Australia and New Zealand. Syst Bot 27:468–484

Li WY, Zhong J, Wang Y-Q (2010) Analysis of amphioxus geographic populations in the West Pacific Ocean based on *COX I* gene. Zool Res 31(4):375–380. doi:10.3724/SP.J.1141.2010.04375

Lobo J, Teixeira MAL, Borges LMS, Ferreira MSG, Hollatz C, Gomes PT, Sousa R et al (2015) Starting a DNA barcode reference library for shallow water polychaetes from the southern European Atlantic coast. Mol Ecol Resour. doi:10.1111/1755-0998.12441

Lowenstein JH, Amato G, Kolokotronis S-O (2009) The real maccoyii: identifying Tuna Sushi with DNA barcodes—contrasting characteristic attributes and genetic distances. PLoS ONE 4(11):e7866. http://dx.doi.org/10.1371/journal.pone.0007866

Lucas C, Thangaradjou T, Papenbrock J (2012) Development of a DNA barcoding systemforseagrasses: successful but not simple. PLoS ONE 7(1):e29987. doi:10.1371/journal.pone.0029987

Maas AE, Blanco-Bercial L, Lawson GL (2013) Reexamination of the species assignment of Diacavolinia pteropods using DNA barcoding. PLoS ONE 8(1):e53889. http://dx.doi.org/10.1371/journal.pone.0053889

Mabragaña E, Díaz de Astarloa JM, Hanner R (2011) DNA barcoding identifies argentine fishes from marine and brackish waters. PLoS ONE 6(12):e28655. http://dx.doi.org/10.1371/journal.pone.0028655

Machida RJ, Tsuda A (2010) Dissimilarity of species and forms of planktonic Neocalanus copepods using mitochondrial COI, 12S, nuclear ITS, and 28S gene sequences. PLoS ONE 5(4):e10278. http://dx.doi.org/10.1371/journal.pone.0010278

Maturana CS, Rodrigo A, Moreno RA, Labra FA, González-Wevar CA, Rozbaczylo N, Carrasco FD, Poulin E (2011) DNA barcoding of marine polychaetes species of southern Patagonian fjords. Revista de Biología Marina y Oceanografía 46(1):35–42

Mecklenburg CW, Moller PR, Steinke D (2011) Biodiversity of arctic marine fishes: taxonomy and zoogeography. Mar Biodiv 41:109–140

Mikkelsen NT, Schander C, Willassen E (2007) Local scale DNA barcoding of bivalves (Mollusca): a case study. Zool Scr 36(5):455–463. http://dx.doi.org/10.1111/j.1463-6409.2007.00289

Mitra A, Trivedi S, Zaman S, Pramanick P, Chakraborty S, Pal N, Fazli P, Banerjee K (2015) Decadal variation of nutrient level in two major estuaries in the Indian subdarbans. Jordan J Biol Sci 8(3):231–236

Molnar JL, Gamboa RL, Revenga C (2008) Assessing the global threat of invasive species to marine biodiversity. Front Ecol Environ 6:485–492

Morrow CC, Picton BE, Erpenbeck D, Boury-Esnault N, Maggs CA, Allcock AL (2012) Congruence between nuclear and mitochondrial genes in Demospongiae: a new hypothesis for relationships within the G4 clade (Porifera: Demospongiae). Mol Phylogenet Evol 62:174–190

Moura CJ, Harris DJ, Cunha MR, Rogers AD (2007) DNA barcoding reveals cryptic diversity in marine hydroids (Cnidaria, Hydrozoa) from coastal and deep-sea environments. Zool Scr 37(1):93–108. http://dx.doi.org/10.1111/j.1463-6409.2007.00312

Nagy ZT, Sonet G, GlawF (2012) First large-scale DNA barcoding assessment of reptiles in the biodiversity hotspot of Madagascar, based on newly designed COI primers. PLoS ONE 7(3):e34506. http://dx.doi.org/10.1371/journal.pone.0034506

Naro-Maciel E, Reid B, Fitzsimmons NN (2009) DNA barcodes for globally threatened marine turtles: a registry approach to documenting biodiversity. Mol Ecol Resour 10:252–263. http://dx.doi.org/10.1111/j.1755-0998.2009.02747.x

Park M-H, Sim C-J, Baek J, Min G-S (2007) Identification of genes suitable for DNA barcoding of morphologically indistinguishable Korean Halichondriidae sponges. Mol Cells 23(2):220–227

Pauls SU, Blahnik RJ, Zhou X, Wardwell CT, Holzenthal RW (2010) DNA barcode data confirm new species and reveal cryptic diversityin Chilean *Smicridea* (*Smicridea*) (Trichoptera: Hydropsychidae). J N Am Benthol Soc 29(3):1058–1074. doi:10.1899/09-108.1
Pires AC, Marinoni L (2010) DNA barcoding and traditional taxonomy unified through Integrative Taxonomy: a view that challenges the debate questioning both methodologies. Biota Neotrop 10(2)
Plaisance L, Caley MJ, Brainard RE, Knowlton N (2011) The diversity of coral reefs: what are we missing? PLoS ONE 6(10):e25026. doi:10.1371/journal.pone.0025026
Pöppe J, Sutcliffe P, Hooper JNA, Wörheide G, Erpenbeck D (2010) COI barcoding reveals new clades and radiation patterns of Indo-Pacific sponges of the family Irciniidae (Demospongiae: Dictyoceratida). PLoS ONE 5(4):e9950. doi:10.1371/journal.pone.0009950
Puillandre N, Strong EE, Bouchet P (2009) Identifying gastropod spawn from DNA barcodes: possible but not yet practicable. Mol Ecol Resour 9:1311–1321. http://dx.doi.org/10.1111/j.1755-0998.2009.02576.x
Raupach MJ, Barco A, Steinke D, Beermann J, Laakmann S, Mohrbeck I et al (2015) The application of DNA barcodes for the identification of marine crustaceans from the North Sea and adjacent regions. PLoS ONE 10(9):e0139421. doi:10.1371/journal.pone.0139421
Razy-Krajka F, Lam K, Wang W, Stolfi A, Joly M, Bonneau R, Christiaen L (2014) Collier/OLF/EBF-dependent transcriptional dynamics control pharyngeal muscle specification from primed cardiopharyngeal progenitors. Dev Cell 29(3):263–276. http://dx.doi.org/10.1016/j.devcel.2014.04.001
Satoh N, Satou Y, Davidson B, Levine M (2003) *Ciona intestinalis*: an emerging model for whole-genome analyses. Trends Genet 19(7):376–381
Schuster A, Erpenbeck D, Pisera A, Hooper J, Bryce M et al (2015) Deceptive desmas: molecular phylogenetics suggests a new classification and uncovers convergent evolution of lithistid-demosponges. PLoS ONE 10(1):e116038. doi:10.1371/journal.pone.011603
Smith KF, Thia J, Gemmill CEC (2012) Barcoding of the cytochrome oxidase I (COI) indicates a recent introduction of *Ciona savignyi* into New Zealand and provides a rapid method for Ciona species discrimination. Aquat Invasions 7(3): 305–313. http://dx.doi.org/10.3391/ai.2012.7.3.002
Sperling EA, Rosengarten RD, Moreno MA, Dellaporta SL (2012). The complete mitochondrial genome of the verongid sponge *Aplysina cauliformis*: implications for DNA barcoding in demosponges. Hydrobiologia 687:61–69 doi:10.1007/s10750-011-0879-x
Steinke D, Zemlak TS, Hebert PDN (2009) Barcoding Nemo: DNA-based identifications for the ornamental fish trade. PLoS ONE 4(7):e6300. http://dx.doi.org/10.1371/journal.pone.0006300
Stolfi A, Christiaen L (2012) Genetic and genomic toolbox of the chordate *Ciona intestinalis*. Genetics 192(1):55–66. http://dx.doi.org/10.1534/genetics.112.140590
Szitenberg A, Becking LE, Vargas S, Fernandez JCC, Santodomingo N, Wörheide G, Ilan M, Kelly M, Huchon D (2013) Phylogeny of Tetillidae (Porifera, Demospongiae, Spirophorida) based on three molecular markers. Mol Phylogenet Evol 67:509–519
Tack JF, Berghe E, Polk PH (1992) Ecomorphology of Crassostrea cucullata (Born, 1778) (Ostreidae) in a mangrove creek (Gazi, Kenya). Hydrobiologia 247:109–117
Tolkin T, Christiaen L (2012) Development and evolution of the ascidian cardiogenic mesoderm. Curr Top Dev Biol 100:107–142. http://dx.doi.org/10.1016/B978-0-12-387786-4.00011-7
Trivedi S, Affan R, Alessa AHA et al (2014) DNA barcoding of Red Sea fishes from Saudi Arabia—the first approach. DNA Barcodes 2:17–20. http://dx.doi.org/10.2478/dna-2014-0003
Trivedi S, Aloufi AA, Ansari AA, Ghosh SK (2015a) Molecular phylogeny of oysters belonging to the genus *Crassostrea* through DNA barcoding. J Entomol Zool Stud 3(1):21–26
Trivedi S, Chakraborty S, Zaman S, Prosenjit P, Fazli P, Amin G, Mitra (2015b) Impact of salinity on the condition factor of commercially important Fin Fish in the lower Gangetic delta. J Environ Sci Comput Sci Eng Technol 4(2):473–480
Trivedi S, Aloufi AA, Rehman H, Saggu S, Ghosh SK (2016a) DNA barcoding: tool for assessing species identification in reptilia. J Entomol Zool Stud 4(1):332–337

Trivedi S, Aloufi AA, Ansari AA, Ghosh SK (2016b) Role of DNA barcoding in marine biodiversity assessment and conservation: an update. Saudi J Biol Sci. http://dx.doi.org/10.1016/j.sjbs.2015.01.00

Trivedi S, Ansari AA, Rehman H, Saggu S, Abbas ZK, Ghosh SK (2016c) DNA barcoding as a molecular tool for the assessment of plant biodiversity. In: Ansari AA, Gill SS (eds) Plant biodiversity: monitoring assessment and conservation. CABI Publications, Wallingford (In press)

Trivedi S, Ueki T, Yamaguchi N, Michibata H (2003) Novel Vanadium-binding proteins (vanabins) identified in cDNA libraries and the genome of the ascidian *Ciona intestinalis*. Biochim Biophys Acta 1630:64–70

Uchimura M, Faye EJ, Shimada S, Inoue T, Nakamura Y (2008) A reassessment of Halophila species (Hydrocharitaceae) diversity with special reference to Japanese representatives. Bot Mar 51:258–268

Uthicke S, Byrne M, Conand C (2010) Genetic barcoding of commercial Bêche-de-mer species (Echinodermata: Holothuroidea). Mol Ecol Resour 10(4)–646. http://dx.doi.org/10.1111/j.1755-0998.2009.02826.x

Vanhove MPM, Tessens B, Schoelinck C, Jondelius U, Littlewood DTJ, Artois T, Huyse T (2013) Problematic barcoding in flatworms: A case-study on monogeneans and rhabdocoels (Platyhelminthes). In: Nagy ZT, Backeljau T, De Meyer M, Jordaens K (eds) DNA barcoding: a practical tool for fundamental and applied biodiversity research. ZooKeys 365:355–379. doi:10.3897/zookeys.365.5776

Vargas S, Erpenbeck D, Göcke C, Hall KA, Hooper JNA, Janussen D, Wörheide G (2013). Molecular phylogeny of Abyssocladia (Cladorhizidae: Poecilosclerida) and Phelloderma (Phellodermidae: Poecilosclerida) suggests a diversification of chelae microscleres in cladorhizid sponges. ZoologicaScripta 42:106–116

Vargas S, Kelly M, Schnabel K, Mills S, Bowden D, Wörheide G (2015) Diversity in a cold hot-spot: DNA-barcoding reveals patterns of evolution among Antarctic demosponges (Class Demospongiae, Phylum Porifera). PLoS ONE 10(6):e0127573. doi:10.1371/journal.pone.0127573

Vargas S, Schuster A, Sacher K, Büttner G, Schätzle S et al (2012) Barcoding sponges: an overview based on comprehensive sampling. PLoS ONE 7(7):e39345. doi:10.1371/journal.pone.003934

Vargas SM, Araújo FCF, Santos FR (2009) DNA barcoding of Brazilian sea turtles (Testudines). Genet Mol Biol 32(3):608–612

Velmurugan S, Prasannakumar C, Manokaran S, Ajithkumar T, Samkamaleson A, Palavesam A (2013) DNA barcodes for marine fungal identification and discovery. Fungal Ecol 6:408–418. http://dx.doi.org/10.1016/j.funeco.2013.05.003

Vences M, Thomas M, Meijden A, Chiari Y, Vieites DR (2005) Comparative performance of the 16S rRNA gene in DNA barcoding of amphibians. Front Zool 2:5. doi:10.1186/1742-9994-2-5

Ward RD, Hanner R, Hebert PDN (2009) The campaign to DNA barcode all fishes. J Fish Biol 74:329–356

Ward RD, Holmes BH, O'hara TD (2008) DNA barcoding discriminates echinoderm species. Mol Ecol Resour 8(6):1202–1211

Ward RD, Zemlak TS, Innes BH (2005) Barcoding Australia's fish species. Phil Trans R Soc Lond B 360:1847–1857

Waycott M, Duarte CM, Carruthers TJB et al (2009) Accelerating loss of seagrasses across the globe threatens coastal ecosystems. Proc Natl Acad Sci USA 106:12377–12381

Waycott M, Freshwater DW, York RA et al (2002) Evolutionary trends in the seagrass genus Halophila (thouars): Insights from molecular phylogeny. Bull Mar Sci 71:1299–1308

Weigt LA, Baldwin CC, Driskell A (2012) Using DNA barcoding to assess Caribbean Reef fish biodiversity: expanding taxonomic and geographic coverage. PLoS ONE 7(7):e41059. http://dx.doi.org/10.1371/journal.pone.0041059

Wörheide G, Erpenbeck D, Menke C (2007) The sponge barcoding project: aiding in the identification and description of poriferan taxa. Porifera Res Biodivers Innov Sustain 2007:123–128

Xiaobo Z, Shaojun P, Tifeng S (2013) Applications of three DNA barcodes in assorting intertidal red Macroalgal Flora in Qingdao, China. J Ocean Univ China 12(1):139–145. http://dx.doi.org/10.1007/s11802-013-2052-9

Zemlak TS, Ward RD, Connell AD (2009) DNA barcoding reveals overlooked marine fishes. Mol Ecol Resour 9(1):237–242. http://dx.doi.org/10.1111/j.1755-0998.2009.02649.x

Zhang H, Zhang Y, Zhang Z, Gao T (2013) DNA barcodes of eight species in genus Sebastes. Biochem Syst Ecol 48:45–50. http://dx.doi.org/10.1016/j.bse.2012.11.012

Zhang J (2011) Species identification of marine fishes in China with DNA barcoding. Evid Based Complement Altern Med. Article ID 978253, 10p. http://dx.doi.org/10.1155/2011/978253

Zhang J-B, Hanner R (2011) DNA barcoding is a useful tool for the identification of marine fishes from Japan. Biochem Syst Ecol 39:31–42

Zhou H, Zhang Z, Chen H, Sun R, Wang H, Guo L, Pan H (2010) Integrating a DNA barcoding project with an ecological survey: a case study on temperate intertidal polychaete communities in Qingdao, China. Chin J Oceanol Limnol 28(4):899–910. doi:10.1007/s00343-010-9131-1

Measurement of a Barcode's Accuracy in Identifying Species

John L. Spouge

Abstract This chapter describes a workflow for measuring a barcode's accuracy when identifying species. First, assemble a database of specimens with their marker sequences and their species binomials. The species binomials provide a "taxonomic gold standard" for species identification and should be as accurate as possible, to avoid penalizing correct species assignment. Second, select a computer algorithm for assigning species to barcode sequences. Only one algorithm (BLAST + P) has improved notably on the simple strategy of assigning specimens to the species of the database sequence(s) nearest under p-distance. Global sequence alignments (e.g., with the Needleman-Wunsch algorithm, or with multiple sequence alignment algorithms) align entire barcode sequences, using all available information, so they sometimes produce more accurate species identifications than local sequence alignments (e.g., with BLAST), particularly when BLAST produces barcode alignments of small subsequences within the sequences. Finally, consensus has settled on "the probability of correct identification" (PCI) as the appropriate measurement of species identification accuracy. The overall PCI for a data set is the average of the species PCIs, taken over all species in the data set. The chapter discusses some variant PCIs, their calculation and the estimation of their statistical sampling errors. It also discusses good practice in incorporating PCR failure and species with singleton representatives into data summaries. For software relevant to this chapter, see http://tinyurl.com/spouge-barcode.

Keywords Barcode · Species identification accuracy · Probability of correct identification

---

J.L. Spouge (✉)
National Center for Biotechnology Information,
National Library of Medicine, National Institutes of Health,
Bethesda, MD 20894, USA
e-mail: spouge@nih.gov

S. Trivedi et al. (eds.), *DNA Barcoding in Marine Perspectives*,
DOI 10.1007/978-3-319-41840-7_2

1 Introduction

The Anthropocene era is the geologic era when human activities started to have a significant effect on the Earth's ecosystems (Zalasiewicz et al. 2000). The Anthropocene is extinguishing species, making the conservation of biodiversity a major challenge. The first step to debate the benefits and costs of preserving biodiversity is to assess the facts, namely, to catalog species throughout the world, and to document changes in their populations, a staggering task requiring more taxonomists than presently exist. Fortunately, the technology of DNA barcodes provides an alternative strategy to field assessment by trained taxonomists, because barcodes require DNA samples and not the immediate identification of specimens. Barcodes have many uses other than taxonomic identification (recognition of novel species, taxonomic classification, and the construction of phylogenetic trees, to name but a few), but applications in biodiversity justify restricting this chapter to the measurement of accuracy in species identification (Hebert et al. 2003a). The ability to measure barcode accuracy when identifying species within a taxon is critical, for it measures the success of the barcode enterprise within that taxon.

In its essence, a barcode is any standardized subset of DNA collected from taxonomic specimens (Floyd et al. 2002). To fix the terminology in the chapter, "marker" connotes any contiguous region of DNA (coding or non-coding), whereas "barcode" is a collective noun connoting the one or more markers comprising the standardized subset of DNA. Presently, all official barcode markers are genes, e.g., CO1 (historically, the first official barcode), matK and rbcL (the barcode markers for plants), and ITS (the barcode marker for fungi). Initial investigations into DNA barcodes, particularly in animals, indicated that when used as a barcode, the cytochrome c oxidase 1 (CO1) gene identified many species correctly (Hebert et al. 2003b), so selection of CO1 as the primary barcode followed naturally (Hajibabaei et al. 2006; Hogg and Hebert 2004; Lorenz et al. 2005; Meyer and Paulay 2005; Saunders 2005; Smith et al. 2005, 2006).

The extension of DNA barcodes to plants (Chase et al. 2005; Cowan et al. 2006; Kress and Erickson 2008; CBOL Plant Working Group 2009) and fungi (Schoch et al. 2012) became problematic with the recognition that the CO1 gene evolved too slowly to identify the corresponding species accurately (Kress et al. 2005). Likewise, CO1 often does not identify insect species accurately (Meier et al. 2006; Huang et al. 2008). The lack of a clear consensus for a barcode in these taxa stimulated interest in an objective, quantitative measurement of barcode accuracy in identifying species. Although consensus about barcodes themselves remains elusive in some taxa, notably plants (Pang et al. 2012; Chen et al. 2015; Ferri et al. 2015), a clear consensus on measuring barcode accuracy has emerged (CBOL Plant Working Group 2009; Meier et al. 2006; Erickson et al. 2008). As a byproduct, the consensus provides a clear standard for measuring barcode accuracy in specific taxa, as well.

Section 2 describes the rationale behind and the methods of evaluating identification accuracy, while Sect. 7 provides a practical chapter summary. The web page http://tinyurl.com/spouge-barcode also provides some practical tools for measuring barcode identification accuracy.

2 Measurement of Species Identification Accuracy

As its ultimate aim, any measurement of the accuracy of species identification should reflect the practical performance of the corresponding barcode database. A barcode database has the following workflow. Users sample a DNA barcode retrieved from a specimen and then query the database with the sample barcode sequence(s); in response, the database software returns a putative species identification. In principle, a species apparently not in the database might evoke the output of "unknown species". Sometimes species boundaries are said to correlate with DNA differences like 2 % (i.e., p-distance of 0.98), but in practice the boundary difference varies with the species. Incomplete lineage sorting can even make the difference vanish. Different attempts to redefine species boundaries with threshold differences (Hebert et al. 2003a; Floyd et al. 2002; Blaxter et al. 2005; Lambert et al. 2005) conflicted heavily with traditional taxonomy (Meier et al. 2006), so no consensus exists on criteria for defining species boundaries with DNA barcodes. Thus, the database output "unknown species" (i.e., "outside every known species boundary") becomes problematic, so for pedagogical clarity, this chapter restricts itself to discussing the identification of known species, i.e., it assumes that every sample for querying the barcode database represents a species already in the database.

3 The Barcode Database

The first step in estimating barcode accuracy is to assemble a database of specimens with their marker sequences and their species binomials. In particular, the species binomials provide a "taxonomic gold standard" for species identification, i.e., they represent a standard for measuring the accuracy of algorithms implemented on a computer. The standard is often imperfect, but it should be the best available.

An investigator should recognize potential defects in a dataset (Spouge and Mariño-Ramírez 2012). To exemplify possible imperfections, the taxonomy in GenBank sequences (unless annotated by the "barcode" keyword) can be noticeably inferior to the taxonomy in databases curated by taxonomists. Imperfect taxonomy degrades measurements of barcode accuracy by penalizing correct species assignments. Moreover, GenBank usually does not provide the original specimen with its entries, rendering it unsuitable for studies of barcodes with more than one marker (e.g., botanic barcode studies using both rbcL and matK), because sequences from different markers are not associated with the single originating specimen). GenBank sequences may also contain hidden sampling biases, particularly if they lack the barcode keyword (Spouge and Mariño-Ramírez 2012). Few methods exist to assess imperfections in a taxonomic gold standard, but investigators can partition available sequences into classes expected to correlate with the accuracy of taxonomic classification (e.g., taxonomically curated and non-curated sequences). Taxonomic inaccuracies can then be inferred from statistical differences between the classes (Suwannasai et al. 2013).

If the aim of the measurement of the identification accuracy is to reflect the performance of the corresponding barcode database (as it should be), the aim constrains the number of specimens sampled within a species and the number of species sampled within genera. Section 5 discusses some nuances of sampling.

The evaluation of barcode accuracy should also account for the PCR failures occurring while obtaining relevant markers from a specimen: clearly, if PCR fails to amplify a marker from a specimen, the marker can contribute little to species identification. Measures of the identification accuracy exist to incorporate the effect of PCR failure directly (Spouge and Mariño-Ramírez 2012). PCR failure rates may diminish rapidly with technological advances, however, so current practice distinguishes PCR failure from identification accuracy by stating the (present) PCR failure rate and then restricting the barcode database to samples with successful PCR amplification (Schoch et al. 2012; Hollingsworth et al. 2009).

4 Algorithms to Assign Species

With an appropriate barcode database in hand, the computer must assign a species to each query (or declare "failure to assign"). The next step, therefore, is to develop a computer method for assigning a species to each specimen and its marker sequence(s). Unless designed for uncommonly narrow purposes, the method needs to handle datasets whose species might have either few or many representative specimens. For future reference, note that like any classification method, species assignment should consider all available information, at least in principle. The discourse below focuses on sequence information, but other types of information (e.g., morphological, geographical, etc.) might be available to influence species assignment.

Species assignment algorithms examining more than a specimen's nearest neighbors (e.g., algorithms building phlylogenetic trees, e.g., parsimony (Farris 1972), neighbor-joining (Saitou and Nei 1987), and Bayesian inference on trees (Munch et al. 2008) have not been noticeably more accurate than the simple strategy of assigning specimens to the species of their nearest neighbor(s) within the barcode database (Austerlitz et al. 2009; Austerlitz 2007; Little and Stevenson 2007). Moreover, many algorithms are too slow for the high-throughput species identifications large barcode databases require (e.g., most tree-building and probabilistic algorithms (Felsenstein 1981, 1988).

Another class of species assignment algorithms ("diagnostic methods") treat differences between aligned sequences as potential taxonomic characters (e.g.,

DNA-BAR (DasGupa et al. 2005), BLOG (Weitschek et al. 2013), CAOS (Sarkar et al. 2008), BRONX (Little 2011), PTIGS-DIdIt (Liu et al. 2011), and Linker (Albu et al. 2011). Diagnostic methods (when properly formalized) are essentially machine-learning methods, which generally require 4 samples per species to be effective (Weitschek et al. 2014). In datasets with noticeably fewer than 4 specimens per species, diagnostic methods may over-fit, reducing their identification accuracy on sparsely represented species.

Sequence distance methods (or related similarity methods, e.g., BLAST and PSI-BLAST (Altschul et al. 1997), BLAST+P (Pang et al. 2012), NN (Austerlitz et al. 2009), and TaxonDNA (Meier et al. 2006) can identify species despite sparse representation. A distance method essentially brings prior knowledge to a species with sparse representation, aiding in its identification.

Most distance methods begin by aligning marker sequences. In contrast, alignment-free distances do not require sequence alignment (Kuksa and Pavlovic 2009), making them simple and fast (Little and Stevenson 2007, Kuksa and Pavlovic 2009). They provide competitive identification accuracy in fish, butterflies, and birds (Kuksa and Pavlovic 2009), but they remain untested in problematic taxa like plants or fungi. So far, therefore, they have not been widely adopted for species identification.

Other distance methods necessitate sequence alignment. Evolutionary distances require a global sequence alignment, i.e., alignment over the full length of the sequences examined. Similarly, sequence distance methods use the full sequence lengths and are in fact equivalent to global sequence similarity methods (Smith et al. 1981), which assess similarity over full sequence lengths. In contrast, local sequence similarity methods assess only the two most similar subsequences in the sequences. (See Fig. 2 in Ref. Spouge and Mariño-Ramírez 2012, for diagrams of local and global alignments.) Local sequence alignment programs (e.g., BLAST) therefore might declare a statistically significant similarity based only on small subsequences displaying convergent evolution (homoplasy) (Wouters and Husain 2001). Local sequence alignment can therefore make distant species appear spuriously close, whereas global alignment always highlights contrasting dissimilarities across the whole sequence. In principle (and therefore if feasible), barcode studies should prefer global alignment (e.g., they should use one of the many tools performing the Needleman-Wunsch Algorithm (Needleman and Wunsch 1970) over local alignment (e.g., they should avoid the Smith-Waterman Algorithm (Smith and Waterman 1981) or BLAST, if possible). Alignments types other than global and local exist (e.g., semi-global alignment), but they did not assign species noticeably better than global alignment, at least in fungi (Schoch et al. 2012; Suwannasai et al. 2013).

In practice, however, nearest neighbor species are critical to species identification. When aligning markers from nearest neighbors, local alignment programs like BLAST often align their full sequence lengths, because local alignment then fuses subsequence alignments by bridging the short gaps between them in closely related sequences. Thus, in the alignments critical to species identification, local alignment programs often perform global alignment anyway. Investigators should be note, however, that local alignment unnecessarily complicates the interpretation of their results, possibly to the point of invalidating them, if the local alignments are over small subsequences of the original sequences. Large gaps in global alignments signal the possibility of local alignment of small subsequences and are a particularly troubling possibility in the intergenic spacers often used as adjunct markers in botanic studies, e.g., trnH-psbA.

The specific alignment or evolutionary distance chosen for a barcode analysis does not influence the nearest neighbor(s) much, so contrary statements notwithstanding (Hebert et al. 2003a), it does not affect the accuracy of most species assignments materially (Suwannasai et al. 2013; Kwong et al. 2012; Fregin et al. 2012; Collins 2012). In a pairwise alignment, therefore, the proportion of nucleotide pairs consisting of different nucleotides ("p-distance") recommends itself as a particularly simple and effective distance (Little and Stevenson 2007).

The Barcode of Life Database (BOLD, http://www.boldsystems.org; Ratnasingham and Hebert 2007) stores sequences in global multiple sequence alignments (MSAs). Many publicly available computer programs (e.g., MUSCLE Edgar 2004), MAFFT (Katoh et al. 2002), or HMMer (Eddy 1995) create MSAs; BOLD uses HMMer to align marker sequences before comparing them. Large barcode databases use MSAs, because MSAs are computationally much faster than pairwise sequence alignments. (For a database of N barcodes, the time for pairwise alignment is approximately proportional to N^2.)

On the other hand, studies with fewer sequences (say, less than 500) have the luxury of more computationally intensive alignment methods. Global alignment performs inconsistently but slightly better overall than MSA, although not statistically beyond sampling errors when used with databases of 500 or fewer sequences (Schoch et al. 2012; Suwannasai et al. 2013; Hollingsworth et al. 2009; Erickson et al. 2008). The BLAST+P method identifies the species of a query by forming an MSA from its 100 top BLAST hits and then assigning species in accord with the closest MSA neighbor under sequence distance. Its identification accuracy noticeably improved on BLAST alone, at least some cases (Pang et al. 2012).

5 Probability of Correct Identification (PCI)

Given a dataset and species assignment algorithm, we now want to measure identification accuracy. In most cases, the measure should scale, so that it estimates the performance of a comparable high-throughput database in identifying a specimen's species correctly. With this aim in mind, consensus has focused on "the probability of correct identification" (PCI) as an appropriate measurement (CBOL Plant Working Group 2009; Meier et al. 2006; Erickson et al. 2008). The definition of PCI is broad enough to accommodate legitimate scientific disagreement about species identification, so in fact the concept generates a class of measures capable of accommodating various scientific needs.

Consider any particular dataset, and for the moment, assume that each species within it generates a known PCI (specified later). The overall PCI for the dataset is the species PCI for each species, averaged over all species in the dataset. If a few data subsets are particularly important (e.g., non-basal angiosperms, basal angiosperms, and gymnosperms within plant taxa (Hollingsworth et al. 2009); or *Pezizomycotina*, *Saccharomycotina*, *Basidiomycota*, and early diverging lineages within fungi (Schoch et al. 2012), the PCIs can be reported separately for each subset. In principle, a weight within the average could reflect a species' under- or over-representation (Pang et al. 2012) or its intrinsic importance within the dataset. Most scientists do not weight averages when calculating overall PCI, however, because usually the weights just represent ephemeral sampling biases. In any case, we need only calculate a species PCI, a probability quantifying identification accuracy in each individual species, to calculate the overall PCI for a dataset.

A leave-one-out procedure ("the jackknife" in statistical language (Efron and Stein 1981) provides the species PCI. Consider a particular species and its representative specimens within the dataset. Remove each representative in turn from the dataset, and use the representative as a query specimen for the dataset. Removed from the dataset, the representative mimics a newly sampled specimen from the species, and the species is "known" if other representatives of the species exist in the dataset. Thus, the leave-one-out procedure is possible only for a species with more than one representative in the dataset. Although a singleton species (i.e., a species with a unique representative) cannot contribute to a species PCI under the leave-one-out, singletons do provide realistic "decoys" within the dataset when considering queries from other species.

Singletons with unique sequences are also a useful ancillary statistic in a barcode study, but the singletons have little relation to the correct identification of the corresponding species in realistic barcode databases (Hollingsworth et al. 2009). For similar reasons, datasets should try to include several species in each genus sampled. In any case, optimistically conflating singleton uniqueness with perfect identification in a heavily sampled species (which is much more demanding) is just plain misleading.

In a non-singleton species, scientists might legitimately disagree over the definition of successful identification of the species. Some scientists might define "success" as a monophyly, where every specimen in the species is closer to all specimens in the species than to any other specimen (CBOL Plant Working Group 2009) Success then is a binary decision, with the species PCI being either 1 or 0. Other scientists might define success pragmatically, by analogy to correct assignment of the species as in a database, where each specimen from the species has as its nearest neighbor(s) only specimens in the species (Meier et al. 2006). Again, species PCI is then either 1 or 0. Additional definitions of success are possible, depending on ties outside the species for a nearest neighbor, assignment of specimens from other species to the species in question, etc.

Some authors have used loose criteria for success (e.g., for some $k > 1$, a specimen's nearest k neighbors must contain at least one other specimen from the same species (Kuksa and Pavlovic 2009). Other authors have experimented with placing additional conditions on "success" as defined above, e.g., the presence of barcode gaps (i.e., p-distance differences between intra- and inter-species comparisons) such as 2 or 3 % (*18*). Although detection of unknown species with p-distance thresholds can be an artificial constraint (Ferguson 2002), any specific choice might be appropriate in different circumstances, depending on the scientific aim.

The following PCI definition relies mostly on averages over individuals. Define the PCI of a query specimen as the fraction of nearest neighbors with the same species as the query, then define the species PCI as the average of the specimen PCI over specimens within a species (Erickson et al. 2008). The definition has little dependence on sudden 0 to 1 changes, as might occur with PCI definitions using barcode gaps. Consequently, its statistical error (discussed briefly in Sect. 6) is smaller than many other definitions of PCI, making the value of the PCI when scaled to large databases more predictable.

In any case, two PCIs are not necessarily comparable just because they both lie between 0 and 1! PCIs are comparable only if their underlying definitions are similar. For example, the definition above using nearest k neighbors obviously produces larger PCIs than the more stringent definition using only the nearest neighbor(s); singleton species if included in a PCI inflate its value unrealistically, etc. Referees should insist on uninflated measures of identification success, to encourage comparability of the resulting PCIs.

6 Statistical Sampling Error

The overall PCI is the (usually unweighted) average of the species PCIs. As a reasonable approximation, assume that species PCIs are independent across species. Every database is a sample of all possible species, so the overall PCI $\hat{p}$ from the database is an estimate of the "true" overall PCI p (i.e., the overall PCI p in a hypothetical database including all organisms on Earth). As such, $\hat{p}$ has a (known) sampling error. The corresponding confidence intervals are often surprisingly broad, making them extremely useful in resolving scientific disagreements.

The binomial distribution provides several possible (but approximately concordant) confidence intervals for $\hat{p}$, of which two at least have appeared in barcode studies. The Wilson score interval (Suwannasai et al. 2013; Little 2011) is described in detail elsewhere (Wilson 1927), whereas the normal approximation was described in a previous review (Spouge and Mariño-Ramírez 2012). In most circumstances, the Wilson score interval is probably preferable, although (as in a t-test) the normal approximation yields the probability that the difference between two PCIs is different from 0.

7 Summary

A consensus has settled on the probability of correct identification (PCI) for the measurement of species identification accuracy with a barcode. The measurement involves several steps.

First, assemble databases corresponding to the markers. The choice of database should receive careful consideration, because it profoundly influences conclusions. Because GenBank taxonomy is often undependable, and because most GenBank sequences do not specify the originating specimen, studies based on GenBank sequences lacking the barcode keyword may have less reproducible conclusions than a carefully controlled taxonomic study. The database should try to include several samples per species and several species per genus.

Second, select a computer algorithm for assigning species to a query specimen's barcode sequences. Only one algorithm has improved noticeably on the following: compute the p-distance between the query and each database marker, and then assign to the query the species of its nearest neighbor(s) in the database. The superior BLAST+P algorithm identifies the species of a query by forming a multiple sequence alignment (MSA) from the query's 100 top BLAST hits in the database and then assigning the query to the species of the closest MSA neighbor under a sequence distance. The essential improvement in BLAST + P might be restricting the MSA to the query's 100 closest neighbors to improve the alignment quality, but definitive conclusions must await further investigation.

Global alignment (e.g., with Needleman-Wunsch algorithm, or with any MSA algorithm) uses all the information in barcode sequences. By contrast, local alignment programs like BLAST might match only small subsequences within two sequences. Thus, barcode investigations using BLAST run the risk of producing an artifact, particularly if the resulting alignments do not extend over entire sequence lengths, and particularly when concluding the inferiority of intergenic markers. As long as sequence alignments use the entire lengths of sequences, algorithms using p-distance (which requires only base-pair counts) identify species just as well as algorithms using other, more complicated distances (e.g., alignment distance and similarity scores, and evolutionary distances like Kimura 2-Parameter Distance, etc.).

Consensus has converged on "the probability of correct identification" (PCI) as a measurement of species identification accuracy in barcode studies. The overall PCI for a dataset is the average of the species PCIs, taken over all species in the dataset. If a dataset contains some distinguished subsets, the investigator can report PCIs for those subsets.

To calculate a species PCI, remove in turn each representative of the species from the database, and consider its distance (e.g., p-distance) from the remaining representatives. Section 5 gives several possible definitions of successful identification within a species. Some were more stringent than others, because scientific purpose makes different definitions of "successful assignment" appropriate to different circumstances. Singletons with unique sequences provide a useful ancillary statistic in barcode investigations, but they bear little on the correct identification of the corresponding species in a realistic barcode database. Referees should discourage the optimistic conflation of singleton uniqueness with perfect identification of a heavily sampled species.

The evaluation of identification accuracy should also assess PCR failure rates. Because the rates may diminish rapidly with technological advances, current practice distinguishes PCR failure from intrinsic identification accuracy by stating the (present) PCR failure rate and then restricting the barcode database for measuring identification accuracy to samples with successful PCR amplification (Schoch et al. 2012; Hollingsworth et al. 2009).

Finally, a dataset provides a statistical sample exemplifying a class of possible datasets. The PCI estimated from a dataset therefore estimates a true overall PCI, and as such, it has a statistical error. The errors are sometimes surprisingly large, and therefore well worth calculating.

For software relevant to this chapter, see http://tinyurl.com/spouge-barcode.

Acknowledgments This research was supported in part by the Intramural Research Program of the NIH, NLM, NCBI.

References

Albu M, Nikbakht H, Hajibabaei M, Hickey DA (2011) The DNA barcode linker. Mol Ecol Resour 11:84–88

Altschul SF, Madden TL, Schaffer AA, Zhang J, Zhang Z, Miller W, Lipman DJ (1997) Gapped BLAST and PSI-BLAST: a new generation of protein database search programs. Nucleic Acids Res 25:3389–3402

Austerlitz F (2007) Comparing phylogenetic and statistical classification methods for DNA barcoding. In: The second international barcode of life conference, Taipei

Austerlitz F, David O, Schaeffer B, Bleakley K, Olteanu M, Leblois R, Veuille M, Laredo C (2009) DNA barcode analysis: a comparison of phylogenetic and statistical classification methods. BMC Bioinform 10:1

Blaxter M, Mann J, Chapman T, Thomas F, Whitton C, Floyd R, Abebe E (2005) Defining operational taxonomic units using DNA barcode data. Philos Trans R Soc B Biol Sci 360:1935–1943

CBOL Plant Working Group (2009) A DNA barcode for land plants. Proc Natl Acad Sci USA 106:12794–12797

Chase MW, Salamin N, Wilkinson M, Dunwell JM, Kesanakurthi RP, Haidar N, Savolainen V (2005) Land plants and DNA barcodes: short-term and long-term goals. Philos Trans R Soc Lond B Biol Sci 360:1889–1895

Chen J, Zhao JT, Erickson DL, Xia NH, Kress WJ (2015) Testing DNA barcodes in closely related species of Curcuma (Zingiberaceae) from Myanmar and China. Mol Ecol Resour 15:337–348

Collins RA (2012) Barcoding's next top model: an evaluation of nucleotide substitution models for specimen identification. Methods Ecol Evol 3:457–465

Cowan RS, Chase MW, Kress JW, Savolainen V (2006) 300,000 species to identify: problems, progress, and prospects in DNA barcoding of land plants. Taxon 55:611–616

DasGupa B, Konwar KM, Mandoiu II, Shvartsman AA (2005) DNA-BAR: distinguisher selection for DNA barcoding. Bioinformatics 21:3424–3426

Eddy SR (1995) Multiple alignment using hidden Markov models. Proc Int Conf Intell Syst Mol Biol 3:114–120

Edgar RC (2004) MUSCLE: a multiple sequence alignment method with reduced time and space complexity. BMC Bioinform 5:113

Efron B, Stein C (1981) The Jackknife estimate of variance. Ann Stat 9:586–596

Erickson DL, Spouge JL, Resch A, Weight LA, Kress JW (2008) DNA barcoding in land plants: developing standards to quantify and maximize success. Taxon 13:1304–1316

Farris JS (1972) Estimating phylogenetic trees from distance matrices. Am Nat 106:645–668

Felsenstein J (1981) Evolutionary trees from DNA sequences: a maximum likelihood approach. J Mol Evol 17:368–376

Felsenstein J (1988) Phylogenies from molecular sequences—inference and reliability. Annu Rev Genet 22:521–565

Ferguson JWH (2002) On the use of genetic divergence for identifying species. Biol J Linn Soc 75:509–516

Ferri G, Corradini B, Ferrari F, Santunione AL, Palazzoli F, Alu M (2015) Forensic botany II, DNA barcode for land plants: which markers after the international agreement? Forensic Sci Int Genet 15:131–136

Floyd R, Abebe E, Papert A, Blaxter M (2002) Molecular barcodes for soil nematode identification. Mol Ecol 11:839–850

Fregin S, Haase M, Olsson U, Alstrom P (2012) Pitfalls in comparisons of genetic distances: a case study of the avian family Acrocephalidae. Mol Phylogenet Evol 62:319–328

Hajibabaei M, Janzen DH, Burns JM, Hallwachs W, Hebert PDN (2006) DNA barcodes distinguish species of tropical Lepidoptera. Proc Natl Acad Sci USA 103:968–971

Hebert PD, Cywinska A, Ball SL, deWaard JR (2003a) Biological identifications through DNA barcodes. Proc Biol Sci 270:313–321

Hebert PD, Ratnasingham S, deWaard JR (2003b) Barcoding animal life: cytochrome c oxidase subunit 1 divergences among closely related species. Proc Biol Sci 270(Suppl 1):S96–S99

Hogg ID, Hebert PDN (2004) Biological identification of springtails (Hexapoda: Collembola) from the Canadian Arctic, using mitochondrial DNA barcodes. Can J Zool (Revue Canadienne De Zoologie) 82:749–754

Hollingsworth PM, Forrest LL, Spouge JL, Hajibabaei M, Ratnasingham S, van der Bank M, Chase MW, Cowan RS, Erickson DL, Fazekas AJ, Graham SW, James KE, Kim K-J, Kress WJ, Schneider H, van AlphenStahl J, Barrett SCH, van den Berg C, Bogarin D, Burgessk KS, Cameron KM, Carine M, Chacón J, Clark A, Clarkson JJ, Conrad F, Devey DS, Ford CS, Hedderson TAJ, Hollingsworth ML, Husband BC, Kellya LJ, Kesanakurti PR, Kim JS, Kim Y-D, Lahaye R, Lee H-L, Long DG, Madriñán S, Maurin O, Meusnier I, Newmaster SG, Park C-W, Percy DM, Petersen G, Richardson JE, Salazar GA, Savolainene V, Seberg O, Wilkinson MJ, Yi D-K, Little DP (2009) A DNA barcode for land plants. Proc Natl Acad Sci USA 106:12794–12797

Huang D, Meier R, Todd PA, Chou LM (2008) Slow mitochondrial COI sequence evolution at the base of the metazoan tree and its implications for DNA barcoding. J Mol Evol 66:167–174

Katoh K, Misawa K, Kuma K, Miyata T (2002) MAFFT: a novel method for rapid multiple sequence alignment based on fast Fourier transform. Nucl Acids Res 30:3059–3066

Kress WJ, Erickson DL (2008) DNA barcodes: genes, genomics, and bioinformatics. Proc Natl Acad Sci USA 105:2761–2762

Kress WJ, Wurdack KJ, Zimmer EA, Weigt LA, Janzen DH (2005) Use of DNA barcodes to identify flowering plants. Proc Natl Acad Sci USA 102:8369–8374

Kuksa P, Pavlovic V (2009) Efficient alignment-free DNA barcode analytics. BMC Bioinform 10: S9

Kwong S, Srivathsan A, Vaidya G, Meier R (2012) Is the COI barcoding gene involved in speciation through intergenomic conflict? Mol Phylogenet Evol 62:1009–1012

Lambert DM, Baker A, Huynen L, Haddrath O, Hebert PDN, Millar CD (2005) Is a large-scale DNA-based inventory of ancient life possible? J Hered 96:279–284

Little DP (2011) DNA barcode sequence identification incorporating taxonomic hierarchy and within taxon variability. PLoS ONE 6(8):e20552

Little DP, Stevenson DW (2007) A comparison of algorithms for the identification of specimens using DNA barcodes: examples from gymnosperms. Cladistics 23:1–27

Liu C, Liang D, Gao T, Pang X, Song J, Yao H, Han J, Liu Z, Guan X, Jiang K, Li H, Chen S (2011) PTIGS-IdIt, a system for species identification by DNA sequences of the psbA-trnH intergenic spacer region. BMC Bioinform 12:1

Lorenz JG, Jackson WE, Beck JC, Hanner R (2005) The problems and promise of DNA barcodes for species diagnosis of primate biomaterials. Philos Trans R Soc B Biol Sci 360:1869–1877

Meier R, Shiyang K, Vaidya G, Ng PK (2006) DNA barcoding and taxonomy in Diptera: a tale of high intraspecific variability and low identification success. Syst Biol 55:715–728

Meyer CP, Paulay G (2005) DNA barcoding: error rates based on comprehensive sampling. PLoS Biol 3:e422

Munch K, Boomsma W, Huelsenbeck JP, Willerslev E, Nielsen R (2008) Statistical assignment of DNA sequences using Bayesian phylogenetics. Syst Biol 57:750–757

Needleman SB, Wunsch CD (1970) A general method applicable to the search for similarities in the amino acid sequence of two proteins. J Mol Biol 48:443–453

Pang X, Liu C, Shi L, Liu R, Liang D, Li H, Cherny SS, Chen S (2012) Utility of the trnH-psbA intergenic spacer region and its combinations as plant DNA barcodes: a meta-analysis. PLoS ONE 7:e48833

Ratnasingham S, Hebert PD (2007) BOLD: the barcode of life data system, Mol Ecol Notes 7:355–364. http://www.barcodinglife.org

Saitou N, Nei M (1987) The neighbor-joining method—a new method for reconstructing phylogenetic trees. Mol Biol Evol 4:406–425

Sarkar IN, Planet PJ, Desalle R (2008) CAOS software for use in character-based DNA barcoding. Mol Ecol Resour 8:1256–1259

Saunders GW (2005) Applying DNA barcoding to red macroalgae: a preliminary appraisal holds promise for future applications. Philos Trans R Soc B Biol Sci 360:1879–1888

Schoch CL, Seifert KA, Huhndorf S, Robert V, Spouge JL, Levesque CA, Chen W, Bolchacova E, Voigt K, Crous PW, Miller AN, Wingfield MJ, Aime MC, An KD, Bai FY, Barreto RW, Begerow D, Bergeron MJ, Blackwell M, Boekhout T, Bogale M, Boonyuen N, Burgaz AR, Buyck B, Cai L, Cai Q, Cardinali G, Chaverri P, Coppins BJ, Crespo A, Cubas PP, Cummings C, Damm U, de Beer ZW, de Hoog GS, Del-Prado R, Dentinger B, Dieguez-Uribeondo J, Divakar PK, Douglas B, Duenas M, Duong TA, Eberhardt U, Edwards JE, Elshahed MS, Fliegerova K, Furtado M, Garcia MA, Ge ZW, Griffith GW, Griffiths K, Groenewald JZ, Groenewald M, Grube M, Gryzenhout M, Guo LD, Hagen F, Hambleton S, Hamelin RC, Hansen K, Harrold P, Heller G, Herrera G, Hirayama, K, Hirooka Y, Ho HM, Hoffmann K, Hofstetter V, Hognabba F, Hollingsworth PM, Hong SB, Hosaka K, Houbraken J, Hughes K, Huhtinen S, Hyde KD, James T, Johnson EM, Johnson JE, Johnston PR, Jones EB, Kelly LJ, Kirk PM, Knapp DG, Koljalg U, Kovacs GM, Kurtzman CP, Landvik S, Leavitt SD, Liggenstoffer AS, Liimatainen K, Lombard L, Luangsa-Ard JJ, Lumbsch HT, Maganti H, Maharachchikumbura SS, Martin MP, May TW, McTaggart AR, Methven AS, Meyer W, Moncalvo JM, Mongkolsamrit S, Nagy LG, Nilsson RH, Niskanen T, Nyilasi I, Okada G, Okane I, Olariaga I, Otte J, Papp T, Park D, Petkovits T, Pino-Bodas R, Quaedvlieg W, Raja HA, Redecker D, Rintoul T, Ruibal C, Sarmiento-Ramirez JM, Schmitt I, Schussler A, Shearer C, Sotome K, Stefani FO, Stenroos S, Stielow B, Stockinger H, Suetrong S, Suh SO, Sung GH, Suzuki M, Tanaka K, Tedersoo L, Telleria MT, Tretter E, Untereiner WA, Urbina H, Vagvolgyi C, Vialle A, Vu TD, Walther G, Wang QM, Wang Y, Weir BS, Weiss M, White MM, Xu J, Yahr R, Yang, ZL, Yurkov A, Zamora JC, Zhang N, Zhuang WY, Schindel D, Fungal Barcoding C (2012) Nuclear ribosomal internal transcribed spacer (ITS) region as a universal DNA barcode marker for Fungi. Proc Natl Acad Sci USA 109:6241–6246

Smith MA, Fisher BL, Hebert PDN (2005) DNA barcoding for effective biodiversity assessment of a hyperdiverse arthropod group: the ants of Madagascar. Philos Trans R Soc B Biol Sci 360:1825–1834

Smith MA, Woodley NE, Janzen DH, Hallwachs W, Hebert PDN (2006) DNA barcodes reveal cryptic host-specificity within the presumed polyphagous members of a genus of parasitoid flies (Diptera: Tachinidae). Proc Natl Acad Sci USA 103:3657–3662

Smith TF, Waterman MS (1981) Identification of common molecular subsequences. J Mol Biol 147:195–197

Smith TF, Waterman MS, Fitch WM (1981) Comparative biosequence metrics. J Mol Evol 18: 38–46

Spouge JL, Mariño-Ramírez L (2012) The practical evaluation of DNA barcode efficacy. Methods Mol Biol 858:365–377

Suwannasai N, Martin MP, Phosri C, Sihanonth P, Whalley AJS, Spouge JL (2013) Fungi in Thailand: a case study of the efficacy of an ITS barcode for automatically identifying species within the Annulohypoxylon and Hypoxylon Genera. PLoS ONE 8:e54529

Weitschek E, Fiscon G, Felici G (2014) Supervised DNA barcodes species classification: analysis, comparisons and results. Biodata Min 7:1

Weitschek E, Van Velzen R, Felici G, Bertolazzi P (2013) BLOG 2.0: a software system for character-based species classification with DNA barcode sequences. What it does, how to use it. Mol Ecol Resour 13:1043–1046

Wilson EB (1927) Probable inference, the law of succession, and statistical inference. J Am Stat Assoc 22:209–212

Wouters MA, Husain A (2001) Changes in zinc ligation promote remodeling of the active site in the zinc hydrolase superfamily. J Mol Biol 314:1191–1207

Zalasiewicz J et al (2000) Are we now living in the Anthropocene? GSA Today 18:4–8

DNA Barcoding: A Tool to Assess and Conserve Marine Biodiversity

Sudakshina Ghosh, Biswabandhu Bankura and Madhusudan Das

Abstract Accurate Species diagnosis is the key element for biodiversity studies and conservation planning. Conventionally, morphological characters are used to identify a species. But, this approach needs a thorough expertise in identifying the external features which often leads to narrowing down of specialization with regard to ascertaining a species within a limited group of taxa. The approach may be particularly valuable for species identification of organisms that are rare, fragile, and/or small, especially when morphological identification is problematic and errors are likely due to simple or evolutionarily conserved body plans. However, each time a new technique has been introduced in science it was accompanied by some debate and distress, and DNA barcoding was no exception. Therefore, more collaborative efforts are needed to explore the potentialities of DNA barcoding in proper species identification across all taxa. At the same time, we need to set a threshold of the genetic variation in species delimitation to find out the cryptic species. It is also an important point to know that the benefits of DNA barcoding are not restricted to taxonomic or systematic research only. The discovery of high-throughput sequencing technologies are going to change the dimension of these techniques in the years to come.

1 Introduction

Accurate Species diagnosis is the key element for biodiversity studies and conservation planning. Conventionally, morphological characters are used to identify a species. However, this approach needs a thorough expertise in identifying the external features which often leads to narrowing down of specialization with regard

S. Ghosh
Department of Zoology, Sister Nibedita Government General Degree College for Girls, Hastings House, 20B, Judges Court Road, Alipore, Kolkata 700027, West Bengal, India

B. Bankura · M. Das (✉)
Department of Zoology, University of Calcutta, 35 Ballygunge Circular Road, Kolkata 700019, West Bengal, India
e-mail: madhuzoo@yahoo.com

S. Trivedi et al. (eds.), *DNA Barcoding in Marine Perspectives*,
DOI 10.1007/978-3-319-41840-7_3

to ascertaining a species within a limited group of taxa. Thus, the regular job of identity establishment of a species not only becomes challenging and laborious but requires in-depth trained professionals. This is more so for larval stages, juveniles and females which are often excluded while describing a species. So assigning a species name to a specimen becomes even more puzzling. Many a times, variation in morphology and phenotypic plasticity may add to the complications in species determination. Also, declines of taxonomists have been observed who could identify and characterize species of many taxa. For evolutionary biologists and ecologists, the main aim is to know how species originate and the factors which make inconsistency in species richness across different biomes of the globe. Sometimes, the entire species diversity of a given region remains unknown, especially for the biodiverse habitats (Purvis and Hector 2000). In these cases, DNA barcode has proved to be an important tool for the detection of cryptic and unidentified species (Smith et al. 2008a). Both ecological field observations as well as DNA barcode markers are essential for finding out new species of insects (Shashank et al. 2014; Bertrand et al. 2014). For instance, DNA barcode sequences identified that the common skipper butterfly, found in Central America, which was initially considered to be a single species as defined by morphological features of the adults actually comprised of numerous species in correspondence to diets and features of the larvae (Hebert et al. 2004a). Likewise, this same technique has been used for cryptic species of hispine beetles (Garcia et al. 2013). Widespread DNA barcode study on Microgastrinae wasps also showed significant understanding of their taxonomy and species discovery (Smith et al. 2012). This technique has also successfully established cryptic diversity among crustaceans, diatoms and fishes (Silva et al. 2014; Hamsher and Saunders 2014; Winterbottom et al. 2014). Thus its use for the finding of new species is developing as a potent tool to clarify the boundaries of species and quantification of their diversity throughout all life stages (Shank et al. 2006; Costion et al. 2011; Bracken-Grissom et al. 2012; Torres et al. 2014).

However, using the data of DNA barcode sequence, reconstruction of a phylogenetic tree to put forward the evolutionary relations among both highly divergent as well as closely related species is a major task. Sometimes, the reconstruction of a phylogenetic tree becomes difficult as short stretches of genetic markers are used as a tool of DNA barcoding. So these community phylogenies can be regarded as a small part of a Tree of Life project, which seeks to gather and rebuild species relations. It has been shown that the problem in having limited nucleotide content can be overcome if species density is high (Smith et al. 2009). Thus as more DNA barcode data are grouped, an increased number of taxa will be woven into increasingly larger phylogenies which eventually shall lead to a well-supported phylogenetic reestablishments of the constituent species from which individual community phylogenies may be cut out for targeted exploration (Kress et al. 2015).

2 DNA Barcoding in Diverse Fields of Study

Barcoding can detect any species from bits of samples. Thus once established, this technique will be able to recognize any unwanted substance in foodstuffs thereby preventing protected species being incorporated and served in commercial goods. By the use of this system, one can discriminate the look alikes. This unmasking can help uncover dangerous organisms camouflaged as harmless ones thereby allowing a more accurate view of biodiversity. The universal barcode of life is provided by ATGC, the four distinct nucleotides that cuts down the ambiguity in identifying an organism. Thus if a standardized library of barcodes can be established, this will enable many more individuals to identify and call by name the marine species they observe. It will ease the identification of not only the abundant and rare species but also the native and invasive ones so that marine biodiversity can get attention at local as well as global scale. Barcoding is therefore such a tool which links biological identification to progressing frontiers of DNA sequencing. Assimilating those links will lead to portable desktop devices and finally to handheld barcoders. This information eventually will enrich the genetic details of marine metazoan species and aid in drawing the tree of life on Earth. Now-a-days it is becoming a regular approach to use DNA barcodes to easily associate diverse life-history stages with different developmental morphologies, such as eggs, larvae, and adults in insects (Pinzon-Navarro et al. 2010).

3 DNA Barcoding in Marine Biodiversity

Majority of our planet being covered by water bodies, the evaluation of marine biodiversity becomes a tough job. The vast marine biome is the home for a large number of plant and animal species. It has already been reported that out of 35 animal phyla, 34 of them have marine representatives while 14 comprise of exclusively marine animals (Briggs 1994; Gray 1997). The vast oceans of the world consist of more than 230,000 known species which represent around 31 metazoan phyla. Possibly another large number of species are yet to be discovered. However, if the rate of extinction of species exceeds that of their discovery, then that becomes a huge cause of concern. This is particularly true in case of diverse marine endangered habitats such as coral reefs (Bucklin et al. 2011). Cryptic species, that are morphologically similar but genetically distinct, are commonly encountered in marine ecosystems. Another difficulty that persists in the marine and estuarine habitat is connecting the larval stages with the respective adult forms by considering only morphology. Both these cases can be effectively addressed by DNA barcoding. Again, invasive species often becomes a threat to the marine biodiversity. This issue can also be resolved globally through DNA barcoding (Molnar et al. 2008). Thus this technique is very much effective for molecular phylogenetic studies, geographical distribution and marine biodiversity conservation (Trivedi et al. 2016b). Many a times, a variety of factors such as biological mechanisms, water dynamics, or historical events become the root of genetic structuring of populations in marine species

(Barbar et al. 2000; Nelson et al. 2000). Recent advancements in deep sea research revealed the existence of extraordinary pharmaceutical properties in a number of deep sea organisms, DNA barcoding of which has gained much attention in the global platform. As such, Census of the Diversity of Abyssal Marine Life (CeDAMar) is dedicated to the barcoding of these deep sea organisms.

4 Biodiversity of Sundarbans Mangroves

Sunderbans, the single largest block of tidal halophytic mangrove forest lying on the delta formed by the rivers Ganges, Brahmaputra and Meghna are listed in the UNESCO world heritage list (http://whc.unesco.org/en/list). They are considered as the world's largest natural nursery. Around 250 marine and estuarine species of fin-fish and shell-fish that are ecologically and commercially important use this place as breeding ground while the juveniles exploit its rich nutrient resources (Trivedi et al.1994). For the culture of tiger prawn (*Penaeus monodon),* the congenial salinity of aquatic sub-subsystem of coastal West Bengal, India is the ideal. However, there is no shrimp hatchery for tiger prawn in and around the Sundarbans mangrove ecosystem due to low salinity. In absence of hatchery, the demand for tiger prawn seeds has become very acute and farmers depend on wild harvest of tiger prawn seeds from estuarine and coastal waters. While doing this, larva and juveniles of non-target species that are caught in the tiny mesh sized nets are thrown away and wasted. This exercise results in great loss of pelagic biodiversity, which might lead to an ecological imbalance (Rao 2004). An earlier report revealed that about 62 species of fin-fish and 22 species of shell-fish that thrive in diverse salinity are wasted during the netting process (Mitra and Banerjee 2005). As they lack a distinct morphology, so it becomes very tough for taxonomists to identify the larva and juveniles. Loss in the demersal fishery sector can be quantified if the taxonomic diversity is confirmed. Hence, it is of utmost importance to identify the juveniles using DNA barcoding, the intent of which is to use large scale screening of reference genes in order to assign unknown individuals to species and to enhance the discovery of new species. Outcome of the study will help to assess the economic loss in the demersal fishery sector and build up a distribution map of larva/juveniles in terms of salinity in and around the Sundarban delta. This will eventually lead to conservation of fin-fish and shell-fish stock at the apex of Bay of Bengal adjacent to Sundarbans.

Mangroves at the juncture of terrestrial, estuarine and near shore marine ecosystem are of enormous ecological and economic significance. It has been reported that the mangrove ecosystem services worldwide are worth at least US$1.6 billion annually (Field et al. 1998; Costanza et al. 1997). But this dynamic and unique ecosystem is getting threatened and depleted. So the conservation of mangroves needs immediate attention. Loss of evolutionary unique species in the mangrove ecosystem has been reported and DNA barcoding provided phylogenetic evidence for creating unified mangrove management plan worldwide (Daru et al. 2013).

5 DNA Barcoding of Marine Microbes

Microbial biodiversity valuation has always been a challenging task. Proper identification of microbes is of utmost importance for preventing the spread of diseases caused by them. DNA barcoding of marine environmental samples has put forward the huge dinoflagellate diversity which serves as primary producers (Stern et al. 2010).

6 DNA Barcoding of Seagrasses, Mangroves and Marine Phytoplanktons

Seagrasses, the submerged flowering plants have much ecological influence on the coastal environment owing to their nutrient recycling ability and high primary productivity. Also, they are known to contain secondary compounds of medicinal value. For example, rosmarinic acid (antioxidant) and zosteric acid (antifouling agent) are obtained from seagrasses. In spite of the diverse distribution of these marine plants worldwide yet, a rapid decline of this species has been reported globally since 1980 (Waycott et al. 2009). Hence, there is a dire need for their conservation. Their identification becomes difficult because of the unavailability of the flower as a distinct morphological trait. In such a case, DNA barcoding can aid in establishing their identity. Several markers have been used for identification of seagrasses such as nuclear ITS for Halophila (Waycott et al. 2002), trnK introns and rbcL for Zostera (Les et al. 2002), ITS1, 5.8S rDNA and ITS2 for Halophila (Uchimura et al. 2008). By using rbcL and matK sequences it was found that DNA barcoding for seagrasses could be developed (Lucas et al. 2012).

7 DNA Barcoding of Marine Algae

Morphological characters are insufficient to differentiate the different species of red marine macro algae. For this, mitochondrial COI gene and UPA (Universal Plastid Amplicon) domain V of the 23S rRNA gene were used as molecular markers for their identification. Results showed that COI was a more sensitive marker to identify a new species Euthoratimburtonii (Clarkston and Saunders 2010). However, owing to lack of universal primers, COI is not effective for the identification of all species. Again, UPA in spite of having universal primers had limitations among closely related species (Xiaobo et al. 2013). Gracilariaceae, a commercially important red algal family, is widely used in biotechnology and microbiology research. Aphycocolloid agar being difficult to identify morphologically, DNA barcoding is helpful in the identification of Gracilaria at the species level (Kim et al. 2010). Thus barcoding can be a suitable technique for keeping a watch on marine and coastal ecosystems for detecting species causing Harmful Algal Bloom (HAB).

8 DNA Barcoding of Marine Zooplanktons

DNA barcoding of zooplanktons provides a solid platform for better understanding of marine ecology. The Census for Marine Zooplanktons (CMarZ) having its DNA Barcoding Centers located in UConn (USA), Bremer-haven (Germany), ORI (Japan), Qingdao (China) and Goa (India) is dedicated to the study of global zooplankton communities.

9 Sponges

The lack of distinct morphological features and high degree of homoplasy have mystified classification within the phylum Porifera (sponges), making DNA barcoding particularly useful in this group. However, the resolving power of COI in sponges differs greatly from their higher metazoan counterparts (Cardenas et al. 2009; Heim et al. 2007; Park et al. 2007; Worheide 2006). The divergence rates of COI are reported to be 10–20 times lower for sponges compared to Bilateria (Shearer et al. 2002; Worheide 2006), and the barcode region typically does not distinguish between species. Several other barcode regions, such as the I3-M11 region of COI (Erpenbeck et al. 2006) and ITS (Park et al. 2007; Poppe et al. 2010) have also been used in other studies.

10 Cnidarians

Very low levels of COI polymorphism in the barcode region of cnidarian class Anthozoa has restriced its usefulness for species discrimination (Shearer et al. 2002). This may be due to the very slow nucleotide substitution rates for mitochondrial genes (Hellberg 2006). Shearer and Coffroth (2008) suggested that the lack of COI divergence among many species indicates the need for different other markers. However, reports showed that COI successfully discriminated four species of gorgonian corals in the Mediterranean region (Calderon et al. 2006). The same marker also differentiated the species of the hexacorallian order Zoantharia (Sinniger et al. 2008). It was observed that genetic distances widely varied among species, genera, and families (Sinniger et al. 2010), which is suggestive of COI to be useful for zoanthid classification. Again, COI evolution in Hydrozoans and Scyphozoans is more typical of other metazoans (Huang et al. 2008). Barcoding based on COI found the presence of several cryptic species among hydrozoans

(Folino-Rorem et al. 2009; Moura et al. 2008). Along with other genes, the species and systematic relationships of Scyphozoa and hydrozoan family Campanulariidae was also resolved using the same marker (Govindarajan et al. 2006).

11 Barnacles

Tsang et al. reported the presence of cryptic species within the acorn barnacle *Chthamalus* (Tsang et al. 2008) and the symbiotic coral barnacle *Wanella* (Tsang et al. 2009) through their barcoding in the western Pacific. The same technique also put forward the fact that genetic divergence was lacking between the species of *Tetraclita* (Tsang et al. 2007).

12 Copepods

To confirm the status of copepod sibling species swarms, many barcoding studies have sought to re-examine morphologically based descriptions. The sibling species of calanoid copepods, *Calanus* (Hill et al. 2001; Unal et al. 2006), *Clausocalanus* (Bucklin and Frost 2009), *Neocalanus* (Machida et al. 2006), and *Pseudocalanus* (Bucklin et al. 2003) have been determined and confirmed by COI barcode region. Bottger-Schnack and Machida examined the cyclopoid copepods (species of Oncaeidae—comprising of more than 100 species and many sibling species) for genetic uniqueness of COI and other genes (Bottger-Schnack and Machida 2010). The copepods although being thought of as taxonomically well known, comprehensive analyses using DNA barcodes revealed their hidden diversity and cryptic speciation across the suborder.

13 Stomatopods

To understand the larval diversity of stomatopods, DNA barcoding has been of much use (Feller and Cronin 2016). This technique has been used by Tang et al. (2010) to recognize stomatopods (mantis shrimp) from the South China Sea. They were successful in linking the unknown larvae to known adult species from that region. In comparison, Barber and Boyce (2006) used barcodes to discriminate larval stomatopods in the Coral Triangle and the Red Sea; however only a fistful could be successfully matched to known adults.

14 Mysids

In a study by Remerie et al. (2006), a marked divergence has been found between Atlantic and Mediterranean populations of a complex of cryptic species within the mysid *Mesopod opsisslabberi*.

15 Amphipods

Amphipoda, one of the most species-rich groups, consists of more than 170 extant families and approximately 10,000 described species till date (Ahyong et al. 2011). For reconstruction of the phylogeny of the genus Amphipod, barcoding may prove to be a suitable technique to ascertain unidentified members inhabiting all parts of the sea, lakes, rivers, sand beaches, and moist habitats on various tropical islands.

16 Euphausiids

Euphausiids are found across all gradients and basins of the World Ocean. They are found over a wide range, from the surface stratum to at least 4000 m depth and from arctic to antarctic waters. In this group, there are 86 known species; although some have geographical variants. They are not only ecologically significant but occasionally may dominate the pelagic communities with regard to abundance and biomass. Barcoding has been of much use for species identification of euphausiids, especially for the larval and juvenile forms. In a study by Bucklin et al. (2007), COI was shown to reliably separate 40 species of euphausiids.

17 Decapods

Adult decapods are clearly identifiable but their larval and juvenile forms are tough to recognize. However, barcoding technique successfully distinguished the species of fiddler crab *Uca*in the North Western Indian Ocean (Shih et al. 2009) and the hermit crab *Clibanarius* of the Ryukyu Archipelago of Japan (Hirose et al. 2010) and also the larvae of *Cancer* crabs in the South Eastern Pacific Ocean (Pardo et al. 2009). On the eastern coastline of Africa, cryptic species of decapod brachyuran mangrove crab *Perisesarma guttatum* were identified using polymorphic DNA sequence of COI (Silva et al. 2010).

18 Molluscs

Barcoding technique has been widely used for marine molluscs which includes chitons (Kelly et al. 2007), gastropods (Hunt et al. 2010; Jennings et al. 2010a; Krug et al. 2007), bivalves (Feng et al. 2011; Lorion et al. 2009; Mikkelsen et al. 2007; Nuryanto et al. 2007; Trivedi et al. 2015), and cephalopods (Allcock et al. 2011; Undheim et al. 2010). By this process, species relationships of the gastropod *Nerita* were shown throughout the South Western tropical Pacific Ocean (Spencer et al. 2007). Again, barcode studies have ruled out the possibility of genetic differentiation among certain species of molluscs. Barcode differences were lacking in the sympatric intertidal limpets (Siphonariidae) off coastal Southeast Africa. This suggested that they are morphotypes of a single species (Teske et al. 2007). Again, 2 clams of the genus *Donax* also showed no noteworthy variation of barcode, thereby representing only one single species (Carstensen et al. 2009). In spite of the barcode data being sparse and taxonomic coverage inclined towards shallow water species, Puillandre et al. (2009) successfully identified gastropod larvae by this method.

19 Annelids

Polychaetes comprise the vast majority of marine annelids with over 12,000 known species. Hardy et al. (2011) carried on a comprehensive barcoding study of polychaetes in the Alaskan and Canadian Arctic which revealed that one-fourth of this species represented two or more discrete genetic lineages. In another study, presence of cryptic species within the cosmopolitan fireworm *Eurythoe complanata* (Barroso et al. 2010) was observed. Likewise, Rice et al. (2008) also found a cryptic species complex within *Polydora cornuta* along the Atlantic and Pacific coasts of North America; by the use of COI, closely related sympatric species of the lugworm *Arenicola* (Luttikhuizen and Dekker 2010) and the oligochaete *Tubificoides* (Erseus and Kvist 2007) were segregated in northeast Atlantic and Scandinavian waters.

20 Echinoderms

One hundred and ninety one species of echinoderms, representing all five classes, were discriminated using DNA barcode (Ward et al. 2008). As a marker, COI has been used in barcoding studies to look at species limits and find out cryptic species of asteroids(Naughton and O'Hara 2009), holothuroids (Uthicke et al. 2010), and crinoids (Helgen and Rouse 2006; Wilson et al. 2007). New procedures and primers for COI sequencing have been produced by synchronized efforts to accelerate barcoding of individuals of this phylum (Hoareau and Boissin 2010).

21 Nematodes

DNA barcoding is a very crucial method for the identification of nematodes as the morphological distinction to describe species is often not possible due to their abundance, small size, and lack of expertise in taxonomy. Nematode diversity in marine sediments has not been explored much; it has been put forward by surveys that diversity of marine nematodes may go beyond 1,000,000 species (Lambshead et al. 2001). As there were complications in developing COI primers, the small-subunit (18S) rRNA gene was used for species identification and diversity (Bhadury et al. 2006; Bhadury and Austen 2010; Floyd et al. 2005). However, some studies have already identified the cryptic diversity of nematodes by the use of COI variation (Derycke et al. 2010; Sanna et al. 2009).

22 Bryozoa

In cosmopolitan bryozoan *Celleporella*, although the phenotypic change was insignificant, yet Gomez-Campo (2007b) found close correspondence between barcode lineages and reproductive isolation of taxa; their divergence into four major clades was also noted (Gomez et al. 2007a).

23 Pycnogonids

From the Antarctic and sub-Antarctic collections, cryptic lineages within the pycnogonid *Colossendeis megalonyx* were found through COI investigation (Krabbe et al. 2010).

24 Chaetognaths

Although subtle morphological differences divide chaetognath species across wide geography, their correct identification is possible using the COI barcode region (Jennings et al. 2010b). Here also, cryptic species was observed by Miyamoto et al. (2010) within the deepsea chaetognath *Caecosagitta macrocephala*. Thus barcoding technology has proved to be an important tool for better understanding of the classification and evolutionary relationships of this phylum.

25 Fishes

In 2005, following barcoding success with fishes (Savolainen et al. 2005; Ward et al. 2005) the global initiative Fish Barcode of Life (FISH-BOL; http://www.fishbol.org) was established. FISH-BOL gathers DNA barcode records for all 30,000 known species of fishes which are used to address commercial fisheries related issues. It also promises to be a potent tool for extending the concept of the natural history and ecological interactions of fish species (Ward et al. 2009). This technique is useful for identifying not only whole fish, fillets, fins, fragments and larvae, but even eggs. Shark fins seized from illegal fishers in northern Australian waters were identified by this process (Holmes et al. 2009). Fish larvae identification by COI barcoding has been quite successful across different parts of the globe as reported from the Great Barrier Reef (Pegg et al. 2006), Caribbean (Victor 2007), and Pacific (Hubert et al. 2010; Paine et al. 2008). A study of fish species done to connect South African and Australian waters revealed that nearly one-third of those species represented two taxa (Zemlak et al. 2009). Likewise, COI barcoding validated a number of new fish species, that includes a goby (Victor 2008), sting ray, Antarctic ray *Bathyraja* (Smith et al. 2008b), handfish *Brachionichthys australis* (Last et al. 2007) and 5 new species of damselfish *Chromis* (Pyle et al. 2008). However, findings from larger-scale barcoding studies of fishes are suggestive of the fact that around 2 % species are overlooked based on the distance metric as proposed by Hebert et al. (2004b). As such, further studies of any barcode that deviates from its putative species by ten times the average divergence within that group should be targeted.

26 Reptiles

The biodiversity hotspot of Madagascar was the first place where large scale DNA barcoding of reptiles (comprising Squamata and Testudines) were conducted with 468 specimens. This study identified 41–48 new species. The technique of DNA barcodes can be useful for conservation and wildlife forensics of sea turtles and other endangered species by identifying their meat and eggs that are traded illegally or the carcasses stranded on beaches (Vargas et al. 2009; Trivedi et al. 2016a). The sea turtles are a group of ancient individuals having a slow mutation rate. In spite of that, all species belonging to this group were successfully recognised and on the basis of genetic distances and character-based methods, no cryptic species were found (Naro-Maciel et al. 2010).

27 Birds

The All Birds Barcoding Initiative (ABBI) was launched in September 2005. Its aim was to barcode around 10,000 known species of birds. Out of 17 sets of species of North American seabirds having overlapping barcodes, 8 species of large white-headed gulls were found by Kerr et al. (2007). Owing to secondary contact and hybridization, these well-formed species are losing their genetic identity. In another study, DNA barcoding analysis was carried on with 387 individuals of 147 species of birds (including seabirds) from the Netherlands (Aliabadian et al. 2013). However given the rapid depletion of mangrove forests, acidification of oceans, increase in salinity and sea level elevation, much more efforts are required for DNA barcoding of sea birds with a view of their conservation.

28 Mammals

Till date, there is a lack of a complete library of DNA barcodes for marine mammals. However, a new species of beaked whales (Ziphiidae), *Mesoplodon perrini*, was discovered from the north Pacific using the barcode technique (Dalebout et al. 2002); *M. traversii* (van Helden et al. 2002), an almost lost species in the Southern Hemisphere was restored in the same way. Likewise, the identity establishment of the tropical bottlenose whale, *Indopacetus pacificus* (Dalebout et al. 2003) was also confirmed.

29 Conclusion

DNA barcoding is an efficient approach of species identification that uses information within a single gene region common across all taxa. The approach may be particularly valuable for species identity establishment of rare, fragile, and/or small organisms, especially when morphological detection is problematic and errors are likely to crop up due to simple or evolutionarily conserved body plans. Whenever a new technique has been introduced in science it was accompanied by some debate and distress, and DNA barcoding was no exception. Therefore, more collaborative efforts are needed to explore the potentialities of DNA barcoding in proper species identification across all taxa. We need to set a threshold of the genetic variation in species delimitation to find out the cryptic species as well. It is also an important point to know that the benefits of DNA barcoding are not restricted to taxonomic or systematic research only. The discovery of high-throughput sequencing technologies are going to change dimension of these techniques in the years to come.

References

Ahyong ST, Lowry JK, Alonso M, Bamber RN, Boxshall GA, Castro P et al (2011) Subphylum Crustacea Brünnich, 1772. Zootaxa 3148:165–191

Aliabadian M, Beentjes KK, Roselaar CS (Kees), Brandwijk HV, Nijman V, Vonk R (2013) DNA barcoding of Dutch birds. Zookeys 2013(365):25–48

Allcock AL, Barratt I, Eleaume M, Linse K, Norman MD, Smith PJ et al (2011) Cryptic speciation and the circumpolarity debate: a case study on endemic Southern Ocean octopuses using the COI barcode of life. Deep Sea Res Part II Topical Stud Oceanogr 58(1–2):242–249

Barber P, Boyce SL (2006) Estimating diversity of Indo-Pacific coral reef stomatopods through DNA barcoding of stomatopod larvae. Proc R Soc Lond B 273:2053–2061

Barber PH, Palumbi SR, Erdmann MV, Moosa MK (2000) Biogeography: a marine Wallace's line? Nature 406:692–693

Barroso R, Klautau M, Solé-Cava A, Paiva P (2010) *Eurythoe complanata* (Polychaeta: Amphinomidae), the 'cosmopolitan' fireworm, consists of at least three cryptic species. Mar Biol 157:69–80

Bertrand C, JanzenD H, Hallwachs W, Burns JM, Gibson JF, Shokralla S, Hajibabaei M (2014) Mitochondrial and nuclear phylogenetic analysis with Sanger and next–generation sequencing shows that, in Area de Conservacion Guanacaste, northwestern Costa Rica, the skipper butterfly named Urbanus belli (family Hesperiidae) comprises three morphologically cryptic species. BMC Evol Biol 14:153

Bhadury P, Austen MC (2010) Barcoding marine nematodes: an improved set of nematode 18S rRNA primers to overcome eukaryotic cointerference. Hydrobiologia 641:245–251

Bhadury P, Austen MC, Bilton DT, Lambshead PJD, Rogers AD, Smerdon GR (2006) Development and evaluation of a DNA-barcoding approach for the rapid identification of nematodes. Mar Ecol Prog Ser 320:1–9

Bottger-Schnack R, Machida R (2010) Comparison of morphological and molecular traits for species identification and taxonomic grouping of oncaeid copepods. Hydrobiologia 2010:1–15

Bracken-Grissom HD, Felder DL, Vollmer NL, Martin JW, Crandall KA (2012) Phylogenetics links monster larvae to deep-sea shrimp. Ecol Evol 2:2367–2373

Briggs JC (1994) Species-diversity—land and sea compared. Syst Biol 43:130–135

Bucklin A, Frost BW (2009) Morphological and molecular phylogenetic analysis of evolutionary lineages within Clausocalanus (Copepoda: Calanoida). J Crust Biol 29:111–120

Bucklin A, Frost BW, Bradford-Grieve J, Allen LD, Copley NJ (2003) Molecular systematic and phylogenetic assesment of 34 calanoid copepod species of the Calanidae and Clausocalanidae. Mar Biol 142:333–343

Bucklin A, Wiebe PH, Smolenack SB, Copley NJ, Beaudet JG, Bonner KG et al (2007) DNA barcodes for species identification of euphausiids (Euphausiacea, Crustacea). J Plankton Res 29:483–493

Bucklin A, Steinke D, Blanco-Bercial L (2011) DNA barcoding of marine metazoa. Ann Rev Mar Sci 3:471–508

Calderon I, Garrabou J, Aurelle D (2006) Evaluation of the utility of COI and ITS markers as tools for population genetic studies of temperate gorgonians. J Exp Mar Biol Ecol 336:184–197

Cardenas P, Menegola C, Rapp HT, Diaz MC (2009) Morphological description and DNA barcodes of shallow-water Tetractinellida (Porifera: Demospongiae) from Bocas del Toro, Panama, with description of a new species. Zootaxa 2009:1–39

Carstensen D, Laudien J, Leese F, Arntz W, Held C (2009) Genetic variability, shell and sperm morphology suggest that the surf clams *Donax marincovichi* and *D. obesulus* are one species. J Mollus Stud 75:381–390

Clarkston BE, Saunders GW (2010) A comparison of two DNA barcode markers for species discrimination in the red algal family Kallymeniaceae (Gigartinales, Florideophyceae), with a description of Euthoratimburtonii sp. Botany 88:119–131

Costanza R, d'Arge R, de Groot R, Farberparallel S, Grasso M, Hannon B et al (1997) The value of the world's ecosystem services and natural capital. Nature 387:253–260

Costion C, Ford A, Cross H, Crayn D, Harrington M, Lowe A (2011) Plant DNA barcodes can accurately estimate species richness in poorly known floras. PLoS ONE 6(11):e26841

Dalebout ML, Mead JG, Baker CS, Baker AN, Helden AL (2002) A new species of beaked whale Mesoplodon perrini sp. n. (Cetacea: Ziphiidae) discovered through phylogenetic analyses of mitochondrial DNA sequences. Mar Mamm Sci 18:577–608

Dalebout ML, Baker CS, Anderson RC, Best PB, Cockcroft VG, Hinsz HL et al (2003) Appearance, distribution, and genetic distinctiveness of longman's beaked whale, *Indopacetus pacificus*. Mar Mamm Sci 19:421–461

Daru BH, Yessoufou K, Mankga LT, Davies TJ (2013) A global trend towards the loss of evolutionarily unique species in mangrove ecosystems. PLoS ONE 8(6):e66686

Derycke S, Ley PD, Ley ITD, Holovachov O, Rigaux A, Moens T (2010) Linking DNA sequences to morphology: cryptic diversity and population genetic structure in the marine nematode *Thoracostoma trachygaster* (Nematoda, Leptosomatidae). Zool Scr 39(3):276–289

Erpenbeck D, Hooper JNA, Worheide G (2006) CO1 phylogenies in diploblasts and the 'Barcoding of Life'—Are we sequencing a suboptimal partition? Mol Ecol Notes 6:550–553

Erseus C, Kvist S (2007) COI variation in Scandinavian marine species of *Tubificoides* (Annelida: Clitellata: Tubificidae). J Mar Biol Assoc UK 87:1121–1126

Feng Y, Li Q, Kong L, Zheng X (2011) DNA barcoding and phylogenetic analysis of Pectinidae (Mollusca: Bivalvia) based on mitochondrial COI and 16S rRNA genes. Mol Biol Rep 38 (1):291–299

Field CB, Osborn JG, Hoffman LL, Polsenberg JF, Ackerly DD, Berry JA et al (1998) Mangrove biodiversity and ecosystem function. Global Ecol Biogeog Let 7:3–14

Floyd RM, Rogers AD, Lambshead PJD, Smith CR (2005) Nematode-specific PCR primers for the 18S small subunit rRNA gene. Mol Ecol Notes 5:611–612

Folino-Rorem N, Darling J, D'Ausilio C (2009) Genetic analysis reveals multiple cryptic invasive species of the hydrozoan genus Cordylophora. Biol Invasions 11:1869–1882

García R, Erin KK, Charles LS, John KW and Terry LE (2013) Using a comprehensive DNA barcode library to detect novel egg and larval host plant associations in a Cephaloleia rolled-leaf beetle (Coleoptera: Chrysomelidae). Biol J Linn Soc 110:189–198

Gomez A, Hughes RN, Wright PJ, Carvalho GR, Lunt DH (2007a) Mitochondrial DNA phylogeography and mating compatibility reveal marked genetic structuring and speciation in the NE Atlantic bryozoan *Celleporella hyalina*. Mol Ecol 16:2173–88

Gomez-Campo C (2007b) Assessing the contribution of genebanks: the case of the UPM seed bank in Madrid. Plant Genet Resour Newslett 151:40–49

Govindarajan AF, Boero F, Halanych KM (2006) Phylogenetic analysis with multiple markers indicates repeated loss of the adult medusa stage in Campanulariidae (Hydrozoa, Cnidaria). Mol Phylogen Evol 38:820–834

Gray JS (1997) Marine biodiversity: patterns, threats and conservationneeds. Biodiv Conserv 6:153–175

Hamsher SE, Saunders GW (2014) A floristic survey of marine tube-forming diatoms reveals unexpected diversity and extensive co-habitation among genetic lines of the Berkeleya rutilans complex (Bacillariophyceae). Eur J Phycol 49(1):47–59

Hardy MS, Carr CM, Hardman M, Steinke D, Corstorphine E, Mah C (2011) Biodiversity and phylogeography of Arctic marine fauna: insights from molecular tools. Mar Biodivers 41 (1):195–210

Hebert PDN, Penton EH, Burns JM, Janzen DH, Hallwachs W (2004a) Ten species in one: DNA barcoding reveals cryptic species in the neotropical skipper butterfly *Astraptes fulgerator*. Proc Natl Acad Sci USA 101:14812–14817

Hebert PDN, Stoeckle MY, Zemlak TS, Francis CM (2004b) Identification of birds through DNA barcodes. PLoS Biol 2:1657–1663

Heim I, Nickel M, Brummer F (2007) Molecular markers for species discrimination in poriferans: a case study on species of the genus Aplysina. In: Custódio MR, Lôbo-Hajdu G, Hajdu E, Muricy G (eds) Porifera research: biodiversity, innovation and sustainability, vol 28. Rio de Janeiro, Brazil, pp 361–371

Helgen LE, Rouse GW (2006) Species delimitation and distribution in *Aporometra* (Crinoidea: Echinodermata): endemic Australian featherstars. Invertebr Syst 20:395–414

Hellberg M (2006) No variation and low synonymous substitution rates in coral mtDNA despite high nuclear variation. BMC Evol Biol 6:24

Hill RS, Allen LD, Bucklin A (2001) Multiplexed species-specific PCR protocol to discriminate four N. Atlantic Calanus species, with amtCOI gene tree for ten Calanus species. Mar Biol 139:279–287

Hirose M, Osawa M, Hirose E (2010) DNA barcoding of hermit crabs of genus *Clibanarius* Dana, 1852 (Anomura: Diogenidae) in the Ryukyu Islands, southwestern Japan. Zootaxa 2414:59–66

Hoareau TB, Boissin E (2010) Design of phylum-specific hybrid primers for DNA barcoding: addressing the need for efficient COI amplification in the Echinodermata. Mol Ecol Resour 10 (6):960–967

Holmes BH, Steinke D, Ward RD (2009) Identification of shark and ray fins using DNA barcoding. Fish Res 95:280–288

Huang D, Meier R, Todd PA, Chou LM (2008) Slow mitochondrial COI sequence evolution at the base of the metazoan tree and its implications for DNA barcoding. J Mol Evol 66:167–174

Hubert N, Delrieu-Trottin E, Irisson J-O, Meyer C, Planes S (2010) Identifying coral reef fish larvae through DNA barcoding: a test case with the families Acanthuridae and Holocentridae. Mol Phylogen Evol 55:1195–1203

Hunt B, Strugnell J, Bednarsek N, Linse K, Nelson RJ, Pakhomov E et al (2010) Poles Apart: The "bipolar" pteropod species Limacina helicina is genetically distinct between the Arctic and Antarctic oceans. PLoS ONE 5(3):e9835

Jennings RM, Bucklin A, Ossenbrugger H, Hopcroft RR (2010a) Species diversity of planktonic gastropods (Pteropoda and Heteropoda) from six ocean regions based on DNA barcode analysis. Deep Sea Res Part II Topical Stud Oceanogr 57(24–26):2199–2210

Jennings RM, Bucklin A, Pierrot-Bults A (2010b) Barcoding of arrow worms (phylum Chaetognatha) from three oceans: genetic diversity and evolution within an enigmatic phylum. PLoS ONE 5(4):e9949

Kathryn FD, Thomas CW (2016) Spectral absorption of visual pigments in stomatopod larval photoreceptors. J Comp Physiol A Neuroethol Sens Neural Behav Physiol 202:215–223

Kelly RP, Sarkar IN, Eernisse DJ, DeSalle ROB (2007) DNA barcoding using chitons (genus *Mopalia*). Mol Ecol Notes 7:177–183

Kerr K, Stoeckle M, Dove C, Weigt L, Francis C, Hebert PD (2007) Comprehensive DNA barcode coverage of North American birds. Mol Ecol Notes 7:535–543

Kim MS, Yang MY, Cho GY (2010) Applying DNA barcoding to Korean Gracilariaceae (Rhodophyta) Cryptogamie. Algologie 31(4):387–401

Krabbe K, Leese F, Mayer C, Tollrian R, Held C (2010) Cryptic mitochondrial lineages in the widespread pycnogonid *Colossendeis megalonyx* Hoek, 1881 from Antarctic and Subantarctic waters. Polar Biol 33:281–292

Kress WJ, García-Robledo C, Uriarte M, Erickson DL (2015) DNA barcodes for ecology, evolution, and conservation. Trends Ecol Evol 30(1):25–35

Krug PJ, Ellingson RA, Burton R, Valdes A (2007) A new *Poecilogonous* species of sea slug (Opisthobranchia: Sacoglossa) from California: comparison with the planktotrophic congener *Alderia modesta*. J Mollus Stud 73:29–38

Lambshead PJD, Tietjen J, Moncrieff CB, Ferrero TJ (2001) North Atlantic latitudinal diversity patterns in deep-sea marine nematode data: a reply to Rex et al. Mar Ecol Prog Ser 210:299–301

Last PR, White WT, Pogonoski JJ (2007) Descriptions of new dogfishes of the genus squalus (Squaloidea: Squalidae). CSIRO, Hobart

Les DH, Moody ML, Jacobs SWL, Bayer RJ (2002) Systematics of seagrasses (Zosteraceae) in Australia and New Zealand. Syst Bot 27:468–484

Lorion J, Duperron Sb, Gros O, Cruaud C, Samadi S (2009) Several deep-sea mussels and their associated symbionts are able to live both on wood and on whale falls. Proc R Soc Lond B 276:177–185

Lucas C, Thangaradjou T, Papenbrock J (2012) Development of a DNA barcoding system for seagrasses: successful but not simple. PLoS ONE 7(1):e29987

Luttikhuizen PC, Dekker R (2010) Pseudo-cryptic species *Arenicola defodiens* and *Arenicola marina* (Polychaeta: Arenicolidae) in Wadden Sea, North Sea and Skagerrak: morphological and molecular variation. J Sea Res 63:17–23

Machida R, Miya M, Nishida M, Nishida S (2006) Molecular phylogeny and evolution of the pelagic copepod genus Neocalanus (Crustacea: Copepoda). Mar Biol 148:1071–1079

Mikkelsen NT, Schander C, Willassen E (2007) Local scale DNA barcoding of bivalves (Mollusca): a case study. Zool Scr 36:455–463

Mitra A, Banerjee K (2005) Living resources of the sea: focus Indian Sundarbans. In: Banerjee Col SR (ed) WWF India. Canning Field Office, 24 Parganas (S), 96 pp

Miyamoto H, Machida R, Nishida S (2010) Genetic diversity and cryptic speciation of the deep sea chaetognath *Caecosagitta macrocephala*. Deep Sea Res Part II Topical Stud Oceanogr 57(24–26):2211–2219
Molnar JL, Gamboa RL, Revenga C (2008) Assessing the global threat of invasive species to marine biodiversity. Front Ecol Environ 6:485–492
Moura CJ, Harris DJ, Cunha MR, Rogers AD (2008) DNA barcoding reveals cryptic diversity in marine hydroids (Cnidaria, Hydrozoa) from coastal and deep-sea environments. Zool Scr 31:93–108
Naro-Maciel E, Reid B, Fitzsimmons NN, Le M, Desalle R, Amato G (2010) DNA barcodes for globally threatened marine turtles: a registry approach to documenting biodiversity. Mol Ecol Resour 10(2):252–263
Naughton KM, O'Hara T (2009) Anew brooding species of the biscuit star *Tosia* (Echinodermata: Asteroidea: Goniasteridae), distinguished by molecular, morphological and larval characters. Invertebr Syst 23:348–366
Nelson JS, Hoddell RJ, Chou LM, Chan WK, Phang VPE (2000) Phylogeographic structure of false clownfish, Amphiprionocellaris, explained by sea level changes on the Sunda shelf. Mar Biol 137(4):727–736
Nuryanto A, Duryadi D, Soedharma D, Blohm D (2007) Molecular phylogeny of giant clams based on mitochondrial DNA Cytochrome C Oxidase I gene. Hayati 14:162–166
Paine MA, McDowell JR, Graves JE (2008) Specific identification using COI sequence analysis of scombrid larvae collected off the Kona coast of Hawaii Island. Ichthyol Res 55:7–16
Pardo LM, Ampuero D, V′eliz D (2009) Using morphological and molecular tools to identify megalopae larvae collected in the field: the case of sympatric *Cancer* crabs. J Mar Biol Assoc UK 89:481–490
Park MH, Sim CJ, Baek J, Min GS (2007) Identification of genes suitable for DNA barcoding of morphologically indistinguishable Korean Halichondriidae sponges. Mol Cells 23:220–227
Pegg GG, Sinclair B, Briskey L, Aspden WJ (2006) Mt DNA barcode identification of fish larvae in the southern Great Barrier Reef, Australia. Sci Mar 70:7–12
Pinzon-Navarro S, Barrios H, Múrria C, Lyal CH, Vogler AP (2010) DNA-based taxonomy of larval stages reveals huge unknown species diversity in neotropical seed weevils (genus Conotrachelus): relevance to evolutionary ecology. Mol Phylogenet Evol 56(1):281–293
Poppe J, Sutcliffe P, Hooper JN, Worheide G, Erpenbeck D (2010) COI barcoding reveals new clades and radiation patterns of Indo-Pacific sponges of the family Irciniidae (Demospongiae: Dictyoceratida). PLoS ONE 5:e9950
Puillandre N, Strong EE, Bouchet P, Boisselier M-C, Couloux A, Samadi S (2009) Identifying gastropod spawn from DNA barcodes: possible but not yet practicable. Mol Ecol Resour 9:1311–1321
Purvis A, Hector A (2000) Getting the measure of biodiversity. Nature 405:212–219
Pyle RL, Earle JL, Greene BD (2008) Five new species of the damselfish genus *Chromis* (Perciformes: Labroidei: Pomacentridae) from deep coral reefs in the tropical western Pacific. Zootaxa 1671:3–31
Rao CAN (2004) Faunal diversity: estuarine ecosystem. Environ Prot Train Res Inst Environ Inform Syst (EPTRI-ENVIS) Newslett 3:8–11
Remerie T, Bourgois T, Peelaers D, Vierstraete A, Vanfleteren J, Vanreusel A (2006) Phylogeographic patterns of the mysid Mesopodopsis slabberi (Crustacea, Mysida) in Western Europe: evidence for high molecular diversity and cryptic speciation. Mar Biol 149:465–481
Rice SA, Karl S, Rice KA (2008) The *Polydora cornuta* complex (Annelida: Polychaeta) contains populations that are reproductively isolated and genetically distinct. Invertebr Biol 127:45–64
Sanna D, Lai T, Francalacci P, Curini-Galletti M, Casu M (2009) Population structure of the *Monocelis lineata* (Proseriata, Monocelididae) species complex assessed by phylogenetic analysis of the mitochondrial Cytochrome c Oxidase subunit I (COI) gene. Genet Mol Biol 32:864–867

Savolainen V, Cowan RS, Vogler AP, Roderick GK, Lane R (2005) Towards writing the encyclopedia of life: an introduction to DNA barcoding. Philos Trans R Soc Lond B 360:1805–1811

Shank TM, Lutz RA, Vrijenhoek RC (2006) Molecular systematics of shrimp (Decapoda: Bresiliidae) from deep-sea hydrothermal vents, I: enigmatic "small orange" shrimp from the Mid-Atlantic Ridge are juvenile *Rimicarisexoculata.* Mol Mar Biol Biotech 7:88–96

Shashank PR, Chakravarthy AK, Raju KR, Bhanu M (2014) DNA barcoding reveals the occurrence of cryptic species in host-associated population of Conogethes punctiferalis (Lepidopt B. R era: Crambidae). Appl Entomol Zool 49:283–295

Shearer TL, van Oppen MJH, Romano SL, Worheide G (2002) Slow mitochondrial DNA sequence evolution in the Anthozoa (Cnidaria). Mol Ecol 11:2475–87

Shearer TL, Coffroth MA (2008) DNA BARCODING: Barcoding corals: limited by interspecific divergence, not intraspecific variation. Mol Ecol Resour 2:247-255

Shih HT, Yeo DCJ, Ng PKL (2009) The collision of the Indian plate with Asia: molecular evidence for its impact on the phylogeny of freshwater crabs (Brachyura: Potamidae). Journal of Biogeography 36(4):703–719

Silva I, Mesquita N, Paula J (2010) Lack of population structure in the fiddler crab *Ucaannulipes* along an East African latitudinal gradient: genetic and morphometric evidence. Mar Biol 157:1113–1126

Silva SE, de Abreu CB, Orlando TC, Wisniewski C, dos Santos-Wisniewski MJ, Hamsher SE et al (2014) A floristic survey of marine tube-forming diatoms reveals unexpected diversity and extensive co-habitation among genetic lines of the Berkeleyarutilans complex (Bacillariophyceae). Eur J Phycol 49:47–59

Sinniger F, Reimer JD, Pawlowski J (2008) Potential of DNA sequences to identify zoanthids (Cnidaria: Zoantharia). Zool Sci 25:1253–1260

Sinniger F, Reimer JD, Pawlowski J (2010) The Parazoanthidae (Hexacorallia: Zoantharia) DNA taxonomy: description of two new genera. Mar Biodivers 40:57–70

Smith PJ, McVeagh SM, Steinke D (2008a) DNA barcoding for the identification of smoked fish products. J Fish Biol 72(2):464–471

Smith PJ, Steinke D, McVeagh MS, Stewart AL, Struthers CD, Roberts CD (2008b) Molecular analysis of Southern Ocean skates (*Bathyraja*) reveals a new species of Antarctic skate. J Fish Biol 73:1170–1182

Smith MA, Fernandez-Triana J, Roughley R, Hebert PD (2009) DNA barcode accumulation curves for understudied taxa and areas. Mol Ecol Resour 9:208–216

Smith AM, Fernández-Triana JL, Eveleigh E, Gómez J, Guclu C, Hallwachs W et al (2012) DNA barcoding and the taxonomy of Microgastrinae wasps (Hymenoptera, Braconidae): impacts after 8 years and nearly 20 000 sequences. Mol Ecol Resour 13(2):168–176

Spencer HG, Waters JM, Eichhorst TE (2007) Taxonomy and nomenclature of black nerites (Gastropoda: Neritimorpha: *Nerita*) from the South Pacific. Invertebr Syst 21:229–237

Stern RF, Horak A, Andrew RL (2010) Environmental barcoding reveals massive dinoflagellate diversity in marine environments. PLoS ONE 5(11):e13991

Tang RWK, Yau C, NG W-C (2010) Identification of stomatopod larvae (Crustacea: Stomatopoda) from Hong Kong waters using DNA barcodes. Mol Ecol Resour 10:439–448

Teske PR, Barker NP, McQuaid CD (2007) Lack of genetic differentiation among four sympatric southeast african intertidal limpets (Siphonariidae): phenotypic plasticity in a single species? J Molluscan Stud 73:223–228

Torres AP, Palero F, dos Santos A, Abello P, Blanco E, Bone A, Guerao G (2014) Larval stages of the deep-sea lobster *Polycheles typhlops* (Decapoda, Polychelidae) identified by DNA analysis: morphology, systematic, distribution and ecology. Helgol Mar Res 68:379–397

Trivedi S, Chaudhuri A, Mitra A, Gupta A, Singh B, Choudhury A (1994) A case study on the loss of biodiversity during prawn seed collection from the Hooghly estuary, India. In: Proceedings of national convention on environment of India – challenges for the 21st century. Institution of Public Health Engineers, Calcutta (India), pp T-IX/32–T-IX/36

Trivedi S, Aloufi AA, Ansari AA, Ghosh SK (2015) Molecular phylogeny of oysters belonging to the genus *Crassostrea* through DNA barcoding. J Entomol Zool Stud 3(1):21–26

Trivedi S, Aloufi AA, Rehman H, Saggu S, Ghosh SK (2016a) DNA barcoding: Tool for assessing species identification in reptilia. J Entomol Zool Stud 4(1):332–337

Trivedi S, Aloufi AA, Ansari AA, Ghosh SK (2016b) Role of DNA barcoding in marine biodiversity assessment and conservation: an update. Saudi J Biolog Sci. http://dx.doi.org/10.1016/j.sjbs.2015.01.00

Tsang L, Chan B, Ma K, Hsu CH, Chu K (2007) Lack of mtDNA and morphological differentiation between two acorn barnacles Tetraclita japonica and T. formosana differing in parietes colours and geographical distribution. Mar Biol 151:147–155

Tsang LM, Chan BK, Ma KY, Chu KH (2008) Genetic differentiation, hybridization and adaptive divergence in two subspecies of the acorn barnacle Tetraclita japonica in the northwestern Pacific. Mol Ecol 17(18):4151–4163

Tsang LM, Chan BKK, Shih F-L, Chu KH, Chen CA (2009) Host-associated speciation in the coral barnacle Wanellamilleporae (Cirripedia: Pyrgomatidae) inhabiting the Millepora coral. Mol Ecol 18:1463–1475

Uchimura M, Faye EJ, Shimada S, Inoue T, Nakamura Y (2008) A reassessment of Halophila species (Hydrocharitaceae) diversity with special reference to Japanese representatives. Bot Mar 51:258–268

Unal E, Frost BW, Armbrust V, Kideys AE (2006) Phylogeography of Calanus helgolandicus and the Black Sea copepod Calanus euxinus, with notes on Pseudocalanus elongatus (Copepoda, Calanoida). Deep Sea R II 53:1961–1975

Undheim EA, Norman JA, Thoen HH, Fry BG (2010) Genetic identification of Southern Ocean octopod samples using mtCOI. C R Biol 333(5):395–404

Uthicke S, Byrne M, Conand C (2010) Genetic barcoding of commercial Beche-de-mer species (Echinodermata: Holothuroidea). Mol Ecol Resour 10:634–646

van Helden AL, Baker AN, Dalebout ML, Reyes JC, Waerebeek KV, Baker CS (2002) Resurrection of *Mesoplodon traversii* (Gray 1874), senior synonym of *M. bahamondi* Reyes, Van Waerebeek, Cárdenas and Yañez, 1995 (Cetacea: Ziphiidae). Mar Mamm Sci 18:609–621

Vargas SM, Araujo FCF, Santos FR (2009) DNA barcoding of Brazilian sea turtles (Testudines). Genet Mol Biol 32:608–612

Victor BC (2007) *Coryphopterus kuna*, a new goby (Perciformes: Gobiidae: Gobiinae) from the western Caribbean, with the identification of the late larval stage and an estimate of the pelagic larval duration. Zootaxa 1526:51–61

Victor BC (2008) Redescription of *Coryphopterus tortugae* (Jordan) and a new allied species *Coryphopterus bol* (Perciformes: Gobiidae: Gobiinae) from the tropical western Atlantic Ocean. J Ocean Sci Found 1:1–19

Ward RD, Zemlak TS, Innes BH, Last PR, Hebert PDN (2005) DNA barcoding Australia's fish species. Philos Trans R Soc Lond B 360:1847–1857

Ward RD, Holmes BH, White WT, Last PR (2008) DNA Barcoding Australasian chondrichthyans: results and potential uses in conservation. Mar Freshwat Res 59:57–71

Ward RD, Hanner R, Hebert PDN (2009) The campaign to DNA barcode all fishes, FISH-BOL. J Fish Biol 73:1–28

Waycott M, Freshwater DW, York RA, Calladine A, Kenworthy WJ (2002) Evolutionary trends in the seagrass genus Halophila (thouars): Insights from molecular phylogeny. Bull Mar Sci 71 (3):1299–1308

Waycott M, Duarte CM, Carruthers TJB, Orth RJ, Dennison WC, Olyarnik S (2009) Accelerating loss of seagrasses across the globe threatens coastal ecosystems. Proc Natl Acad Sci USA 106 (30):12377–12381

Wilson N, Hunter R, Lockhart S, Halanych K (2007) Multiple lineages and absence of panmixia in the "circumpolar" crinoid *Promachocrinus kerguelensis* from the Atlantic sector of Antarctica. Mar Biol 152:895–904

Winterbottom R, Hanner RH, Burridge M, Zur M (2014) A cornucopia of cryptic species: a DNA barcode analysis of the gobiid fish genus Trimma (Percomorpha, Gobiiformes). Zookeys 381:79–111

Worheide G (2006) Low variation in partial cytochrome oxidase subunit I (COI) mitochondrial sequences in the coralline demosponge Astrosclera willeyana across the Indo-Pacific. Mar Biol 148:907–912

Xiaobo Z, Shaojun P, Tifeng S (2013) Applications of three DNA barcodes in assorting intertidal red Macroalgal Flora in Qingdao. China. J Ocean Univ China 12(1):139–145

Zemlak TS, Ward RD, Connell AD, Holmes BH, Hebert PDN (2009) DNA barcoding reveals overlooked marine fishes. Mol Ecol Resour 9:237–242

Safety Assessment and Authentication of Seafood Through DNA Barcoding

Shalini Saggu, Hasibur Rehman, Subrata Trivedi, Al Thbiani Aziz and Jayda G. Eldiasty

Abstract Consumers of today have more apprehension on food safety ever before and high value markets such as EU and USA market are place much attention on it. Quality of food is a very important aspect of human life and people become more and more anxious about nutrition, food safety and environmental issues that determine their acceptance of food products. Food traceability is a central issue for the identification of improper labeling of processed food and feed and there are rules aimed to protect consumers and producers against fraudulent but the tools available are not always appropriate. Recently, DNA barcoding has gained support as a rapid, cost-effective and broadly applicable molecular diagnostic technique for seafood authentication. DNA barcoding methodologies are progressively more applied not only for scientific purposes but also for diverse real-life uses. The present chapter brings out a comprehensive review on the seafood authentication and fraud. Developing country markets are placing poor attention on food safety and sanitation measures and rising levels of market information, education and awareness programmes will help to minimize the gap.

Keywords COI · DNA barcoding · Seafood · Safety · Security

1 Introduction

Seafood products are a common consumer choice and a variety of cooking methods are used in seafood preparation. The increased awareness of the value of food quality is reflected in an increased request by consumers of assurances regarding the origin and content of the food they buy. At the same time, companies must be able to confirm the authenticity of their products, to comply with the regulations and to protect their products from global competition (Scarano and Rao 2014). For these

S. Saggu (✉) · H. Rehman · S. Trivedi · A.T. Aziz · J.G. Eldiasty
Faculty of Sciences, Department of Biology, University of Tabuk,
Tabuk 71491, Saudi Arabia
e-mail: ssaggu@ut.edu.sa; shalinisaggu05@gmail.com

S. Trivedi et al. (eds.), *DNA Barcoding in Marine Perspectives*,
DOI 10.1007/978-3-319-41840-7_4

reasons, the identification of the origin of the ingredients present in food or feed and the characterization of materials entering and exiting the food chains are of particular relevance.

DNA barcoding shows enormous promise for the rapid identification of organisms at the species level. There has been much recent debate, however, about the need for longer barcode sequences, especially when these sequences are used to construct molecular phylogenies. All food products must comply with the description provided by the manufacturers or processors, with reference to the origin of the ingredients, details of the transformation process, the geographic origin and the identity of the species, breeds or varieties used. The partial or complete replacement with cheaper components is one of the most common frauds to the consumers and producers.

Nowadays DNA barcoding is an established technique that involves sequencing gene segments and comparing the results with orthologous reference sequences in public databases (Hebert et al. 2003). The objective of DNA barcoding is to identify the sample species by sequencing a single gene that is universally amplified across metazoans using as template genetic materials isolated from a small portion of organisms in any stage of their life history; theoretically, all species are delineated by their unique barcode sequence, or by a tight cluster of very similar sequences (Ward et al. 2005). The core assumption of DNA barcoding is that variation in the nucleotide sequence is lower within a species than between different species (Meyer and Paulay 2005). Typically, genes within the mitochondrial genome are used for DNA barcoding of animals. Mitochondrial genome lacks introns, pseudogenes and repetitive sequences in mitochondrial genome facilitate sequence alignments of the amplified genes (Lin et al. 2005). Complete mtDNA genome sequences are publicly available; primers can therefore be designed to amplify and sequence any species that has a published mtDNA genome (Simmon and Weller 2001; Lin and Hwang 2008).

The cox1 gene, which encodes for the cytochrome oxidase subunit I, was originally proposed to be a specific mitochondrial marker for animal DNA barcoding. Molecular identification based on cox1 sequencing has provided reliable results for several animal groups that have been tested (Hebert et al. 2004a, b) and has also been expanded to include high resolution at the species level for fish (Hogg and Hebert 2004; Dawnay et al. 2007; Ivanova et al. 2007; Hubert et al. 2008). Based on these encouraging results, the barcoding community has established a Fish-BOL (Fish Barcode of Life) initiative that seeks to assemble a comprehensive reference system based on cox1 sequencing for all of the estimated 20,000 marine and 15,000 fresh-water fish species (http://www.fishbol.org/index.php). The main goal of this project is to help manage fish biodiversity and develop the Catalogue of Life (http://www.catalogueoflife.org/search.php) DNA barcoding can be applied to several fields, including biodiversity monitoring (e.g. taxonomic, ecological and conservation studies) and forensic science. Additionally, DNA barcoding can be used to identify organisms that lack distinctive morphological features (i.e., are in the larval stage or because of homoplasy and phenotypic plasticity of a given diagnostic character to environmental factors (Vences et al. 2005). Application of DNA barcoding for food authentication has recently gained attention because of food safety concerns, including incorrect food labeling, food substitutions (Marko et al. 2004) or recent food contamination. According to U.S. Food and Drug Administration, fish substitutions in seafood derivatives are becoming increasingly common; thus, analytical methods to verify food labeling are needed (http://www.fda.gov/Food/FoodSafety/ProductSpecificInformation/Seafood/RegulatoryFishEncyclopediaRFE/ucm071528.htm). In this chapter we also tried to go through the literature available on the seafood authentication and sea food safety by using our search filter in PubMed (Table 1).

2 DNA-Barcoding a Reliable Method for Seafood Authentication

To detect and prevent species substitution on the commercial market, a number of methods have been developed based on the unique protein or deoxyribonucleic acid (DNA) profiles found in different species. Protein-based methods are generally reliable for testing fresh or lightly processed seafoods, but become impractical in heavily processed foods, where proteins are degraded (Rasmussen and Morrissey 2008). Although some studies have reported the use of enzyme-linked immunosorbent assays with heat-treated seafood products (Carrera et al. 1997; Asensio et al. 2003) this method does not work well with closely related species and requires the development of species-specific antibodies. In contrast, DNA-based methods have numerous advantages over protein-based methods, including a higher information content, greater resistance to degradation, increased specificity and sensitivity, and presence in all cell types (Rasmussen and Morrissey 2008).

There are three main steps in DNA-based identification: DNA isolation, polymerase chain reaction (PCR) amplification, and detection of species based on unique DNA profiles. This review will primarily discuss the PCR amplification and species identification steps, but some advances in DNA isolation techniques will also be highlighted. Some important characteristics to consider when selecting a DNA-based method for use in seafood species identification include sample processing time and costs, equipment and startup costs, reproducibility, reliability, range of target species, and ability to recover and identify DNA from processed products, complex food matrices, and mixed-species samples. Further, for large-scale screening purposes, methods that can be applied in automated and high-throughput settings are desirable. Numerous DNA-based detection methods have been used for seafood species identification, including PCR sequencing, PCR–restriction fragment length polymorphism (RFLP), species-specific PCR, random amplified polymorphic DNA (RAPD), and single-stranded conformational polymorphism (SSCP) (Rasmussen and Morrissey 2008; Teletchea 2009). Each of these methods has some advantages and disadvantages, as outlined in Rasmussen and Morrissey. For example, PCR sequencing, PCR–RFLP, and species-specific PCR all exhibit high reproducibility. However, traditional PCR sequencing can be costly

Table 1 Search method used and the number of articles available on PUBMED

Search criteria/keywords	Articles available on PUBMED
Seafood safety	1416
Seafood safety and fraud	06
Seafood safety and DNA barcoding	07
Seafood and DNA barcoding	24
Seafood fraud	20
Seafood fraud and DNA barcoding	05

and time-consuming and cannot be used to identify multiple species in one sample. PCR–RFLP and species-specific PCR are less costly and can generally be used with mixed-species samples, but they are vulnerable to errors from intraspecies variation and they do not provide the high level of information acquired with PCR sequencing. In addition, RAPD and SSCP do not require prior knowledge of DNA sequence information, but they exhibit reduced reproducibility compared with other methods, and RAPD is vulnerable to DNA degradation.

3 DNA Barcoding of Seafood and Other Food Commodities

DNA barcoding is predominantly successful when applied to seafood because: (i) in comparison to other animal sources (e.g. cattle, sheep, goat, horse) the number of species is higher, so the effectiveness of the technique is enhanced; (ii) classical identification approaches are not useful in many cases (following industrial processing, morphological characteristics are often lost and classical identification processes are no longer effective) and (iii) identification can often proceed beyond species level, allowing the identification of local varieties and hence the origin of the product. The technique has been used to identify commercial fraud, e.g. the illegal and dangerous substitution of the toxic puffer-fish mislabelled as monkfish. Despite its proven effectiveness, few studies on the application of DNA barcoding to certain categories of seafood (e.g. crabs and lobsters) have been conducted. Therefore, more extensive studies are required to confirm the potential use of this technique on all kinds of seafood, as a reliable traceability tool (Table 2).

The applicability of DNA barcoding for the identification and traceability of mammalian (e.g. beef, pork, lamb, venison, horse) and avian (e.g. chicken, turkey) meat is also identified. However, there are several cases of species or breeds with the same DNA profile. In this case DNA barcoding would not be able to return a correct identification, thereby making it impossible to track some meat products. This occurrence is common in livestock. An example is cattle where many breeds are derived from hybridisation events.

Table 2 Example of the some food product's with DNA barcoding studies

Food product	References
Spices	De Mattia et al. (2011)
Meat	Teletchea et al. (2008), Cai et al. (2011)
Cod	Di Pinto et al. (2013)
Fish	Filonzi et al. (2010)
Anchovy	Jérôme et al. (2008)
Shark	Barbuto et al. (2010)

There are no technical limitations to the application of DNA barcoding to the traceability of plant raw materials. However, at cultivar level, the reduced genetic diversity often requires analysis of large portions of the genome. This has cost implications and is contrary to the basic DNA barcoding methodology, which requires the analysis of short and universal DNA regions only. Regarding dairy products, the authors pointed out that although no studies based on a strict DNA barcoding approach have been conducted, the use of molecular tools to characterize and trace dairy products is gaining large acceptance.

Regarding processed foods, DNA barcoding has been used to identify commercial tea, fruit species in yogurt, and fruit residues in juices, purees, chocolates, cookies etc. However, certain challenges were highlighted. During processing, the DNA structure of many ingredients (e.g. seeds, fruits, plants and animal parts) can be transformed as a result of physical (i.e. heating, boiling, and UV radiation) or chemical (i.e. addition of food preservative, artificial sweeteners) treatments. For this reason, the application of DNA barcoding on transformed commodities can be ineffective.

4 Seafood Deception

Information on a food product is indispensable for consumers to let them choosing one food product over another. The choice can reflect everyday life, holy concerns or health concerns (e.g. absence of peanuts, lactose intolerance or gluten for individuals with particular allergies). For these reasons erroneous depiction and mislabeling of a food product are unlawful, particularly if the food has been processed removing the ability to distinguish the components. All studies that have investigated seafood fraud have found it and there was not a single study which has reported 0 % fraud overall. The huge majority (91 %) of studies focused their sampling at the retail end of the supply chain. The few studies that sampled mid-chain were split between landings, distributors, processors, and wholesalers.

The studies includes mostly surveys of seafood species sold at retail, but also covers instances of fraud such as government investigations of illegal practices in the industry and other observations by scientists, governments, and consumers. To date, studies and investigations have been conducted in 29 countries and on all continents except Antarctica. The United States has the highest number of studies on seafood fraud at 39, followed by Spain at 14 and Italy at 11. Although the percentage of seafood fraud found in the studies ranges from 1.5 to 100 %, the average level of mislabeling found in these studies is 22 %.

Seafood fraud is not new and has been in the press for nearly a century. The earliest account of seafood fraud identified thus far comes from a news report from 1915, while the most recent was published in 2014. This increase can be attributed to rapid advances in DNA-based technology, the species authentication method used by most researchers today. DNA-based technology has simplified the process of species identification and made it more cost-effective than the previous technique called isoelectric focusing.

5 Conclusions

As demonstrated in this chapter, DNA-based techniques for species identification are rapidly evolving. The three most commonly used methods, PCR sequencing, PCR–RFLP, and species-specific PCR, are all experiencing ongoing technological advances leading to increased speed and automation potential. Emerging techniques for species identification, such as micro-arrays and NGS technologies have provided extremely high-throughput, automated platforms for DNA analysis that could be applied to regulatory screening of seafood products or large-scale sequencing operations. Eventually, the choice of method(s) is dependent on a combination of factors, including the desired scale of automation, speed, and throughput; the range and degree of processing of target species; and the required level of information content and specificity.

In recent years, the DNA barcoding has been exploited for both phylogenetic (research) and commercial (legal) purposes in many taxa, not only of fishes but also of crustaceans and mollusks. In fact, if it is true that DNA barcoding was initially exploited almost exclusively for biodiversity studies in marine organisms, it is also true that more recently DNA barcoding has also been proposed for genetic traceability of seafood products (Aquino et al. 2011; Lakra et al. 2011; Zhang et al. 2011; Trivedi et al. 2014, 2016).

Conventional DNA barcoding is an efficient tool that can be used to identify food components and thus validate label information contents. Molecular authentication of fresh and processed food based on DNA markers offers a valuable contribution for the identification of genetic material along the production chains since, in principle, DNA can be extracted from any food product. Finally, the application of DNA markers for the identification, characterization, and traceability of food component can improve the transparency of food production systems, helping honest producers in the protection of their food productions.

References

Aquino LM, Tango JM, Canoy RJ, Fontanilla IK, Basiao ZU, Ong PS, Quilang JP (2011) DNA barcoding of fishes of Laguna de Bay, Philippines. Mitochondrial DNA 22:143–153

Asensio L, González I, Rodríguez MA, Hernández PE, García T, Martín R (2003) Development of a monoclonal antibody for grouper (*Epinephelusmarginatus*) and wreck fish (*Polyprionamericanus*) authentication using an indirect ELISA. J Food Sci 68(6):1900–1903. doi:10.1111/j.1365-2621.2003.tb06990.x

Barbuto M, Galimberti A, Ferri E, Labra M, Malandra R, Galli P et al (2010) DNA barcoding reveals fraudulent substitutions in shark seafood products: the Italian case of "palombo" (*Mustelus spp.*). Food Res Int 43(1):376–381

Cai YS, Zhang LA, Shen FJ, Zhang WP, Hou R, Yue BS et al (2011) DNA barcoding of 18 species of Bovidae. Chin Sci Bull 56:164–168

Carrera E, Garcia T, Cespedes A, Gonzalez I, Sanz B, Hernandez P, Martin R (1997) Immunostick colorimetric ELISA assay for the identification of smoked salmon (*Salmosalar*), trout (*Oncorhynchusmykiss*) and bream (*Bramaraii*). J Sci Food Agric 74:547e550

Dawnay N, Ogden R, Mc Ewing R, Carvalho GR, Thorpe RS (2007) Validation of the barcoding gene COI for use in forensic genetic species identification. Forensic Sci Int 173:1–6

De Mattia F, Bruni I, Galimberti A, Cattaneo F, Casiraghi M, Labra M (2011) A comparative study of different DNA barcoding markers for the identification of some members of *Lamiacaea*. Food Res Int 44(3):693–702

Di Pinto A, Di Pinto P, Terio V, Bozzo G, Bonerba E, Ceci E, Tantillo G (2013) DNA barcoding for detecting market substitution in salted cod fillets and battered cod chunks. Food Chem 141 (3):1757–1762

Filonzi L, Chiesa S, Vaghi M, Marzano FN (2010) Molecular barcoding reveals mislabelling of commercial fish products in Italy. Food Res Int 43(5):1383–1388

Hebert PDN, Cywinska A, Ball SL, deWaard JR (2003) Biological identifications through DNA barcodes. Proc R Soc B 270:313–321

Hebert PDN, Penton EH, Burns JM, Janzen DH, Hallwachs W (2004a) Ten species in one: DNA barcoding reveals cryptic species in the neotropical skipper butterfly *Astraptesfulgerator*. PNAS 101(41):14812–14817

Hebert PDN, Stoeckle MY, Zemlak TS, Francis CM (2004b) Identification of birds through DNA barcodes. PLoS Biol 2(10):e312. doi:10.1371/journal.pbio.0020312

Hogg ID, Hebert PDN (2004) Biological identification of springtails (Hexapoda: Collembola) from the Canadian Arctic, using mitochondrial DNA barcodes. Can J Zool 82:749–754

Hubert N, Hanner R, Holm E, Mandrak NE, Taylor E, Burridge M et al (2008) Identifying Canadian freshwater fishes through DNA barcodes. PLoS ONE 3(6):e2490. doi:10.1371/journal.pone.0002490

Ivanova NV, Zemlak TS, Hanner R, Hebert PDN (2007) Universal primer cocktails for fish DNA barcoding. Mol Ecol Notes 7:544–548. doi:10.1111/j.1471-8286.2007.01748.x

Jérôme M, Martinsohn JT, Ortega D, Carreau P, Verrez-Bagnis V, Mouchel O (2008) Toward fish and seafood traceability: anchovy species determination in fish products by molecular markers and support through a public domain database. J Agric Food Chem 56(10):3460–3469

Lakra WS, Verma MS, Goswami M, Lal KK, Mohindra V, Punia P, Gopalakrishnan A, Singh KV, Ward RD, Hebert P (2011) DNA barcoding Indian marine fishes. Mol Ecol Resour 11:60–71

Lin WF, Hwang DF (2008) Application of species-specific PCR for the identification of dried bonito product (Katsuobushi). Food Chem 106:390–396

Lin WF, Shiau CY, Hwang DF (2005) Identification of four Thunnus tuna species using mitochondrial cytochrome b gene sequence and PCR-RFLP analysis. J Food Drug Anal 13:382–387

Marko PB, Lee SC, Rice AM, Gramling JM, Fitzhenry TM, McAlister JS, Harper GR, Moran AL (2004) Mislabelling of a depleted reef fish. Nature 430:309–310

Meyer CP, Paulay G (2005) DNA barcoding: error rates based on comprehensive sampling. PLoS Biol 3(12):e422. doi:10.1371/journal.pbio.0030422

Rasmussen RS, Morrissey MT (2008) DNA-based methods for the identification of commercial fish and seafood species. Compr Rev Food Sci Food Saf 7(3):280–295

Scarano D, Rao R (2014) DNA markers for food products authentication. Diversity 6:579–596. doi:10.3390/d6030579

Simmon RB, Weller SJ (2001) Utility and evolution of cytochrome b in insects. Mol Phylogen Evol 20(2):196–210

Teletchea F (2009) Molecular identification methods of fish species: reassessment and possible applications. Rev Fish Biol Fish 19:265e293

Teletchea F, Bernillon J, Duffraisse M, Laudet V, Hänni C (2008) Molecular identification of vertebrate species by oligonucleotide microarray in food and forensic samples. J Appl Ecol 45 (3):967–975

Trivedi S, Affan R, Alessa AHA et al (2014) DNA barcoding of Red Sea fishes from Saudi Arabia—the first approach. DNA Barcodes 2:17–20. http://dx.doi.org/10.2478/dna-2014-0003

Trivedi S, Aloufi AA, Ansari AA, Ghosh SK (2016) Role of DNA barcoding in marine biodiversity assessment and conservation: an update. Saudi J Biol Sci. http://dx.doi.org/10.1016/j.sjbs.2015.01.00

Vences M, Thoma M, Bonett RM, Vieites DR (2005) Deciphering amphibian diversity through DNA barcoding: chances and challenges. Philos Trans R Soc B 360:1859–1868

Ward RD, Zemlak TS, Innes BH, Last PR, Hebert PDN (2005) DNA barcoding Australia's fish species. Philos Trans R Soc B Biol Sci 360(1462):1847–1857

Zhang H, Zhang Y, Gao T, Li P, Xu H (2011) Genetic identification of two species of Pleuronichthys by DNA barcoding. Chin J Oceanol Limnol 29:967–972

Bioinformatics Tools in Marine DNA Barcoding

Pradosh Mahadani, Subrata Trivedi, Hasibur Rehman and Shalini Saggu

Abstract Information science has been applied to manage the information generated in molecular biology to produce the field called Bioinformatics. Application of bioinformatics in several ways is an integral part of DNA barcode research. Since the origin of the DNA barcode concept, three major criteria were set up to test the efficiency of barcode regions. These are (i) Universal Primers to amplify the barcode region (ii) Calculation of Barcode gap (intra and interspecies distance) (iii) Species resolution power. For evaluation of these criteria, DNA barcode community is applying bioinformatics tools and algorithms in primer design, distance calculation, phylogenetic analysis, etc. DNA Barcoding produces a huge amount of Cytochrome c oxidase subunit I (COI) sequences information for species identification from animal kingdom of marine biodiversity. Similarly, Consortium of Barcode of Life (CBOL) plant group proposed plastid genes *rbcL* and *matK* either singly or in combination as the standard DNA barcode for plants (CBOL Plant Working Group 2009). ITS of the nuclear ribosomal RNA standardized as the universal barcode marker for fungi. Bioinformatics play a major role in storing of DNA barcode information, and it is easy to retrieve from the database. Software-based sequence quality assurance is main starting and check point for the production of barcode sequences. In this chapter, we discussed different tools and methods of bioinformatics and their proper utilization.

Keywords Bioinformatics · DNA barcoding · COI · BLAST

P. Mahadani
Department of Genetics, ICAR-National Research Center for Orchids,
Pakyong, Sikkim, India

S. Trivedi (✉) · H. Rehman · S. Saggu
Faculty of Sciences, Department of Biology,
University of Tabuk, Tabuk 71491, Saudi Arabia
e-mail: strivedi@ut.edu.sa; trivedi.subrata@gmail.com

S. Trivedi et al. (eds.), *DNA Barcoding in Marine Perspectives*,
DOI 10.1007/978-3-319-41840-7_5

1 Introduction

1.1 DNA Barcode Repository

A database is a collection of information stored on a computer in a systematic way, such that a computer program can access it easily. Databases that are available via the web also became an in disable tool for biological research. The store data need to be accessed in a meaningful way, and often contents of several databanks or databases have to access simultaneously and correlated with each other. Bioinformatic's databases are publicly available and are designed, developed and maintained by different organization located across countries in the world. These databases must have developed using different database management system e.g. MySQL, Oracle, etc. or stored as spread sheets, flat files and simple text files on different hardware flatforms. The Barcode of Life Data System (BOLD) has evolved into primary resources for the DNA barocding community as well as NCBI or its sister genomic repositories, DNA Data Bank of Japan (DDBJ), European Molecular Biology Laboratory (EMBL).

1.2 The Barcode of Life Data System (BOLD)

In 2005, online database The Barcode of Life Data Systems (www.barcodinglife.org) established for the collection and management of specimen, distributional, and molecular data as well as analytical tools to support their validation. BOLD is freely available to any researcher with interest in DNA barcoding (Ratnasingham and Hebert 2007). On drawback of prior version, BOLD version 3.6 was released in October 2013 to address the usability gaps. BOLD divided into four major sections. These are Database, Taxonomy, Identification, and Workbench (Fig. 1).

- **Database**: These database resources are four different parts.

  I. **Public data portal**: contain the entire DNA barcode sequences on BOLD. This database can be used to access and download the associated specimen data and sequences.

  II. **BIN database**: Barcode Index Numbers (BINs) are an interim taxonomic system for animals. Barcodes are clustered algorithmically, generating a web page for each cluster which is deposited in this database.

  III. **Primer database**: A searchable database of barcode primers, which includes primer statistics.

  IV. **Publication database**: A searchable, community maintained database of barcode papers linked to published datasets. Search by title, abstract or author keywords (Fig. 2).

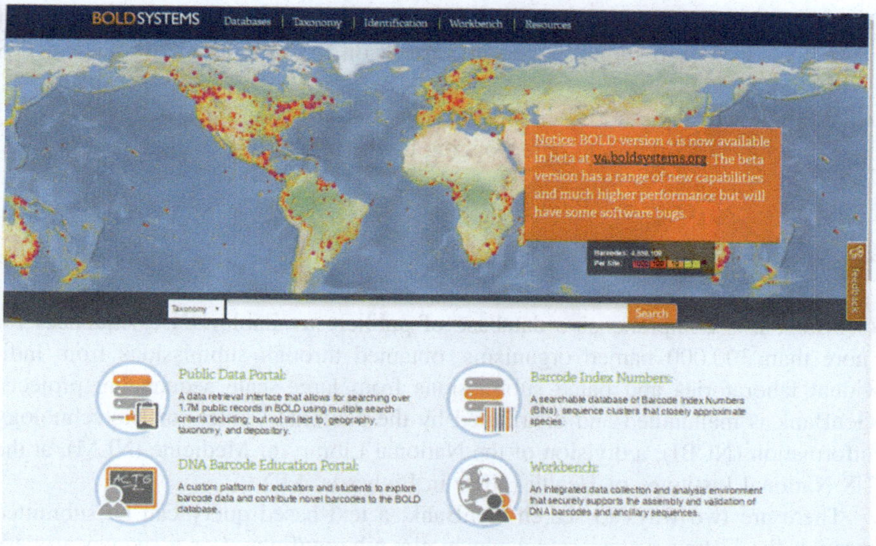

Fig. 1 Barcode of life webpage portal

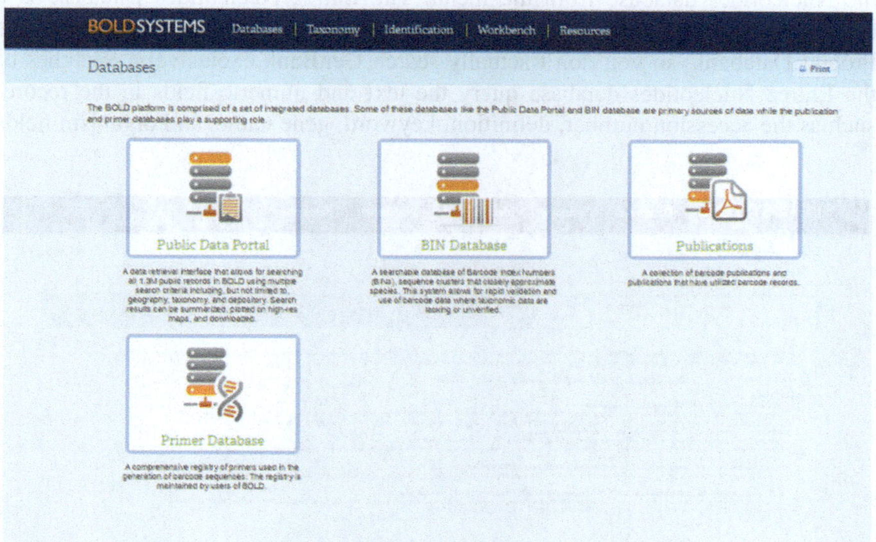

Fig. 2 Barcode of life database page

- **Taxonomy**: This link provides access to the taxonomy browser, a public resource which contains a page that displays the images, distribution map and other details for each taxon on BOLD. User can easily examine the progress of DNA barcoding of different taxonomic hierarchy.

- **Identification**: Through this section identification of unknown sequences of animal, plant and fungal is possible based on the COI, matK, rbcL, and ITS genes (Fig. 3).
- **Workbench**: The workbench link provides access to the BOLD data analysis and management workbench. After logging in, the initial page is the User Console.

2 Genbank

GenBank is a comprehensive database of publicly available DNA sequences for more than 300,000 named organisms, obtained through submissions from individual laboratories and batch submissions from large-scale sequencing projects. GenBank is maintained and distributed by the National Center for Biotechnology Information (NCBI), a division of the National Library of Medicine (NLM), at the US National Institutes of Health (NIH) in Bethesda, MD.

There are two ways to search GenBank: a text-based query can be submitted through the Entrez system at www.ncbi.nlm.nih.gov/Entrez/ or a sequence query can be submitted through the BLAST family of programs (see http://www.ncbl.nlm.nih.gov/BLAST/). To search GenBank through the Entrez system you would select the Nucleotides database from the menu. The Entrez Nucleotides Database is a collection of sequences from several sources, including GenBank, RefSeq, and the Protein Databank, so you don't actually search GenBank exclusively. Searches of the Entrez Nucleotides database query the text and numeric fields in the record, such as the accession number, definition, keyword, gene name, and organism fields

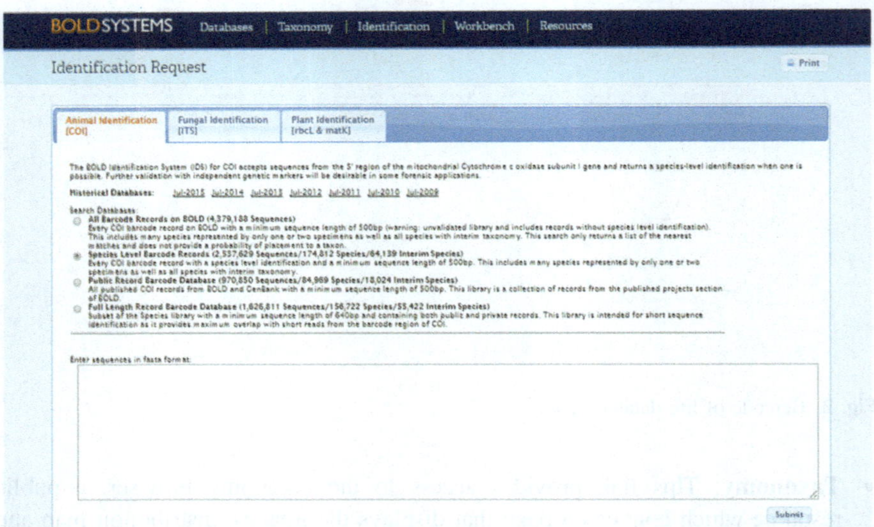

Fig. 3 Barcode of life identification page

to name just a few. Nucleotide sequence records in the Nucleotides database are linked to the PubMed citation of the article in which the sequences were published. Protein sequence records are linked to the nucleotide sequence from which the protein was translated. If you have obtained a record through a text-based Entrez Nucleotides Database search you can read the nucleotide sequence in the record. However, most researchers wish to submit a nucleotide sequence of interest to find the sequences that are most similar to theirs. This is done using the BLAST (Basic Local Alignment Search tool) programs. You select the BLAST program you wish to use depending upon the type of comparison you are doing (nucleotide to nucleotide, or nucleotide to protein sequence, etc.) and then you select the database to run the query in (any of several nucleotide or protein databases).

2.1 *Database Search and Sequence Alignment*

Database sequence similarity search is an important methodology in DNA barcoding which reveals biologically significant sequence relationships, and it suggests future investigation strategies. Sequence alignment is a common and a powerful tool to compare novel sequences (DNA, RNA or amino acids) with previously characterized gene in the database. Sequence alignment is the procedure of comparing two (pairwise) or more (multiple) sequences by searching for a series of character patterns that are the same order in the alignments correspond to mutation, and gap corresponds to insertion and deletions. Gaps are also introduced to more similar characters between the sequences involve.

Sequence alignment's tools are classified as global and local alignments. Global alignments cover the entire length of the specified sequences. All the characters in both sequences are compared in the alignments. The algorithm used in global alignments is the Needlmen and Wunch algorithms. In local alignment, stretches of sequences with the high density of matches are aligned. Sequences that differ in length or sequence that share a common domain are suitable candidates for this. Local alignment is first described by the Smith-Waterman alignments to find out alignment against a database.

2.2 *BLAST*

Bioinformaticians have developed so called 'heuristic' algorithms, which allow searching a database in considerably less time. The most popular one is Basic Local Alignment Search Tool (BLAST) Percent similarity of the resulting DNA or protein sequences was analyzed through BLAST (Altschul et al. 1990; http://www.ncbi.nlm.nih.gov/blast/), a choice is offered between the different BLAST programs through different hyperlinks (nucleotide blast, protein blast, blastx, tblastn, tblastx etc.) (Fig. 4).

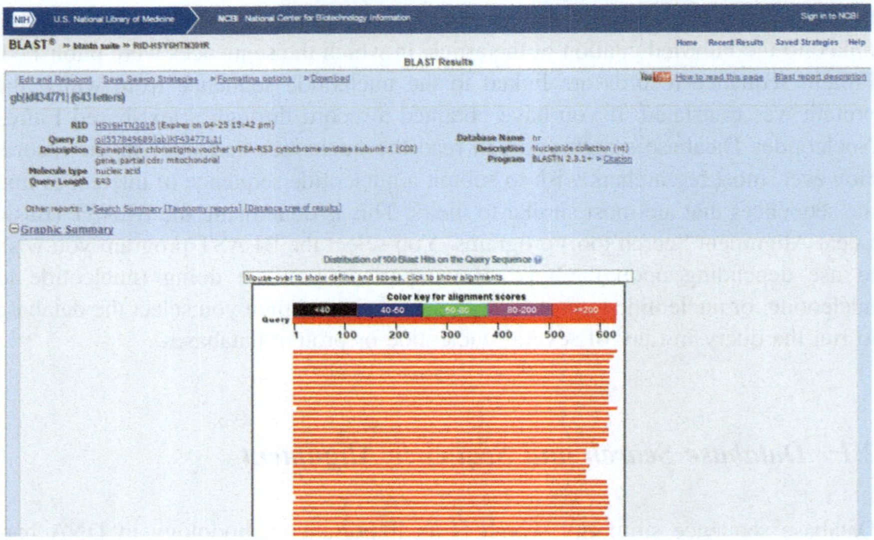

Fig. 4 Showing the BLAST result on http://blast.ncbi.nlm.nih.gov/Blast.cgi

As query for barcode sequences in nucleotide BLAST (BLASTN), were used in FASTA format or Accession No. We specified the database because our sequences were from mitochondria or chloroplast DNA. So, we choice the "others" database option. We also selected megablast under the program selection header, which optimized the search for highly similar sequences and clicked on the Blast button to initiate the search. The outputs of the megablast search contain a table with sequencing producing significant alignments. We also used specialized BLAST in bl2seq for the alignment two (or more) sequence and primer blast to make specific primer.

BLAST uses statistical method to evaluate hits for their significance. BLAST program identifies sequences having share common words of a pre-set size (K-tuple) in the database sequences and these matching words are extended only if they score higher than pre-defined threshold. The default threshold for the E-value is 10 and default word size is 11. The E-value is the probability that the query match is due to randomness. The lower the E-value is more significant the match. The score (in bits) is a value attributed to the alignment but is independent of the scoring matrix used, while E-values of 10^{-3} and below are often considered indicative of statistically significant results.

2.3 *Multiple Sequence Alignment*

A multiple sequence alignment (MSA) arranges a set of sequences in a scheme where positions believed to be homologous are written a common column. The gap represent a deletion, an insertion in the sequences that do not have a gap, or a combination of insertion and deletions. MSA gives biologist the ability to extract biological important but perhaps widely dispersed sequence similarity that can give biologist hints about the evolutionary history of certain sequences. The MSA, homology search algorithm is some time called 'many-against-eachother' search because the input is a small defined set of sequences which are compared only against each other, not against an entire database. There are several approaches, one of the most popular being the progressive alignment strategy used by the clustal family of programs.

2.4 CLUSLAT

The most commonly used software for progressive alignment is CLUSTALW and CLUSTALX. This program are identical to each other in term of alignment method but offer either a simple text interface (ClustalW) suitable for high-throughput tasks or a graphical interface (ClustalX). ClustalX and ClustalW will take a set of input sequences and carry out the entire progressive alignment procedure automatically. The sequences are aligned in pairs in order to generate a distance matrix that can be used to generate a distance matrix that can be used to make a simple initial tree of the sequence. Finally, the multiple sequence alignment is carried out using the progressive approach (Thompson et al. 1997).

ClustalW and ClustalX are both freely available and can be downloaded from the EMBL/EBI file server (ftp://ftp.ebi.ac.uk/pub/software/) or from ICGEB in Strasbourg, France (ftp://ftp-igbmc.u-strasbg.fr/pub/ClustalW/ and ftp://ftp-igbmc.u-strasbg.fr/pub/ClustalX/). In each case, ClustalX (X stands for X windows) provides a graphical user interface with colorful display of alignments. Open ClustalX and open the sequence file using File → Load Sequences. The graphical display allows the user to slide over the unaligned protein sequences. Select Do complete Alignment from the Alignment menu. ClustalX performs the progressive alignment (progress can be followed up in the lower left corner), and creates an output guide tree file and an output alignment file in the default Clustal format. It is, however, possible to choose a different format in the Output Format Options from the Alignment menu. ClustalX also allows the user to change the alignment parameters (from Alignment Parameters in the Alignment menu). If an alignment shows, for example, too many large gaps, the user can try to increase the gap-opening penalty and redo the alignment. ClustalX indicates the degree of conservation at the bottom of the aligned sequences, which can be used to evaluate a given alignment (Fig. 5).

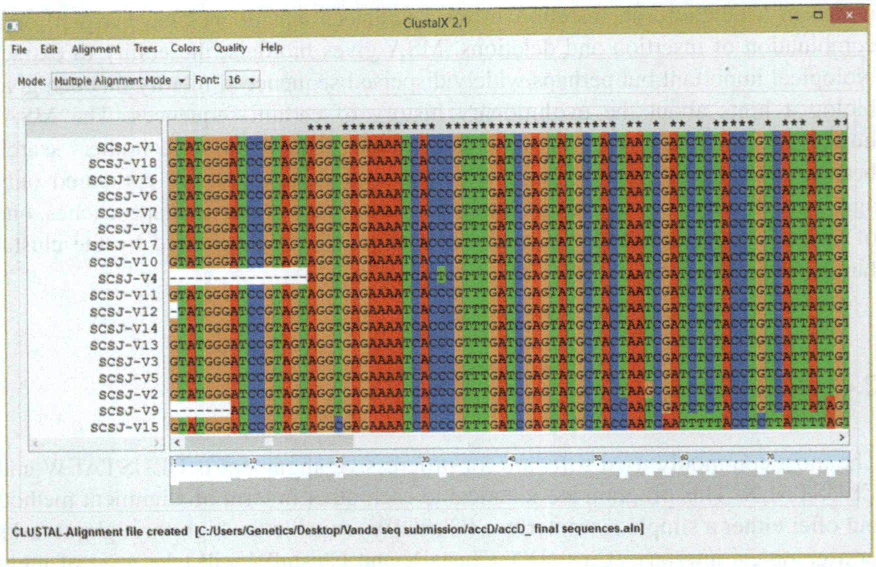

Fig. 5 Result showing alignment though CLUSTAL

3 Quality Control of DNA Barcode Data

Quality checking is one of the crucial steps for the generation of DNA barcode sequences. The purpose of bidirectional sequencing is to increase the confidence of sequence quality. There are two check points one is trimming from raw trace file and second, 3′ and 5′ terminals were clipped to generate consensus sequences and checking of open reading frame for the protein coded sequences. Trace file are assemble in sequence editing software (Applied Biosystems Sequence Scanner v1.0, BioEdit) and sequence with greater than 2 % ambiguous bases were discarded, using quality Value of 40 for bidirectional reads. Manual editing of raw traces and subsequent alignments of forward and reverse sequences enabled us to assign edited sequences for most species. In some cases of discrepancy, both the sequences were reviewed and quality values of the sequences were considered to determine the most likely nucleotide. Then the 3′ and 5′ terminals were clipped to generate consensus sequences for each sample. Finally, each of the sequences was compared in NCBI through BLASTN. The sequences were translated using the online software ORF finder (http://www.ncbi.nlm.nih.gov/gorf/gorf.html) and aligned through BLASTP to examine whether the partial amino acid codes were coherent with the particular gene frame and without any stop codon.

3.1 *BioEdit*

BioEdit is a mouse-driven, easy-to-use sequence alignment editor and sequence analysis program. BioEdit is intended to supply a single program that can handle most simple sequence and alignment editing and manipulation functions that researchers are likely to do on a daily basis, as well as a few basic sequences analyses.

3.1.1 Sequence Manipulations Suit

The Sequence Manipulation Suite (http://www.bioinformatics.org/sms2/) is written in JavaScript 1.5, which is a lightweight, cross-platform, object-oriented scripting language. JavaScript is now standardized by the ECMA (European Computer Manufacturers Association). The first version of the ECMA standard is documented in the ECMA-262 specification. The ECMA-262 standard is also approved by the ISO (International Organization for Standards) as ISO-16262. JavaScript 1.5 is fully compatible with ECMA-262, Edition 3. Sequences submitted to the Sequence Manipulation Suite instead manipulated by the web browser, which executes the JavaScript. The Sequence Manipulation Suite was written by Paul Stothard (University of Alberta, Canada). Short descriptions of the programs which were used in this study.

3.1.2 Reverse Complement

Converted a DNA sequence into its reverse, complement, or reverse-complement counterpart. The entire IUPAC DNA alphabet was supported, and the case of each input sequence character was maintained.

3.1.3 ORF Finder

Searched for open reading frames (ORFs) in the DNA sequence. The program returned the range of each ORF, along with its protein translation. ORF Finder supports the entire IUPAC alphabet and several genetic codes. ORF Finder was used to search newly sequenced DNA for potential protein encoding segments.

4 Primer Design

Before starting a barcode project on any new marine group, it is essential to test the performance of existing primers on fresh specimens from a range of species in the target group. Bioinformatics tools are very useful to test the efficiency of newly design primer or existing primers. Proper primer design is actually one of the most important factors/steps in successful DNA sequencing. Efficacy and sensitivity of

PCR largely depend on the efficiency of primers. The primers which are unique for the target sequence to be amplified should fulfill certain criteria such as primer length, GC %, annealing and melting temperature, 5′ end stability, 3′ end specificity etc. PCR is capable of amplifying a single target DNA fragment out of a complex mixture of DNA. This ability depends on the specificity of the primers. Primers are short single-stranded oligonucleotides which anneal to template DNA and serve as a "primer" for DNA synthesis. In order to achieve the geometric amplification of a DNA fragment, there must be two primers, one flanking each end of the target DNA. It is essential that the primers have a sequence that is complementary to the target DNA. The forward primer which will be complementary to the lower strand and must run 5′–3′. The reverse primer which will be complementary to the upper strand and must run 3′–5′. However, we always write DNA sequences in the 5′–3′ direction so the reverse primer would be written as reverse complementary form. We take primers sequences from consensus sequences of multiple sequence alignment and slightly modified some nucleotide of the primer sequences on the basis of criteria of primers (Table 1).

5 Phylogenetic Analysis

Phylogenetics is the science of estimating the evolutionary past, in the case of molecular phylogeny, based on the comparison of DNA or protein sequences. DNA barcodes are used both to identify species and to draw attention to overlooked and new species; they can help identify candidate exemplar taxa for a comprehensive phylogenetic study. Barcode of Life projects create a perfect taxonomic sampling environment for conducting phylogenetic studies on different branches of the Tree of Life. Consequently, phylogenies that are constructed on top of barcode libraries,

Table 1 Softwares available and their features for primer design

Name of the software/program	Features	Available
OligoCalc	Web-accessible, client-based computational engine to calculate molecular weight, solution concentration, melting temperature, etc.	http://basic.northwestern.edu/biotools/OligoCalc.html
PCR products	The program searches for perfectly matching primer annealing sites that can generate a PCR product	(http://www.bioinformatics.org/sms2/pcr_products.html)
Primer blast	It uses Primer3 (http://bioinfo.ut.ee/primer3-0.4.0/primer3/input.htm) to design PCR primers and then submits them to BLAST search against user-selected database	http://www.ncbi.nlm.nih.gov/tools/primer-blast/
Prime3 (http://frodo.wi.mit.edu/primer3/input.htm)	This software is very powerful and control over the size of product desired, primer size and Tm range, and presence/absence of a 3′–GC clamp etc.	http://biotools.umassmed.edu/bioapps/primer3_www.cgi

Table 2 Softwares available for phylogenetic analysis

Software/programe	References
Phylip	http://evolution.genetics.washington.edu/phylip.html
MEGA	http://www.megasoftware.net/
PAUP	http://paup.csit.fsu.edu/

in a given taxonomic group, are less likely to be influenced by insufficient taxon sampling. Additionally, barcoding aids in pinpointing and subsequent replacement of taxa with attributes—such as exceptionally elevated rates of evolution or nucleotide compositional biases—that can mislead the recovery of phylogenetic trees. Barcode sequence data can also provide a shared genomic cornerstone for the variable repertoire of genes that can be used to build the phylogenetic tree. It can be used as a link between the deeper branches of the tree to its shallow, species-level branches. Barcode sequences have been analyzed mainly by using phylogenetic tree reconstruction methods such as NJ, these barcode-based trees should not be interpreted as phylogenetic trees (Table 2).

6 Methods of Tree Construction

The methods for calculating phylogenetic trees fall into two general categories. These are distance-matrix methods, also known as clustering or algorithmic methods (e.g. UPGMA, neighbour-joining, Fitch Margoliash), and discrete data methods, also known as tree searching methods (e.g. parsimony, maximum likelihood, Bayesian methods. Distance is relatively simple and straightforward—a single statistic, the distance (roughly, the percent sequence difference), is calculated for all pairwise combinations of OTUs, and then the distances are assembled into a tree. Discrete data methods examine each column of the alignment separately and look for the tree that best accommodates all of this information. Unsurprisingly, distance methods are much faster than discrete data methods. However, a distance analysis yields little information other than the tree. Discrete data analyses, however, are information rich; there is an hypothesis for every column in the alignment, so you can trace the evolution at specific sites in the molecule. Barcode sequences have been analyzed mainly by using phylogenetic tree reconstruction methods such as NJ, these barcode-based trees should not be interpreted as phylogenetic trees.

6.1 *Neighbor-Joining Method*

Neighbor-joining (*NJ*) a heuristic method for estimating the ***minimum evolution*** tree originally developed by Saitou and Nei (1987) and modified by Studier and Keppler (1988). NJ is conceptually related to clustering, but does not require the data to be

ultrametric. The principle of NJ is to find pairs of operational taxonomic units (OTUs) that minimize the total branch length at each stage of clustering of OTUs starting with a star-like tree. The neighbor-joining method is therefore a special case of the ***star decomposition*** method. The ***Neighbor-joining*** (***NJ***) method constructs a tree by sequentially finding pairs of neighbors, which are the pairs of OTUs connected by a single interior node. This algorithm does not attempt to cluster the most closely related OTUs, but rather minimizes the length of all internal branches and thus the length of the entire tree. The NJ algorithm starts by assuming a star-like tree that has no internal branches. In the first step, it introduces the first internal branch and calculates the length of the resulting tree. The algorithm sequentially connects every possible OTU pair and finally joins the OTU pair that yields the shortest tree. The length of a branch joining a pair of neighbors, X and Y to their adjacent node is based on the average distance between all OTUs and X for the branch to X, and all OTUs and Y for the branch to Y, subtracting the average distances of all remaining OTU pairs. This process is then repeated, always joining two OTUs (neighbors) by introducing the shortest possible internal branch. The *Fitch–Margoliash* method is a distance-matrix method that evaluates all possible trees to find the tree that minimizes the differences between the pairwise genetic distances and the distance represented by the sum of branch lengths for each pair of taxa in the tree. NJ has the advantage of being very fast, which allows the construction of large trees including hundreds of sequences; this significant difference in speed of execution compared to other distance methods has undoubtedly accounted for the popularity of the method. Distance methods are implemented in many different software packages, including Phylip, Mega5, Paup*4 and many more.

6.1.1 Models

This is all complicated by the fact that molecular evolution is ancient history, a kind of molecular archaeology where we are trying to recover the past by extrapolating backward from a small set of surviving clues. If little evolution has occurred, this is fairly straightforward. However, and quite rapidly, the true evolutionary difference between two sequences becomes obscured by multiple mutations (changes on top of changes), especially at the more rapidly evolving sites. In these cases, a simple count of the differences between two sequences will underestimate how much evolution has actually occurred. Various models (corrections) have been developed to try to estimate the true difference between sequences based on their present states, such as amino acid substitution matrices (e.g. Dayoff, Blossom, etc.) or gamma corrections (giving more weight to changes at slowly evolving sites), etc. However, it is beyond the scope of this chapter to explain these, and the interested reader should consult one of several excellent texts on molecular evolution for further detail.

6.1.2 K2P Model

The Kimura model (Kimura 1980) is an extension of the Jukes and Cantor (JC) basic model (Tamura et al. 2007). This model distinguishes between two types of substitutions: transitions, where a purine is replaced by another purine (A<–>G) or a pyrimidine is replaced by another pyrimidine (C<–>T), and transversions, where a purine is replaced by a pyrimidine or vice versa (A or G<–>C or T). The model assumes that the rate of transitions is different from the rate of transversions. For the species-level analysis, nucleotide sequence divergences were calculated using the Kimura-2-Parameter (K2P) model, the best metric when distances are low as in DNA barcode sequence.

6.1.3 Bootstrap Value

Bootstrap analysis is a widely used sampling technique for estimating the statistical error in situations in which the underlying ***sampling distribution*** is either unknown or difficult to derive analytically. The bootstrap method offers a useful way to approximate the underlying distribution by resampling from the original data set. Felsenstein (1993) first applied this technique to the estimation of confidence intervals for phylogenies inferred from sequence data. First, the sequence data are bootstrapped, which means that a new alignment is obtained from the original by randomly choosing columns from it with replacements. Each column in the alignment can be selected more than once or not at all until a new set of sequences, a *bootstrap replicate*, the same length as the original one has been constructed. Therefore, in this resampling process, some characters will not be included at all in a given bootstrap replicate and others will be included once, twice, or more. Second, for each reproduced (i.e. artificial) data set, a tree is constructed, and the proportion of each clade among all the bootstrap replicates is computed. This proportion is taken as the statistical confidence supporting the monophyly of the subset. Bootstrap values should be displayed as percentages, not raw values. This makes the tree easier to read and to compare with other trees. By convention, only bootstrap values of 50 % or higher are reported; lower values mean that the node in question was found in less than half of the bootstrap replicates.

7 Incorporation of Indel Information in DNA Barcode

Insertion and deletion (indels) attract increasing interest because they play an important role in genomic evolution. Hardly a few studies incorporated the indel information in phylogenetic analysis and tried to treat different indel as separate binary characters or fifth state characters. Simmons and Ochoterena (2000) proposed simple indel coding (SIC) and the complex indel coding (CIC) procedures to treat indels with six rules. Muller (2006) developed a Modified Complex Indel coding (MCIC) based on state transformation cost. Simple indel coding method coded all indels characters by using Gap Coder and FastGap and IndelCoder menu of SeqState program deals with MCIC. Simple indel coding is conservative and easy to apply in indels contain sequences. After sequence alignment, for each gap a number is assigned and each gap position is also reported. In the distance matrix, 1 represents presence of each gap, and 0 represent the absence of each gap But, this method is does not utilized the all the available information and CIC rules are very difficult to translate into a clearly formulate algorithm. Modified Complex Indel Coding (MCIC) method primarily derived from the Complex indel coding. In MCIC method, first aligned sequences are numbered and converted into binary pattern. Ogden and Rosenberg (2007) used gap information during tree construction under the maximum parsimony principle and concluded that all the three gap coding methods perform equally well in topological accuracy. However, maximum likelihood method based analysis (e.g. RAxML, PAUP, etc.) is statistically inconsistent when sequences evolved with indels. Muñoz-Pajares (2013) first time developed a software package, Substitution and Indel Distances to Infer Evolutionary Relationships (SIDIER), on R language. SIDIER combines both the gap distance and substitutions distance to infer evolutionary relationships. This combine distance can be used across a wider range of phylogenetic problems and also useful for barcode gap calculation (Muñoz-Pajares 2013). SIDIER is promising software for intra and inter specific calculation in DNA barcode studies as well as to infer phylogenetic relationships. We suggested that the indel-rich loci may be valuable for phylogenetic but careful attention should be require for selecting alignment algorithm and indel coding methods (Table 3).

Table 3 Softwares available for selecting alignment algorithm and indel coding methods

Software name	Method used	Reference and availability
GapCoder	SIC	(Young and Healy 2003) and (http://www.home.duq.edu/~youngnd/GapCoder)
FastGap	SIC	(Borchsenius 2007) and (http://192.38.46.42/aubot/fb/FastGap_home.htm)
SeqState	MCIC	(http://www.nees.uni-boon.de/downloads/)
PAUP*	Missing data/fifth character	(http://paup.csit.fsu.edu)
SPInDel	Diverse statistical methods	(http://www.portugene.com/SPInDel/SPInDel_webworkbench.html)
SIDIER	SIC/MCIC/fifth characters + substitution	(http://cran.r-project.org/web/packages/sidier/index.html)

References

Altschul SF, Gish W, Miller W, Myers EW, Lipman DJ (1990) Basic local alignment search tool. J Mol Biol 215:403–410

Borchsenius F (2007) FastGap 1.0.8. Software. http://192.38.46.42/aubot/fb/FastGap_home.htm

CBOL Plant Working Group (2009) A DNA barcode for land plants. Proc Natl Acad Sci USA 106:12794–12797

Felsenstein J (1993) Phylogeny Inference Package (PHYLIP) version 3.5. University of Washington, Seattle

Kimura M (1980) A simple method for estimating evolutionary rates of base substitutions through comparative studies of nucleotide sequences. J Mol Evol 16:111–120

Muller K (2006) Incorporating information from length-mutational events into phylogenetic analysis. Mol Phylogenet Evol 38:667–676

Muñoz-Pajares AJ (2013) SIDIER: substitution and indel distances to infer evolutionary relationships. Methods Ecol Evol 4:1195–1200

Ogden TH, Rosenberg MS (2007) How should gaps be treated in parsimony? A comparison of approaches using simulation. Mol Phylogenet Evol 42:817–826

Ratnasingham S, Hebert PDN (2007) BOLD: the barcode of life data system. Mol Ecol Notes 355–364. www.barcodinglife.org

Saitou N, Nei M (1987) The neighbor-joining method—a new method for reconstructing phylogenetic trees. Mol Biol Evol 4:406–425

Simmons MP, Ochoterena H (2000) Gaps as characters in sequence-based phylogenetic analyses. Syst Biol 49:369–381

Studier JA, Keppler KJ (1988) A note the neighbor-joining algorithm of Satou and Nei. Mol Biol Evol 5(6):729–731

Tamura K, Dudley J, Nei M, Kumar S (2007) MEGA4: Molecular Evolutionary Genetics Analysis (MEGA) software version 4.0. Mol Biol Evol 24:1596–1599

Thompson JD, Gibson TJ, Plewniak F, Jeanmougin F, Higgins DG (1997) The CLUSTAL_X windows interface: flexible strategies for multiple sequence alignment aided by quality analysis tools. Nucleic Acids Res 25:4876–4882

Young ND, Healy J (2003) GapCoder automates the use of indel characters in phylogenetic analysis. BMC Bioinform 4

Part II
DNA Barcoding of Marine Invertebrates

Morphological and COI Sequence Based Characterisation of Marine Polychaete Species from Great Nicobar Island, India

V. Sekar, R. Rajasekaran, C. Prasannakumar, R. Sankar, R. Sridhar and V. Sachithanandam

Abstract DNA barcoding has proved to be a powerful alternative method to traditional morphological approaches, allowing to complement identification techniques for living organisms. In this study we assess intraspecific and interspecific genetic divergence Among the 6 genera marine polychaetes from Great Nicobar Island of Souther part of the Andaman and Nicobar Island. The present study results suggested that high level of interspecific genetic variation was observed between *Lysidice collaris* and *Terebella ehrenbergi* (0.727). The minimum genetic distance (0.316) was observed between genera *Phyllodoce fristedti* and *Ceratonereis mirabilis*. Morphological identification of the polychaetes in this study was supported by the molecular data, as shown by the congruence and high similarity between the sequences produced in the present study and those available in GenBank. This study presents the first information on DNA barcoding for polychaetes species in the Great Nicobar Island, and it establishes the effectiveness of DNA barcoding for identification of marine polychaetes species from Andaman and Nicobar Island, thus making it available to a much broader range of scientists.

Keywords Polychaetes · Cytochrome c oxidase subunit I · Great Nicobar Island · Andaman and Nicobar Island

V. Sekar (✉) · R. Rajasekaran · C. Prasannakumar · R. Sankar
Faculty of Marine Sciences, Centre of Advanced in Marine Biology,
Annamalai University, Parangipettai Campus, Chidambaram
608502, Tamil Nadu, India
e-mail: sekarveera15@gmail.com

V. Sachithanandam
Department of Ocean Studies and Marine Biology, Pondicherry University,
Andaman and Nicobar Island 744112, India
e-mail: pondiunisachin@gmail.com

Present Address:
V. Sekar · R. Sankar · R. Sridhar · V. Sachithanandam
National Centre for Sustainable Coastal Management,
Ministry of Environment, Forest and Climate Change, Chennai 600025, India

S. Trivedi et al. (eds.), *DNA Barcoding in Marine Perspectives*,
DOI 10.1007/978-3-319-41840-7_6

1 Introduction

The world is facing a global biodiversity crisis (Novacek and Cleland 2001; Bellwood et al. 2004). The rapid loss of marine and terrestrial biodiversity has prompted efforts to catalogue this biodiversity in Census of Marine Life (CoML). Nowadays morphologically taxonomical identifications are declining in number the reason behind this being the decline of taxonomic skills (Hopkins and Freckleton 2002) and insufficient funding. Besides, the upcoming scientists focus on the advance techniques of molecular methods.

Annelida, one of the most successful animal phyla, exhibits an amazing variety of morphological forms. Disparity between some of the forms is so great that until molecular tools were used. Some annelid lineages were not commonly recognized as belonging to the group (Halanych and Janosik 2006). Generally polychaetes are the most dominant and diverse macrofaunal groups in marine benthic habitats. Usually they characterized up to 80 percentages of high species richness and diverse group in marine sediments plays fundamental ecological roles in the benthic communities (Hutchings 1998). Among annelid workers there have been working hypotheses for the interrelationships of major annelid lineages related to the molecular techniques. The relationships within polychaete groups have been difficult to discern and molecular data only partly corroborate classifications done on morphological grounds.

In the last decade a molecular revolution in taxonomy has take place in parallel with morphological taxonomy. DNA barcoding has the potential to be a practical method for identification of ongoing research program for the creation of a long term data base on caterpillar towards identification of cryptic species. DNA barcoding will have broad scientific applications. It will be of great utility in conservation biology, including biodiversity surveys. These approaches to describe, catalogue and identify the diversity have been increasingly adopted in biodiversity studies, (Hebert et al. 2003; Blaxter 2004). Hebert et al. (2010) proposed the molecular technique DNA bar-coding using a short DNA sequence has been increasingly adopted in taxonomic identification studies. Currently, these methods are being applied to a wide range of taxonomic groups (Floyd et al. 2002; Hebert et al. 2004a, b; Ward et al. 2005; Meyer et al. 2008). The main goal of these DNA bar-coding is to characterize the biological diversity using a short DNA sequence that will facilitate and expedite taxonomic identification. This initiative has helped in the discovery of new species many of which have been shown to be morphologically cryptic, thereby considerably improving biodiversity assessment in poorly studied benthic fauna (Olivares et al. 2001; Blaxter et al. 2004; Derycke et al. 2005; Leasi and Todaro 2009). Careful perusal of literature revealed that many works have been undertaken using DNA Bar-coding molecular tools and worldwide important and monumental works as a sole source to identify many species and sibling species.

Folmer et al. (1994) studied DNA primers for amplification of mitochondrial cytochrome *c* oxidase subunit 1 from diverse metazoan invertebrates. Chotteau et al. (1994) made an attempt to discover two related genes in a marine worm, the polychaete of *Nereis diversicolor*. Gibson et al. (1999) carried out the non-developmental genetic variation of molecular and morphological evidence of a single species of spionidae, *Boccardia proboscidea*, with multiple development modes using RAPD-PCR. Christer et al. (2000) traced out the phylogeny of the tubificidae, and of most of its subfamilies and some of its genera revisited, on the basis of sequences of 18S ribosomal DNA in a selection of species. McHugh (2000) studied molecular systematic of polychaetes in various development can occurred for the couple of years. The testing of sister-group relationships is being incorporated into the most recent studies and numbers of analyses were increased relationships within polychaete. The researches on polychaetes have a relative short history and the achievements in improving the understanding of disputed taxonomic and phylogenetic problems of different polychaete groups are rather scarce and sporadic.

DNA barcode can be of significant help for taxonomical, ecological and biological studies, mainly for specific identification of biological communities and biodiversity in general (Valentini et al. 2009). The main objective of using the DNA barcode is the creation of reference databases with nucleotide sequences of currently known species (Ratnasingham and Hebert 2007).

Mitochondrial genome in most of the animals is 15,000 bp and holds phylogenetic information that can be examined as gene rearrangement of amino acid or nucleotide data. Up to 2006, only four complete annelid genomes were available in GenBank, but now recognizing that traditional taxonomic approaches often overlook polychaete species (Schmidt 2003), recently examined variation in mitochondrial DNA (mtDNA) sequences and demonstrated that such analysis is valuable for the discrimination of closely related polychaete species studies (Nygren and Pleijel 2010; Ratnasingham and Hebert 2007).

Cytochrome c oxidase subunit 1 gene, is the part of electron transport chain, is the most commonly sequenced mitochondrial gene for invertebrate animals. In contrast to the ribosomal genes like 18S and 28S, protein coding genes usually capture phylogenetic signal at two levels. Nucleotide substitutions accumulate most quickly in the third, or wobble position of the codon. The initial target for DNA barcoding described in this protocol is mitochondrial cytochrome c oxidase subunit 1 (CO1). Selection of an appropriate gene is a critical strategic and practical decision, with significant consequences for the overall success of this project.

Hebert et al. (2003) had suggested a section of the mitochondrial DNA gene CO1 for species identification. Once sequenced, this gene fragment could be used as a 'barcode' to distinguish between species. CO1 is the best candidate for this taxonomic tool, as it has a high degree of conservation and insertions and deletions are rare (Moritz and Cicero 2004). It also has many rapidly evolving nucleotide sites, which will allow differentiation between even recently evolved species (Nylander et al. 1999). Compared to the nuclear genome the mitochondrial genome lacks introns, has had restricted exposure to recombination and has a haploid mode of inheritance (Saccone et al. 1999). Hebert et al. (2003) demonstrated that the presence of high level of diversity between species sequences allowed for the successful assignment of 98 % of samples of cryptic lepidopteran species. Mitochondrial DNA sequence divergences have also been successfully used to distinguish between species of North American birds (Hebert et al. 2004a), spiders (Hebert and Barrett 2005), cryptic species of butterflies (Hebert et al. 2004b), leeches (Siddall and Budinoff 2005), springtails (Stevens and Hogg 2003; Hogg and Hebert 2004), beetles (Monaghan et al. 2005), oligochaetes (Nylander et al. 1999), worms (Bely and Wray 2004), extinct moas (Lambert et al. 2005), and various other species of vertebrates and invertebrates (Saccone et al. 1999; Hebert et al. 2003).

Recent day's annelid systematics and the in group relationships of polychaetes are matter of ongoing debates. In the last decade, a number of molecular studies revealed several cryptic species of polychaetes within divergent genera, such as *Perinereis* (Scaps et al. 2000), Syllis (Maltagliati et al. 2000), Dipolydora (Manchenko and Radashevsky 2002), Neanthes and Hediste sp. (Breton et al. 2003) and Ophelina (Maltagliati et al. 2004). In India limited studies have been done in Capitellidae (Samidurai 2010) and Nereidae (Satheeshkumar and Jagadeesan 2010). Brett (2006) made an attempt to test the mtDNA COI gene in polychaete worms. Pop et al. (2007), Bleidorn et al. (2006), Kvist et al. (2010), Carr et al. (2011), Maturana et al. (2011), Canales-Aguirre et al. (2011) evaluated the utility of COI of the mitochondrial DNA for the taxonomic determination and suggested that CO1 gene is a useful molecular marker for fast and accurate taxonomic determination of benthic polychaetes.

Fernando and Rajasekaran (2007) have erected a new species of Namalycastis through morphological taxonomy. However, only explorations of new habitats like creeks and Islands, especially deep-sea, may result the discovery of new species. Cryptic species that are morphologically similar but genetically distinct were shown to be common in marine systems (Knowlton 2001). Consequently, a more careful look at the world oceans might show even by numbers, that in island ecosystems many of the organisms are highly modified to other terrestrial and aquatic ecosystems.

The present to elucidate polychaete identification through morphological and COI gene sequences based analysis carried out from Nicobar Island. This is first ever documentation from this remote part of country.

2 Materials and Methods

2.1 Specimen Collection

In the present study 6 specimens (namely P1, P2, P3, P4, P5 and P6) were collected from the dead coral patches and specimens list are given in Table 1 Small pieces of tissue were taken from each individual were preserved in 95 % ethanol and stored in −20 °C until further analysis could be done in the laboratory.

Table 1 Shows for list of specimen collection site and GenBank accession numbers

Family	Sample/species	Position lat/long	Vocher. no. CSMB	Assess. no.
Terebellidae	P1-*T. ehrenbergi*	06° 54.606′ N; 93° 55.770′ E	VNR6	JX966313
Eunicidae	P2-*L. collaris*	06° 52.993′ N; 93° 55.990′ E	LXN1	KC208488
Nereidae	P3-*P.n. brevicirrus*	06° 55.962′ N; 93° 55.896′ E	CMPB2	JX966314
Phyllodocidae	P4-*P. fristedti*	06° 54.606′ N; 93° 55.770′ E	VNR7	JX966312
Spionidae	P5-*S. squamata*	06° 55.962′ N; 93° 55.896′ E	B-Qu.1	KC208487
Nereidae	P6-*C. mirabilis*	06° 52.993′ N; 93° 55.990′ E	LXN5	KC208486

2.2 *DNA Extraction*

Standard DNA extraction protocol of Prasannakumar et al. (2011) was adopted. Briefly, small pieces of tissue from each individual was placed in 1.5 mL eppendorf tube and 500 µL of solution I (0.1 M NaCl, 10 mM Tris-HCl, pH 8.0 and 1 mM EDTA pH 8.0 and 2 % SDS) was added. The tissue was homogenised using a sterile homogenizer following which 5 µL of proteinase K (1 mg/mL) was added and vortexed for 5 min. The sample was incubated at 55 °C in a water bath for 2 h. After incubation the samples was chilled over ice for 10 min and solution II (6 m NaCl) was added and mixed by inverting the tubes several times. The tube was chilled on ice for 5 min and centrifuged at 8000 rpm for 15 min. About 500 µL of the supernatant was carefully collected into a new tube and two volume of (i.e., 1 mL) 100 % ethanol was added to precipitate the DNA. The precipitate was pellet down at 8000 rpm for 5 min and the supernatant was removed without touching the pellet. The DNA pellet was washed with 500 µL of ice cold ethanol and centrifuged at 11,000 rpm for 5 min. The supernatant was carefully removed and the excess liquid was decanted. The pellet was air dried and re-suspended with 50 µL of MilliQ water.

2.3 *PCR Amplification*

Primer pairs of Folmer et al. (1994) were used for amplification. It is been found that not a single pair of primer was effective in amplifying all the specimens collected. Corresponding primers used in amplifying specific species is listed in Table 2.

Primers were synthesised in a commercial company Bioserve biotechnologies, Pvt. Ltd. (India). Polymerase chain reaction (PCR) was conducted with final concentrations of 1 µM primers, 3 mM MgCl2, 0.4 mM each dNTP and 2.5 units of Taq polymerase (Merck, India). The thermocycle profile for amplifying barcode region consisted of warm up at 94 °C for 1 min, 5 cycles at 94 °C for 30 s, annealing at 45–50 °C for 40 s, and extension at 72 °C for 1 min, followed by

Table 2 Folmer's primers used for amplifying COI sequences of polychaetes

Sample	Best primer pairs	Primer sequences
P1, P5, P6	dgLCO/dgHCO	5′-GGTCAACAAATCATAAAGAYATYGG-3′/5′-TAAACTTCAGGGTGACCAAARAAYCA-3′
P2	LCO/dgHCO	5′-GGTCAACAAATCATAAAGATATTGG-3′/5′-TAAACTTCAGGGTGACCAAARAAYCA-3′
P3, P4	LCO/HCO	5′-GGTCAACAAATCATAAAGATATTGG-3′/5′-TAAACTTCAGGGTGACCAAAAAATCA-3′

30–35 cycles of 94 °C for 30 s, 51–54 °C for 40 s, and final extension at 72 °C for 10 min.

PCR amplicons of COI gene was gel checked using 1.2 % China agarose prepared in 1X TAE buffer. About 3 μl Ethidium Bromide was used as staining dye and 5 μl of Bromothymol blue was used as tracking dye. About 100 V DC was maintained between the electrodes of gel apparatus for 20 min. Following electrophoresis the gel was analyzed and pictured in gel doc system (Bioserve Biotechnologies, Pvt. Ltd.).

2.4 DNA Sequencing

QIAGEN QIAquick™ kit was used for sequencing reaction. The sequencing PCR was done to amplify one strand of barcode gene employing the primer LCO1490 only under standard PCR conditions. The samples were precipitated and suspended in 40 μl of loading solution provided with the kit. Sequencing was done with MegaBace sequencer at Bioserve Biotechnologies, India.

2.5 DNA Sequence Analysis and BLAST

COI gene sequences produced in the study were subjected to BLAST analysis through BLASTN 2.2.26 (Zhang et al. 2000). The sequence chromatograms are read manually and double checked for miscall and base spacing using Chromas Pro (Ver.1.5) (www.technelysium.com.au/ChromasPro.html).

The DNA sequences were aligned using ClustalX (ver.2.0), (Larkin et al. 2007). The phylogenetic trees are constructed using MEGA 4.0 (Molecular Evolutionary Genetic Analysis) software. MEGA is an integrated tool for conducting automatic and manual sequence alignment, inferring phylogenetic trees, mining the web base data bases, estimating the rates of molecular evolution, and testing evolutionary hypothesis.

2.6 DNA Sequences and GenBank Accession Numbers

The CO1 sequences produced in the present study could be accessed through accession numbers (Table 2).

3 Results

3.1 Systematic Account

3.1.1 Order Terebellida; Family Terebellidae; *Terebella Ehrenbergi* (Grube 1870)

Description
Body 35–40 mm in length with long and coiled tentacles, tentacles are filamentous, numerous and slender. Three pairs of arborescent gills are present on segments 2, 3 and 4 (Fig. 1a). Lateral lobes are lacking. The peristomium has eye spots. Notosetae are first present from the fourth segment and continue posteriorly but are absent from the last 40 segments. Setae are very slender and distally serrated (Fig. 1b). Uncini are in single rows on segments 5 to 10 and the last segment and in double rows on other segment. Each uncinus has 3 to 5 large teeth (Fig. 1c).

Material Observed: Nicobar Island.

Habitat: Soft tube forming on dead and live corals at 1 m water depth.

Distribution: Japan, China Sea, Burma and Red Sea. **India**: Gulf of Mannar, Andaman and Nicobar Islands, Mahanadi estuary and Gujarat.

Remarks: The present material agrees well with the original description.

3.1.2 Order Eunicida; Family Eunicidae *Lysidice Collaris* (Grube 1870; Gravier 1900; Day 1967; Rao and Soota 1977)

Description
Prostomium is distinctly bilobed in front and has two reniform eyes located near the outer base of the paired antennae (Fig. 2a). The 3 prostomial antennae are slender, second dental plate with three heavy teeth. In anterior segments the dorsal cirri are slenderer than ventral ones (Fig. 2b). In posterior segments the dorsal cirri become shorter. Setae include capillary setae (Fig. 2d) bidentate composite falcigers, comb setae and bidentate subacicular hooks are first present at setiger 21 and continue posteriorly (Fig. 2e).

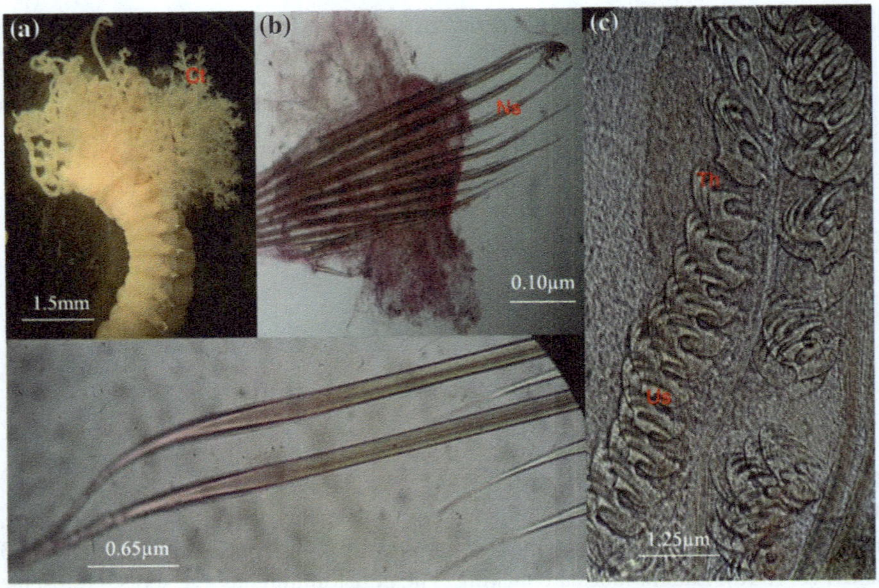

Cl - Coiled tentacles, Ns - Notosetae, Th - Teeth, Us - Uncinus

Fig. 1 *Terebella ehrenbergi* **a** Anterior end, **b** Notopodial seta, **c** Thoracic uncinus

Material: 114 specimens collected from St. 1–8, 10 and 11.

Distribution: Indian Ocean, Pacific Ocean, Persian Gulf, Red Sea. **India**: Andaman and Nicobar Islands, Kilakarai, Pamban, Gujarat coast and Gulf of Mannar.

Habitat: Boring in dead corals live on cervices of dead corals.

Remarks: Present materials agree well with the Day (1967) descriptions. In earlier examination 32 specimens collected from St. 1, 2, 5, 7, 8, 11 and 13.

3.1.3 Order Phyllodocida; Family Nereididae; *Perinereis Nuntia Brevicirrus* (Grube 1876; Fauvel 1932; Parulekar 1971)

Description

Maximum length specimen is 100 mm long and 6 mm wide, 108 setigers. Prostomium is pyriform, pairs of eyes in trapezoidal arrangement situated on the posterior part of prostomium (Fig. 3a). Tentacles are short and small, distally slender the palps large, especially the basalia expanded the terminalia very small button shaped. The longest peristomial cirrus extends back to setigers 7.

The paragnaths on proboscis have the following arrangement: I = 3 cones, II = 12–15 cones in 3 oblique rows; III = 13 cones in 3 longitudinal rows; IV a

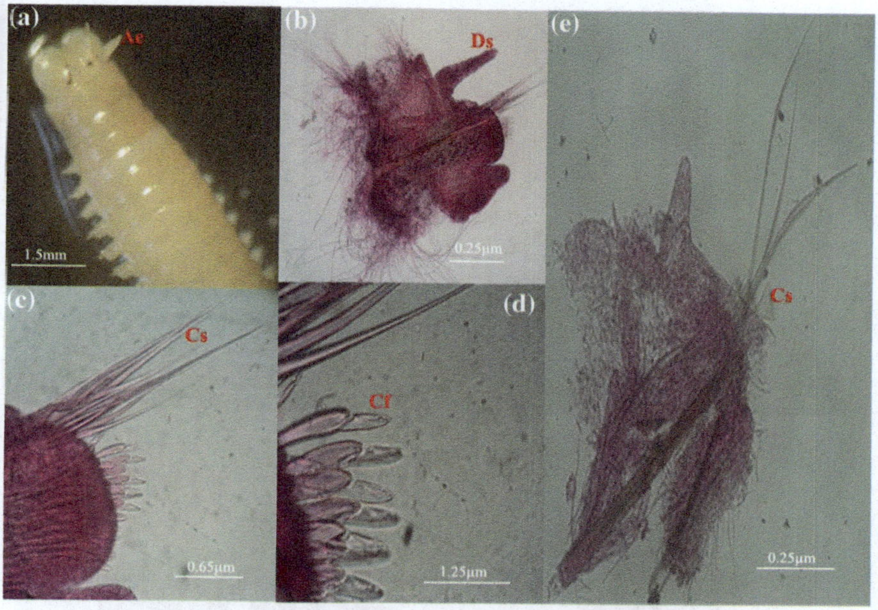

Ae - Antennae, Ds- Dorsal cirri, Cs- Capillary setae, Cf- Composite falcigers, Cs- Comb setae

Fig. 2 *Lysidice collaris* **a** Anterior end, **b** Anterior foot, **c** Posterior foot, **d** Limbate capillary, **e** Falciger and Comb setae

dense triangular group; V = 3 cones in a triangle; VI a transverse row of 5 to 8 flattened broad paragnaths; VII and VIII = 30–40 cones in 3 irregular rows.

Typical parapodia have all ligules conical with the dorsal ones the longest (Fig. 3b). Dorsal cirri are slender and extend distally somewhat beyond the tips of dorsal ligules (Fig. 3c). The anterior setigers, more than 10 in a live specimen are blue—black or green—black, the posterior region pale—brown. Notosetae with homogomph spinigers and neurosetae with hetrogomph falcigers (Fig. 3d).

Distribution: Japan, Australia, New Zealand, New Caledonia, Malay Archipelago, Indian Ocean, Saint Paul Island, Red Sea. **India**: Gulf of Mannar, Tuticorin, Cape Comorin, Andaman and Nicobar Islands, Maharashtra and Goa Coast.

Material:14 specimens collected from St. 2, 3, 5, 6, 7 and 11.

Habitat: Found among barnacles and oysters and in dead coral crevices at low tide.

Remarks: The present materials agree well with the Fauvel (1953) description.

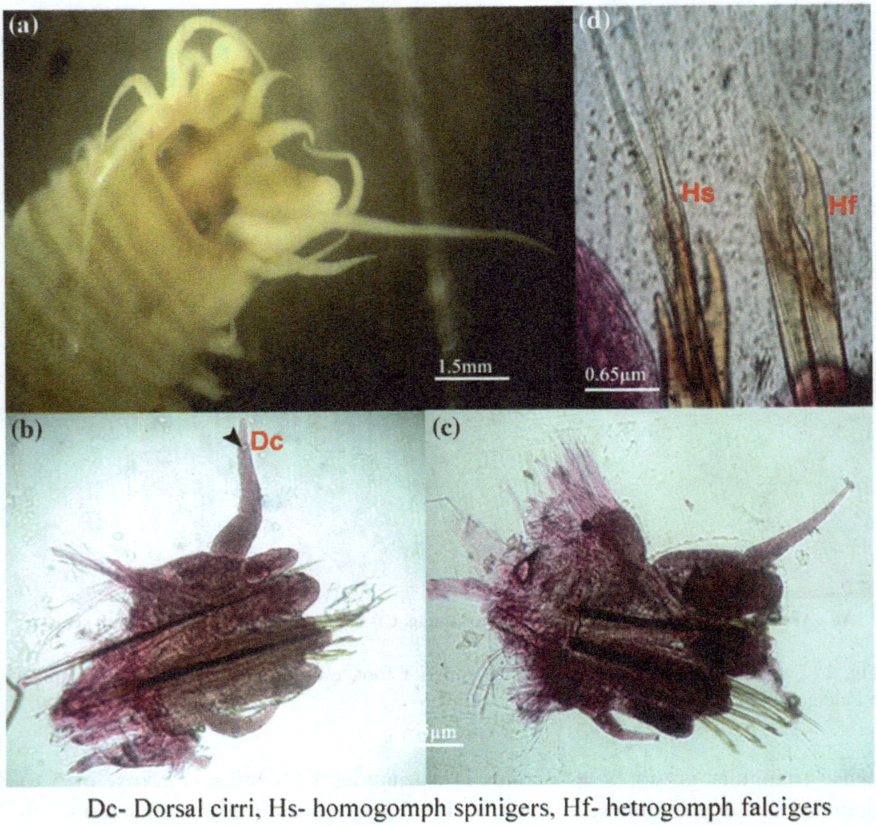

Fig. 3 *Perinereis nuntia brevicirrus* **a** Anterior end, **b** Anterior foot, **c** Posterior foot, **d** setae structure

3.1.4 Order Phyllodocida; Family Phyllodocidae; *Phyllodoce Fristedti* (Bergstrom 1914; Day 1962; Tampi and Rangarajan 1964)

Description

Body long, slender with numerous segments. Prostomium heart-shaped with a pair of prominent black eyes (Fig. 4a). In posterior—margin of prostomium notched and a small occipital tentacle present. Four short, subulate tentacles. Longest tentacular cirri reach back to 7th setiger. Numerous irregular rows of short papillae at the base of the long proboscis. Feet are uniramous. Dorsal and ventral cirri are foliaceous, lanceolate nearly twice as long as broad. Ventral cirri small and broad (Fig. 4c). Compound setae minutely serrated (Fig. 4d).

Distribution: Indian Oecean, Ceylon; **India**: Andaman and Nicobar Islands.

Material: 2 specimens collected from St.1and 4.

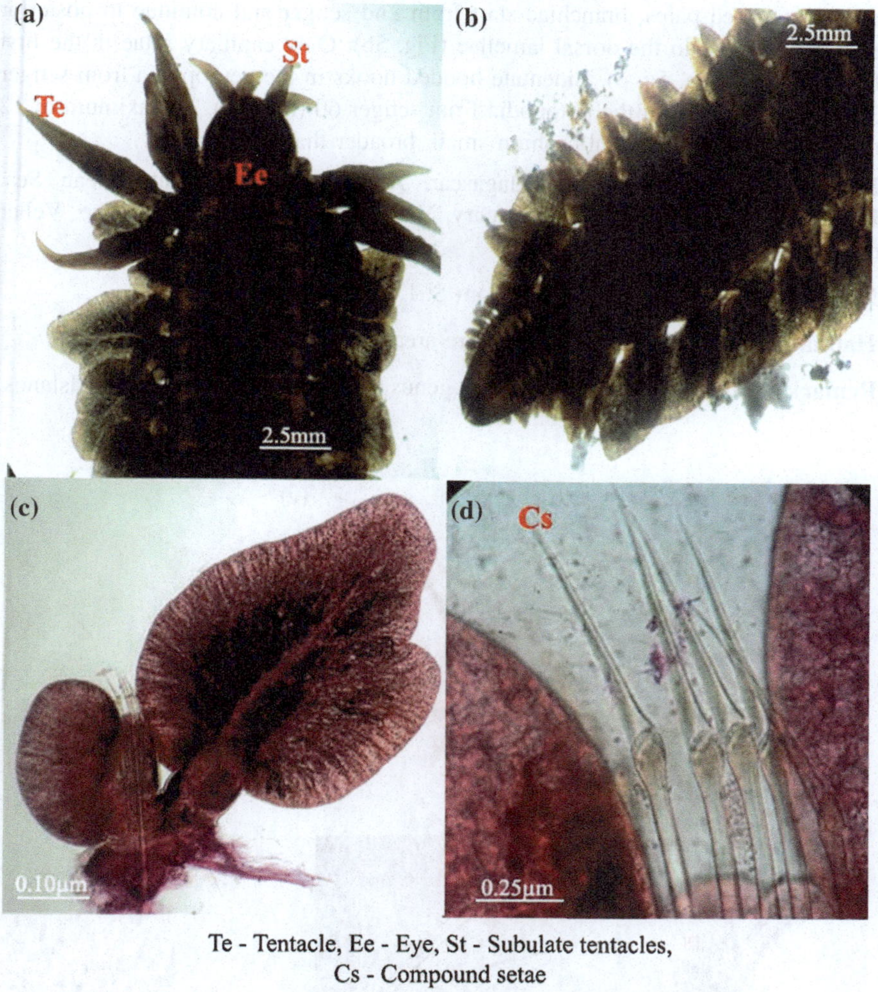

Fig. 4 *Phyllodoce fristedti* **a** Anterior end, **b** Anterior foot, **c** Posterior foot, **d** Setae

Habitat: Crevices of dead corals and beach rocks in intertidal zone.

Remarks: The present material agrees well with the descriptions of Day (1967).

3.1.5 Order Spionida; Family; Spionidae; *Scolelepis Squamata* (Muller 1806; Delle Chiaje 1825; Fauvel 1927; Day 1967; Misra and Chakraborty 1991; Rao 2001)

Description

Body 55–60 mm long, prostomium pointed anteriorly with 4–5 pairs of eyes in a row and a well-marked occipital keel reaching 2nd setiger (Fig. 5a). A pair of long and stout coiled palps, branchiae start from 2nd setiger and continue to posterior end and attached to the dorsal lamellae (Fig. 5b). Only capillary setae in the first few segments (Fig. 5c, d). Bidentate hooded hooks in the neuropodia from setiger 30–35 onwards and in the notopodia from setiger 60 (Fig. 5e). A maximum of 12 neuropodial hooks pygidial cushion small, broader than long.

Distribution: Mosambique, Madagascar, Atlantic Ocean, Mediterranean Sea. **India**: Orissa coast, Rushikulya estuary, Visakhapatam coast, Pulicat lake, Vellar estuary and Godavary estuary.

Material: 13 specimens collected from St.1, 2, 3, 9 and 11.

Habitat: Silty sediments in sandy shore areas.

Remarks: This is the first record of the genus from Andaman and Nicobar Islands.

Cl- Coiled palps, Pr- Prostomium, Dl - Dorsal lamellae, Cs- Capillary setae, Nh - Neuropodial hooks

Fig. 5 *Scolelepis squamata* **a** Anterior end, **b** Anterior foot, **c** Posterior foot, **d** Hooded hook, **e** Notosetae

3.1.6 Order Phyllodocida; Family Nereididae; *Ceratonereis Mirabilis* (Kinberg 1866; Gravier 1901; Fauvel 1953; Day 1967; Misra and Chakraborty 1991)

Description

The prostomium is broad, more than twice as wide as long and has a deep cleft between the antennae (Fig. 6a). The basalia of palps is quite long and terminalia button shaped. 2 pairs of eyes in rectangular arrangement, the longest peristomialcirus extend back to setiger 17. Prostomium and dorsum of palps are light green, dorsum of segment has distinct light green or green-brown band, which becomes lighter towards posterior. The other parts of the body are white. Paragnaths are present only on maxillary ring of the proboscis: I 0; II 10–13 cones in 2 oblique clusters; III 7–9 cones in one cluster; IV 10–14 cones, lateral teeth of the jaw are indistinct (Fig. 6e). The first two pairs of parapodia are uniramous, the rest biramous. The dorsal cirrus is very long, 3 times as long as notoligule, neuroligule is slightly shorter but thicker. The dorsal and ventral cirri are digitate; acicular lobes are very small, only as a projection, shorter than ventral cirrus. The dorsal cirrus of anterior parapodia is 5 times long as notoligule (Fig. 6b) and notoligulesdigitate, while supra-notoligules thicker.

The acicular lobes of neuropodium are short, distally obtuse; neuroligule short but slightly longer than neuro-acicular lobe. The dorsal segments of middle and posterior cirrus are rather long (Fig. 6c). Anterior notoseate are homogomph spinigers. Indistinct hetrogomph falcigers appear from middle parapodia, end of terminal piece is beaked. Some posterior setigers bear homogomph falcigers in which the end of terminal piece bifid. Notopodial falcigers homogomph, neuropodial falcigers homogomph (Fig. 6d).

Distribution: Red Sea, Persian Gulf, Indian and Atlantic oceans, Japan, New Caledonia, New Zealand, Honolulu, Australia, Brazil, West Indies **India**: Lakshadweep, Andaman and Nicobar Islands, Krusadai island, Pamban, Kilakarai, Maharashtra and Goa Coast.

Material: 25 specimens collected from station 1 to 10.

Habitat: Silty sand substratum under coral rubbles and surface of dead coral.

Remarks: The species is characterised by its cleft prostomium and the presence of notopodial falcigers on posterior setigers.

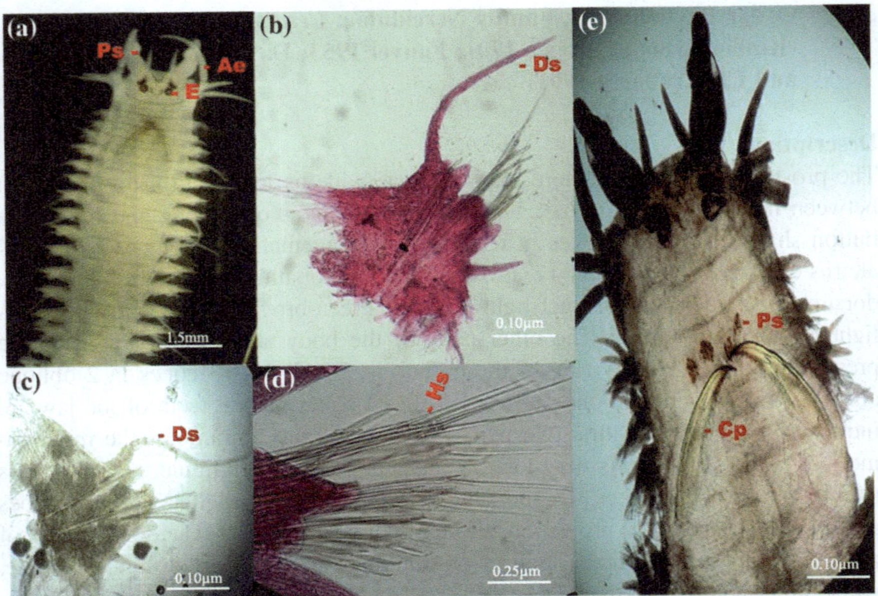

Ps - Palps, E - Eye, Ae - Antennae, Ds - Dorsal cirrus, Hs - Homogomph spiniger, Hs - Homogomph falciger, Nf - Neuropodial falciger(Long) and Short, Ps - Paragnaths, Cp - Chitinous paragnath

Fig. 6 *Ceratonereis mirabilis* **a** Anterior end, **b** Anterior foot, **c** Posterior foot, **d** Setae, **e** Dorsal view of proboscis

4 DNA Barcoding Analysis Though COI Gene Sequences

Precise location of sample collection and specimen identification along with its CO1 accession numbers were tabulated (Table 1).

Successful amplification and sequencing of CO1 gene was achieved from 6 polychaete genera from Nicobar Islands with majority of taxa possessing unique sequences. A distinct pattern of variable region was observed in the COI gene sequences among all the taxa. The COI molecule exhibiting variable regions were tested for molecule barcoding. Almost all the sequences showed a similarity between 85 and 98 % when compared with the polychaete sequences available online in Genbank and EMBL data base (Table 3). The threshold limit for differentiating polychaete species based on COI sequence similarity was lacking. Sequences generated in this study agree with morphological taxon.

In the present study it was observed that the polychaete species *Lysidice collaris* (P2) recorded shared 98 % similarity (2 % variation). Similarly the species *Scolelepis squamata* (P5) shared 98 % similarity (2 % variation sequenced from USA) with *Scolelepis squamata* previously sequenced from Canada. The other specimens *Terebella ehrenbergi* (P1), *Perinereis nuntia* (P3), *Phyllodoce maculate* (P4) and *Ceratonereis mirabilis* (P6) sequenced for first time.

Table 3 Similarity between the COI sequences produced in the present study and the reference sequences from GenBank database with its geographical identity

Strain name	Biological name	Similarity (%)	Reference strain name and accession number	Country of occurrence of reference strain
P1	*T. ehrenbergi*	93	*Terebella tantabiddycreekensis* and EU835671	Australia
P2	*L. collaris*	98	*Lysidice collaris* and GQ497557	USA
P3	*P. nuntia*	92	*Perinereis vallata* and HQ705192	Chile
P4	*P. fristedti*	93	*Phyllodoce maculate* and GU670818	Canada
P5	*S. squamata*	98	*Scolelepis squamata* and HM473680	Canada
P6	*C. mirabilis*	85	*Diplocirrus longisetosus* and HQ024296	Canada

5 Mitochondrial DNA Sequencing Analysis

Among the CO1 sequence of the 6 polychaetes, 403 variable positions were observed which results in 6 different haplotyps. The mean number of pair wise differences between two polychaete individuals was 0.219 ± 0.1. The phylogenitic analysis of CO1 dataset is presented as a NS dendogram with bootstrap analysis (Fig. 7) with closest blast hits showing a grouping of all species into 6 clades, supported by high bootstrap values. The Mp analysis yielded 32 most paramonious trees, each with a length of 891 steps, a contriteness index (CI) of 0.66 and a retention index of 0.84. The phylogenetic tree showed distinct clade for the 6 genera as determined from morphology.

Six specimens formed 6 distinct clades in the constructed phylogram. The clade 1 the one in the top of the phylogram (Fig. 7) contained *Lydice collaris* CASMB LXN1 along with *Lydice collaris* DQ470469, DQ470468 and GQ497557 with 98.2 % identity. Clade 2 consists of *Scolelepis squamata* CASMB B-Quarry 1, which showed 96.7 % similarity with *Scolelepis squamata* Bioug-BAMPOL0179 (HM473680) and Bioug-BAMPOL0017 (HM473679). Clade 3 has *Perinereis nuntia brevicirrus* CASMB CMP2 with a 91.2 and 90.4 %, sequence identity towards *Perinereis valdata* HQ705193 and HQ705196 respectively. Clade 4 contained *Terebella ehrenbergi* CASMB VNR6 and clade 5 *Phyllodoce fristedti* CASMB VNR7 shows 91.1 and 94.4 % identify with the members of the family Terebellidae and Phyllodocidae.

Phylogenetic relationships between the polychaetes of Andaman were shown (Fig. 7). The voucher name CASMB indicates the sequences of present study and other voucher names in the phylogram represents the reference sequences extracted from Genbank database. Since most of the sequences (n = 4) were sequenced for

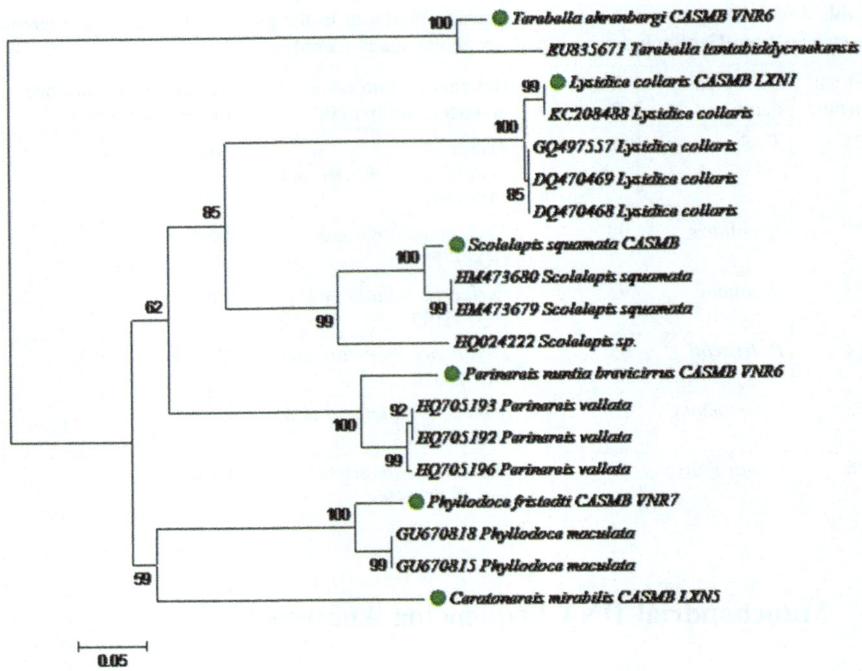

Fig. 7 Cluster analysis of the COI sequences (K2P) for 6 polychaetes species from Nicobar Islands

the first time, Genbank did not contained similar sequences for CASMB-LXN5, hence no similar sequences were used. CASMB-LXN5 formed the out-group in the constructed phylogram.

Members of same species clustered in the same clade proving the reliability of COI gene sequences in identifying polychaetes. *Ceratonereis mirabilis* CASMB LXN5 was found to be a distant relative to all other genera and none of the closest similarity was observed during the blast due to the lack of close sequences in the database it was transparently placed outside of the cluster.

6 Genetic Distance

Inter-generic pair-wise distance between polychaete genera showed high levels of intraspecific gene flow among the Nicobar Islands sampling sites with all 6 lineages being highly divergent with mean distance of 0.48. The genetic distance between the genera of polychaets was calculated using pair wise distance analysis via maximum likelihood method (Table 4). Among the 6 genera analysed, maximum distance value was observed between *Lysidice collaris* CASMB LXN1 and

Table 4 Variations among the COI pair-wise distances within the polychaetes

	T. ehrenbergi	*L. collaris*	*P. n. brevicirrus*	*P. fristedti*	*S. squamata*	*C, mirabilis*
T. ehrenbergi						
L. collaris	0.727					
P.n. brevicirrus	0.647	0.375				
P. fristedti	0.623	0.425	0.366			
S. squamata	0.725	0.364	0.350	0.361		
C. mirabilis	0.665	0.446	0.378	0.316	0.440	

Terebella ehrenbergi CASMB VNR6 (0.727). The minimum genetic distance (0.316) was observed between genera *Phyllodoce fristedti* CASMB VNR7 and *Ceratonereis mirabilis* CASMB LXN5.

7 Discussion

In the marine realm, annelid are one of the most dominant animal phyla exhibiting an amazing variety of morphological forms with each species having specific characters, yet they are less studied compared to other taxa of similar ecological importance and complicated morphological evidence (Fauchald 1977). Morphological identifications of polychaete species is time consuming and somewhat inaccurate leading to possible misidentifications and morphological traits are complex due to the high levels of homoplasy (Eklof 2010). Although morphological identification has been performed following the early description in monographs of Day (1967), Fauvel (1930, 1932) and Fauchald (1977).

DNA barcoding is a well accepted taxonomic method which uses a short genetic marker to facilitate identification of a particular species even by non-specialist. DNA barcoding can reliably assign unknown specimens to known species, also flagging potential cryptic species and genetically distant populations (Radulovici et al. 2010). The popularity of COI DNA 'barcoding' is increasing rapidly, with mass amounts of invertebrates and vertebrates collected in the field inevitably becoming a mass of data to be added. With so much data needing to be processed, 'taxonomic impediment' exists just as much for molecular data as it does for traditional collections (Brower 2006). Hebert et al. (2004a, b) proposed DNA barcoding works under the principle that interspecies variations are greater than the intra-species variations allowing one to distinguish the species using nucleotide sequences. CO1 has been accepted as universal barcode to delineate animal life. Sequence variation in a segment of the mitochondrial cytochrome c oxidase I gene was employed to compare morphological versus molecular diversity estimates. The phylogenetic relationships of these polychaete taxa are matter of ongoing debates in recent papers on annelid morphology (Westheide et al. 1999). Based on multiple studies Hebert et al. (2003) suggested an approximately 650 bp of the CO1 gene which is relatively easy to amplify with standard primers and is sufficient enough to obtain resolution on all levels between species and phylum for majority of the groups albeit with some exceptions (Ward 2009). CO1 has been proposed as the principle gene for barcoding organisms. Sequence variation in a segment of the mitochondrial cytochrome c oxidase I gene was employed to compare morphological versus molecular diversity.

Morphological identification of the polychaetes in this study was supported by the molecular data (n = 2), as shown by the congruence and high similarity between the sequences produced in the present study and those available in GenBank. Discrepancies between the morphological identifications and the 3 genera closest matches in the molecular database are most likely the result of the limited taxonomic coverage of COI sequences available in GenBank relative to the vast diversity of marine polychaetes. The likely applicability of a COI identification system to new animal groups and geographical settings suggests the feasibility of creating an identification system for animals-at-large scale.

Carr et al. (2011) demonstrated the effectiveness of DNA barcoding as a tool for species identification in polychaetes. The clustering pattern of COI barcodes flagged misidentifications, guided taxonomic decisions, and facilitated the detection of diversity overlooked by the current taxonomic system. In Indian waters several author have barcoded various flora and fauna (Persis et al. 2009; Lakra et al. 2009; Ajmal Khan et al. 2010, 2011; Akbar John et al. 2010; Prasannakumar et al. 2011; Sachithanandam et al. 2012) but no study have compared morphological and molecular data. Such comparisons have been previously published for various fauna (Bleidorn et al. 2003; McHugh 2000; Rota et al. 2001; Struck et al. 2002), whereas in polychaetes has been published in NCBI (Satheeshkumar and Jagadeesan 2010; Samidurai 2010).

Phylogenetic relationships among different organisms are of fundamental importance in biology and one of the prime objectives of DNA sequence analysis is phylogeny reconstruction for understanding evolutionary history of organisms (Chaudhri and Das 2001). Molecular phylogenetic analyses of closely related species provide insights into their relationships, allowing us to verify their morphological taxonomic classification. Sometimes, such studies indicate that the previously assumed classification is wrong or not sufficient (Timm et al. 2008). In the present study six species of polychaetes belonging to four different families was barcoded. Among these 4 barcodes was not available in NCBI earlier. It is the first new information for Genbank database that will act as a benchmark and reference data for identifying respective polychaetes species around the world in near the future.

In accordance with previous analyses most hypotheses on relationships among polychaete families are only weakly supported (Bleidorn et al. 2003; Brown et al. 1999; Kojima 1998), Orbinidae are closely related to the Parergodrilidae a result supporting the analysis of Struck et al. (2002). Interestingly, this relationship was also found in the cladistic analyses by Rouse and Fauchald (1997). Irrespective of the method the analysis suggests a probable paraphyly of the Orbiniidae with regard to Questa (Bleidorn et al. 2003). Among the 6 species barcoded in the present study, members of the same species clustered in the same clade proving the reliability of COI gene sequences in identifying polychaetes. *Ceratonereis mirabilis* CASMB LXN5 was found to be a distant relative to all other genera and no close similarity was observed during the blast analysis in GenBank due to the lack of close sequences in the database it was transparently placed outside of the cluster. This indicates the necessity to expand the polychaete barcodes data in the Genbank. Maturana et al. (2011) assessed intraspecific and interspecific genetic divergence among marine polychaetes of 13 polychaetes species and identified high levels of interspecific variation among 31 analyzed sequences. Mean pairwise sequence distance comparisons ranged from 0.2 to 0.4 % in previous study, but the in the present study interspecific comparisons were much higher and ranged between 18 and 47 %.

Lower intraspecific divergences were found within Nicobar polychaetes, which showed an average of 26.4 % ranging from 13.8 to 36.8 %. The smaller intraspecific divergences are unlikely to confound species boundaries. This is similar to what was observed between two morphologically similar species (Sato 1999; Ball et al. 2005; Monaghan et al. 2005). Other studies such as spider barcoding that uses CO1 sequences has found much lower defining divergence levels of <2 % (Hebert and Barrett 2005). Limited sample size is found to be an obstacle in performing such analysis. However among 6 genera analysed in the present study maximum distance value was observed between *Lysidice collaris* and *Terebella ehrenbergi* (72.7 %). The minimum genetic distance (31.6 %) was observed between genera *Phyllodoce fristedti* and *Ceratonereis mirabilis*. Among 6 sequences produced in the present study, only 2 have been identified precisely. Other 4 sequences have been identified only to family level. Hence the sequences serve as a bench mark data besides highlighting the importance in expanding the polychaete barcode database. This study presents the first information on DNA barcoding for polychaetes species in the Great Nicobar Island, and it establishes the effectiveness of DNA barcoding for identification of marine polychaetes species from Andaman and Nicobar Island, thus making it available to a much broader range of scientists. In Future, Indian water need more studies on marine polychaete though COI gene sequence approach, will contribute increase biodiversity.

References

Ajmal Khan S, Lyla PS, Akbar John B, Prasanna Kumar C, Murugan S (2011) Identifying marine fin fishes using DNA barcodes. Curr Sci 101:9–12

Ajmal Khan S, Lyla PS, Akbar Jhon B, Prasanna Kumar C, Murugan S, Jala KCA (2010) DNA, barcoding of *Stolephorus indicus* and *Stolephorus commersonni*. Biotechnology 9(3):373–377

Akbar John B, Prasannakuma C, Lyla PS, Ajmal Khan S, Jalal KCA (2010) DNA barcoding of *Lates calcarifer* (Bloch, 1970). Res J Biol Sci 5(6):414–419

Ball SL, Hebert PDN, Burian SK, Webb JM (2005) Biological identifications of mayflies (Ephemeroptera) using DNA barcodes. J N Am Benthol Soc 24(3):508–524

Bellwood DR, Hughes TP, Folke C, Nystrom M (2004) Confronting the coral reef crisis. Nature 429:827–833

Bely AE, Wray GA (2004) Molecular phylogeny of naidid worms (Annelida: Clitellata) based on cytochrome oxidase *I*. Mol Phylogenet Evol 30:50–63

Blaxter ML (2004). The promise of DNA taxonomy. Philos Trans R Soc Lond Ser B Biol Sci. 359:669–679

Blaxter M, Elsworth B, Daub J (2004) DNA taxonomy of a neglected animal phylum: an unexpected diversity of tardigrades. Proc Roy Soc Lond Biol Sci 271:S189–S192

Bleidorn C, Podsiadlowski L, Bartolomaeus T (2006) The complete mitochondrial genome of the orbiniid polychaete Orbinia latreillii (Annelida Orbiniidae)—A novel gene order for Annelida and implications for annelid phylogeny. Gene 370:96–103

Bleidorn C, Vogt L, Bartolomaeus T (2003) New insights into polychaete phylogeny (Annelida) inferred from 18S rDNA sequences. Mol Phylogenet Evol 29:279–288

Breton S, Dufresne F, Desrosiers G, Blier PU (2003) Population structure of two northern hemisphere polychaetes, *Neanthes virens* and *Hediste diversicolor* (Nereididae) with different life history traits. Mar Biol 142:707–715

Brett CD (2006) Testing the effectiveness of the mt DNA Cytochrome c oxidase subunit 1 (COI) gene locus for identifying species of Polychaete worm (Polychaeta: Annelida) in New Zealand

Brower AVZ (2006) Problems with DNA barcodes for species delimitation: "ten species" of *Astraptes fulgerator* reassessed (Lepidoptera: Hesperiidae). Syst Biol 4:127–132

Brown S, Rouse G, Hutchings P, Colgan D (1999) Assessing the usefulness of histone H3, U2 snRNA and 28rDNA in analyses of polychaete relationships. Aust J Zool 47:499–516

Canales-Aguirre CB, Rozbaczylo N, Hernandez CE (2011) Genetic identification of benthic polychaetes in a biodiversity hotspot in the southeast Pacific. Revista de Biologia Marina y Oceanografia 46:89–94

Carr CM, Hardy SM, Brown TM, Macdonald TA, Hebert PDN (2011) A Tri-Oceanic perspective: DNA barcoding reveals geographic structure and cryptic diversity in Canadian polychaetes. PLoS ONE 6(7):222–232

Chotteau LA, Laudet V, Flourens A, Begue A, Leprince D, Fontaine F (1994) Identification of two ets related genes in a marine worm, the polychaete annelid Nereis diversicolor. FEBS Lett 354:62–66

Chaudhri P, Das S (2001) Statistical analysis of large DNA sequences using distribution of DNA words. Curr Sci 80(9):1161–1166

Christer E, Prestegaard T, Mari K (2000) Phylogenetic analysis of Tubificidae (Annelida: Clitellata) based on 18S rDNA sequences. Mol Phylogenet Evol 15(3):381–389

Day JH (1962) Polychaeta from several localities in the Western Indian Ocean. Proc zool Soc Lond 139(4):627–656

Day JH (1967) A monograph on the Polychaeta of Southern Africa. Pts I and II Brit Mus Nat Hist 656:1–878

Delle Chiaje S (1825) Memorie sulla storia e notomia degli animali senza vertebre del regno di Napoli. Naples 3:1–232

Derycke S, Remerie T, Vierstraete A, Backeljau T, Vanfleteren J, Vincx M, Moens T (2005) Mitochondrial DNA variation and cryptic speciation within the free-living marine nematode Pellioditis marina. Mar Ecol Prog Ser 300:91–103

Eklof J (2010) Taxonomy and phylogeny of polychaetes. Intellecta Infolog, Goteborg, 33 pp

Fauchald K (1977) The Polychaete worms. Definition and keys to the orders, families and genera. Nat His Mus Ang Sci Ser 28:1–188

Fauvel P (1927) Polychetes sendentaries. Addenda aux Errantes, Archiannelida, Myzoztomaires. Fauna Fr 16:1–494

Fauvel P (1932) Polychaeta of Indian Museum. Calcutta Mem Ind Mus 12:1–262

Fauvel P (1953) Annelida Polychaeta. In: The Fauna of India including Pakistan, Ceylon, Burma and Malaya. The Indian Press Ltd, Allahabad, 507 pp

Fauvel P (1930) The littoral fauna of Krusadai Island in the Gulf of Mannar. Bull Madras Gov Mus (New Series—Natural History Section) I(2):1–72

Fernando OJ, Rajasekaran R (2007) A new species of Namanereidinae: Namalycastis glasby sp. nov. from Indian waters. J Bombay Nat Hist Soc 104:1

Folmer O, Black M, Hoeh W, Lutz R, Vrijenhoek R (1994) DNA primers for amplification of mitochondrial cytochrome c oxidase subunit I from diverse metazoan invertebrates. Mol Mar Bio Biotechnol 3(5):294–299

Floyd R, Abebe E, Papert A, Blaxter M (2002) Molecular barcodes for soil nematode identification. Mol Ecol 11:839–850

Gibson G, Paterson IG, Taylor H, Woolridge B (1999) Molecular and morphological evidence of a single species, Boccardia proboscidea (Polychaeta: Spionidae), with multiple development modes. Mar Biol 134:743–751

Gravier C (1900) Contribution to the study of polychaete annelids. I partie. Nouv Archs Mus Hist Nat 3:137–282

Gravier C (1901) Contribution to the study of polychaete annilides Red Sea. Nouv Arch Mus Hist Nat Paris (Ser. 4) 3:147–268

Grube AE (1870) Beschreibung neuer oder wenig beakannten von Heron Ehrenberg gesammalter Anneliden aus den Rothen Meeres. Mber Akad Wiss Berl 1870:484–521

Grube AE (1876) Bemerkungen uber die Familie der Aphroditeen, (Gruppe Polynoina, Acoeta, Polylepidea). Jber Schles Ges Vaterl Kult 53:46–72

Halanych KM, Janosik AM (2006) A review of molecular markers used for Annelid phylogenetics. Integr Comp Biol 46:533–543

Hebert PDN, Cywinska A, Ball SL, Dewaard JR (2003) Biological identifications through DNA barcodes. Proc R Soc Lond B Biol Sci 270:313–322

Hebert PDN, Stoeckle MY, Zemlak TS, Francis CM (2004a) Identification of birds through DNA Barcodes. PLoS Biology 2(10):1657–1663

Hebert PDN, Penton EH, Burns JM, Janzen DH, Hallwachs W (2004b) Ten species in one: DNA barcoding reveals cryptic species in the neotropical skipper butterfly *Astraptes fulgerator*. Proc Natl Acad Sci USA 101:14812–14817

Hebert PDN, Barrett RDH (2005) Identifying spiders through DNA barcodes. Can J Zool 83:481–491

Hebert PDN, Cywinska AA, Ball SL, deWaard JR (2010) Biological identifications through DNA barcodes. Proc R Soc Lond 270:313–321

Hogg ID, Hebert PDN (2004) Biological identification of springtails (Hexapoda: Collembola) from the Canadian Arctic, using mitochondrial DNA barcodes. Can J Zool 82(5):749–754

Hopkins GW, Freckleton RP (2002) Declines in the numbers of amateur and professional taxonomists: implications for conservation. Ani Cons 5:245–249

Hutchings P (1998) Biodiversity and functioning of polychaetes in benthic sediments. Biodivers Conserv 7:1133–1145

Kinberg JGH (1866) Annulata nova. Ofvers K Vetensk Akad Forh 22:167–169, 239–258

Knowlton N (2001) Molecular genetic analyses of species boundaries in the sea. Hydrobiologia 420:73–90

Kojima S (1998) Paraphyletic status of Polychaeta suggested by phylogenetic analysis based on the amino acid sequences of elongation factor-1 alpha. Mol Phylogenet Evol 9:255–261

Lambert DM, Baker A, Huynen L, Haddrath O, Hebert PDN, Millar CD (2005) Is a large-scale DNA-based inventory of ancient life possible? J Hered 96(3):279–284

Larkin M, Blackshields G, Brown N, Chenna R, McGettigan P, McWilliam H (2007) Clustal W and Clustal X version 2.0. Bioinformatics 23(21):2947–2948

Lakra WS, Verma MS, Goswami M, Lal KK, Mohindra V, Punia P, Gopalakrishnan A, Singh KV, Ward RD, Hebert P (2009) DNA barcoding Indian marine fishes. Mol Ecol Resour 11(1):60–71

Leasi F, Todaro MA (2009) Meiofaunal cryptic species revealed by confocal microscopy: the case of *Xenotrichula intermedia* (Gastrotricha). Mar Biol 156:1335–1346

Manchenko GP, Radashevsky VI (2002) Genetic divergences between two sibling sympatric Dipolydora species (Polychaeta: Spionidae) from the Sea of Japan, and a new species description. J Mar Biol Assoc UK 82:193–199

Maltagliati F, Peru AP, Casu M, Rossi F, Lardicci C, Curini-Galletti M, Castelli A (2000) Is *Syllis gracilis* (Polychaeta: Syllidae) a species complex? An allozyme perspective. Mar Biol 136:871–879

Maltagliati F, Casu M, Castelli A (2004) Morphological and genetic evidence supports the existence of two species in the genus Ophelia (Annelida, Polychaeta) from the Western Mediterranean. Biol J Linn Soc 83:101–113

Maturana CS, Moreno RA, Labra FA (2011) DNA barcoding of marine polychaetes species of southern Patagonian fjords. Revista De Biología Mar Oceano 46:35–42

McHugh D (2000) Molecular phylogeny of the Annelida. Can J Zool 78:1873–1884

Monaghan MT, Balke M, Ryan-Gregory T, Vogler AP (2005) DNA-based species delineation in tropical beetles using mitochondrial and nuclear markers. Philos Trans R Soc Lond B 360:1925–1933

Meyer M, Briggs AW, Maricic T, Hober B, Hoffner B, Krause J, Weihmann A, Paabo S, Hofreiter M (2008) From micrograms to picograms: quantitative PCR reduces the material demands of high-throughput sequencing. Nucleic Acids Res 36:e5

Misra A, Chakraborty RK (1991) Polychaetes from Lakshadweep. Fauna of Lakshadweep. Zool Surv India (State Fauna Series) 2:137–165

Moritz C, Cicero C (2004) DNA barcoding: promise and pitfalls. PLoS Biol 2(10):1529–1531

Muller OF (1806) Zoologia Danica Seu Animalium Daniae et Norvegiae rariorum ac minus notorum, Descriptiones et Historia. Havniae 160 pp

Novacek M, Cleland EE (2001) The current biodiversity extinction event: scenarios for mitigation and recovery. Proc Natl Acad Sci USA 98:5466–5470

Nygren A, Pleijel F (2010) From one to ten in a single stroke—resolving the European *Eumida sanguinea* (Phyllodocidae: Annelida) species complex. Mol Phyl Evol 58:132–141

Nylander JAA, Erseus C, Källersjo M (1999) A test of monophyly of the gutless Phallodrilinae (Oligochaeta, Tubificidae) and the use of a 573-bp region of the mitochondrial cytochrome oxidase I gene in analysis of annelid phylogeny. Zool Scr 28:305–313

Olivares RA, Fleeger JW, Foltz DW (2001) Decoupling of molecular and morphological evolution in deep lineages of a meiobenthic harpacticoid copepod. Mol Biol Evol 18:1088–1102

Pop DD, Zdrenghea LM, Procociuc A Popal (2007) Gene polymorphism of angiotensin-converting enzyme and angiotensin II type 1 receptor in patients with congestive heart failure. Rome J Intern Med 45:349–354

Prasannakumar C, Akbar John B, Khan SA, Lyla PS, Murugan S, Rozihan M, Jalal KCA (2011) Efficiency of Universal Barcode Gene (Goxi) on morphologically cryptic mugillidae fishes delineation. Trends Appl Sci Res 4:1819–3579

Parulekar AH (1971) Polychaete from Maharashtra and Goa. J Bombay Nat Hist Soc 68 (3):726–749

Persis M, Reddy ACS, Rao LM, Khedkar GD, Ravinder K, Nasruddin K (2009) CO1 (Cytochrome oxidase–I) sequence based studies of Carangid fishes from Kakinada coast. India Mol Biol Rep 36:1733–1740

Radulovici AE, Archambault P, Dufresne F (2010) DNA barcodes for marine biodiversity: moving fast forward? (February). Diversity 2(4):450–472

Rao CAN (2001) Polychaete fauna of the Godavari Estuary Zool Surv India Esturine Ecosystem Series 4:21–32

Ratnasingham S, Hebert PDN (2007) The barcode of life data system. Mol Ecol Notes 7:355–364. http://www.barcodinglite.org

Rao CAN, Soota TD (1977) On the occurrence of *Poecilochaetus serpens* Allen (Polychaeta: Poecilochaetidae) in the Andaman and Nicobar Islands. Newsl Zool Surv India 3(6):346–347

Rota E, Martin P, Erseus C (2001) Soil-dwelling polychaetes: enigmatic as ever? Some hints on their phylogenetic relationship as suggested by a maximum parsimony analysis of 18S rRNA gene sequences. Contrib Zool 70:127–38

Rouse GW, Fauchald K (1997) Cladistics and polychaetes. Zoolog Scr 26:139–204

Sachithanandam V, Mohan PM, Muruganandam N, Chaaithanya IK, Dhivya P, Baskaran R (2012) DNA barcoding, phylogenetic study of *Epinephelus* spp. from Andaman coastal region, India. Indian J Geo-Mar Sci 41(3):203–211

Saccone C, DeCarla G, Gissi C, Pesole G, Reynes A (1999) Evolutionary genomics in the Metazoa: the mitochondrial DNA as a model system. Gene 238:195–210

Samidurai S (2010) Studies on macrobenthos of different mangrove habitats of Tamil Nadu, India: ecology, molecular taxonomy and mosquitocidal properties. Ph.D. thesis. Annamalai University, p. 174

Satheeshkumar P, Jagadeesan L (2010) Phylogenetic position and genetic diversity of Neridae—polychaeta based on molecular data from 16S r RNA sequences. Middle East J Sci Res 6 (6):550–555

Sato OW (1999) Polydorid species (Polychaeta: Spionidae) in Japan, with descriptions of morphology, ecology and burrow structure. Boring Species J Mar Biolo Assoc UK 79:831–848

Scaps P, Rouabadh A, Lepretre A (2000) Morphological and biochemical evidence that *Perinereis cultrifera* (Polychaeta: Nereididae) is a complex of species. J Mar Biol Assoc UK 80:735–736

Schmidt WWH (2003) Cosmopolitan versus cryptic meiofaunal polychaete species: an approach to a molecular taxonomy. Helgol Mar Res 57:1–6

Kvist S, Sarkar IN, Erseus C (2010) Genetic variation and phylogeny of the cosmopolitan marine genus Tubificoides (Annelida: Clitellata: Naididae: Tubificinae) J Mol Evol 57:687–702

Siddall ME, Budinoff RB (2005) DNA-barcoding evidence for widespread introductions of a leech from South American *Helobdella triserialis* complex. Conserv Genet 6:467–472

Stevens MI, Hogg ID (2003) Long-term isolation and recent range expansion from glacial refugia revealed for the endemic springtail *Gomphiocephalus hodgsoni* from Victoria Land. Antarctica Mol Ecol 12:2357–2369

Struck TH, Westheide W, Purschke G (2002) Progenesis in Eunicida (Polychaeta: Annelida) separate evolutionary events? Evidence from molecular data. Molec Phylogen Evol 25:190–199

Tampi PRS, Rangarajan K (1964) Some polychaetous annelids from the Andaman waters. J Mar Biol Ass India 6(1):98–123

Timm J, Figiel M, Kochzius M (2008) Contrasting patterns in species boundaries and evolution of anemonefishes (Amphiprioninae, Pomacentridae) in the centre of marine biodiversity. Mol Phyl Evol 49:268–276

Valentini A, Pompanon F, Taberlet P (2009) DNA barcoding for ecologist. Trends Ecol Evol 24:110–117

Ward RD (2009) DNA barcode divergence among species and genera of birds and fishes. Mol Ecol Resour 9:1077–1085

Ward RD, Zemlak TS, Innes BH, Last PR, Hebert PDN (2005) DNA barcoding Australia's fish species. Phil Trans R Soc B 360:1847–1857

Westheide W, McHugh D, Purschke G, Rouse GW (1999) Systematization of the Annelida: different approaches. Hydrobiologia 402:291–307

Zhang Z, Schwartz S, Wagner L, Miller W (2000) A greedy algorithm for aligning DNA sequences. J Comput Biol 7:203–214

Revised Phylogeny of Extant Xiphosurans (Horseshoe Crabs)

B. Akbar John, Hassan I. Sheikh, K.C.A. Jalal, K. Zaleha and B.Y. Kamaruzzaman

Abstract An attempt was made to revise the molecular phylogeny of extant xiphosurans (Horseshoe crabs) using universal barcode gene cytochrome oxidase C subunit 1. All four extant horseshoe crab species namely *Limulus polyphemus* (American horseshoe crab), *Tachypleus gigas, T. tridentatus* and *Carcinoscorpius rotundicauda* (Asian conspecifics) together with predicted ancestral lineages (insects, scorpions and common crabs) were considered for phylogram construction using distance matrix methods. Genetic distance (GD) data analysis revealed the distant genetic relatedness of *L. polyphemus* with Asian conspecifics. More interestingly, the monophyletic origin of *Tachypleus gigas* and *Tachypleus tridentatus* was quite evident in the phylogram which other molecular markers failed to address. Close genetic relatedness of horseshoe crabs with insects showed that they might have evolved from ancient aquatic insects. The efficiency of cytochrome oxydase C subunit 1 gene in species level identification among the horseshoe crab genome was clear in both the phylogram together with the precise identification of the differential developmental stages to the species level.

Keywords Horseshoe crabs · Living fossil · Xiphosuran · Genetic lineage · Malaysia

---

B. Akbar John (✉) · K. Zaleha
INOCEM Research Station (IRS), Kulliyyah of Science, International Islamic University Malaysia, Kuantan, Pahang, Malaysia
e-mail: akbarjohn50@gmail.com

H.I. Sheikh
Department of Biotechnology, Kulliyyah of Science, International Islamic University Malaysia, Kuantan, Malaysia

K.C.A. Jalal · B.Y. Kamaruzzaman
Department of Marine Science, Kulliyyah of Science, International Islamic University Malaysia, Kuantan, Malaysia

S. Trivedi et al. (eds.), *DNA Barcoding in Marine Perspectives*,
DOI 10.1007/978-3-319-41840-7_7

1 Introduction

Among the bewildering array of animal taxa, horseshoe crabs are unique in their genetic makeup and are commonly known as "Living Fossil". An Intriguing characteristic of horseshoe crabs is that they are morphologically similar in look and having virtually unchanged genetic makeup which helped them in withstanding various environmental stresses for the past 150 million years (Rudkin and Young 2009; Kamaruzzaman et al. 2011). It is interesting to note that, within the impressive diversity and ecological range of extant chelicerate arthropods, only the xiphosurid horseshoe crabs retain a primitively obligate aquatic habit harking back to their distant genealogical roots. It was noted that the fossil record of the basic xiphosurid horseshoe crab body plan has been extended back to the late Ordovician Period, about 445 million years ago, demonstrating an origin that lies outside of the paraphyletic 'synziphosurines' (Obst et al. 2012).

Global distribution pattern of four extant species of horseshoe crabs showed restricted inhabitation of *Limulus polyphemus* along the American coastline especially in Gulf of Mexico, while other three Asian conspecifics such as *Tachypleus tridentatus, T. gigas, Carcinoscorpius rotundicauda* are inhabiting Indo-china coastal waters from Bay of Bengal up to South Philippines. In Malaysia, their distribution was noted in both east and west coast of Peninsular Malaysia with the restricted distribution of *T. tridentatus* in Borneo Island (Sabah and Sarawak) (Akbar John et al. 2011, 2012). During spawning time adult horseshoe crabs migrate from the offshore continental shelf to spawn on intertidal sandy beaches (in case of *T. gigas*) and sandy mud beaches and mangrove area (in case of *C. rotundicauda*) during full and new moon days (Zaleha et al. 2010).

Detailed morphological descriptions of the four species of horseshoe crabs were given by Mikkelsen (1988) and the comparisons of morphological differences between species were presented by Chiu and Morton (2003). Morphological characters, including the shape of the prosoma and telson, the shapes and numbers of claspers in the male and the shapes of marginal spines in the female, have been considered important in the classification of horseshoe crabs (Sekiguchi and Nakamura 1979; Mikkelsen 1988). Chiu and Morton (2003) had extensively discussed morphological variations between *Tachypleus tridentatus* and *Carcinoscorpius rotundicauda* using conventional morphometric approach. Yamasaki et al. (1988) had studied geographic variations in body sizes (maximum prosomal width) of the four extant horseshoe crab species and he proved significant variation in prosomal width of *T. tridentatus* and C. *rotundicauda* collected from different countries. These little morphological differentiations among horseshoe crab lineages have resulted in substantial controversy concerning the phylogenetic relationship among the extant species of horseshoe crabs, especially among the three species in the Indo-Pacific region. Earlier studies suggested that the three species constitute a phylogenetically irresolvable trichotomy (Xia 2000).

These discrepancies have attracted various researchers to concentrate on their genomic structure to differentiate the species at the gene level. Hence, the present study was aimed to revise the existing molecular phylogeny of horseshoe crabs and to predict their possible ancestry using universal barcode gene (mitochondrial Cytochrome Oxidase C subunit 1 gene) as a benchmark reference.

2 Materials and Methods

2.1 Sample Collection, Preparation and Larval Rearing

Matured horseshoe crabs (*Tachypleus gigas*) were collected from Balok and Pekan nesting grounds and a female *T. gigas* sample was collected from Pulau Gaya, Sabah, Eastern Malaysia. Mangrove horseshoe crab (*Carcinoscorpius rotundicauda*) samples were collected from Sitieu mangrove forest (Terengganu, East coast of Malaysia) during May and August 2010. All the samples were identified to the species level using conventional taxonomic keys (Yamasaki et al. 1988; Chiu and Morton 2003) Fertilized eggs of *T. gigas* were sampled from Pekan and immediately transported to Institute of Aquatrop (University Malaysia Terengganu) in an aerated condition. Eggs were kept in filtered sea water (salinity 33 ± 2 ppt) under aerated condition in a larval rearing tank for 30 days. After a month, Pretrilobite stage of *T. gigas* swimming in the amniotic fluid were sampled and preserved in 95 % ethanol for DNA sequencing. In next 2 weeks, free swimming trilobite stage was sampled from larval rearing tank and preserved in 95 % ethanol as mentioned above. Precautions were taken to avoid fungal attack on developing eggs by constantly changing the filtered sea water in every 3 days. Simultaneously matured female crabs were dissected out using sterilized scissors and forceps to collect Apodeme tissue and Immatured eggs. Samples such as immatured egg, matured egg, flesh (Apodeme), pretrilobite larvae and trilobite larvae were collected in 1.5 ml Eppendorf tube containing 95 % ethanol for DNA isolation. Prior to this, all the samples were photographed for future reference. The sample details and coordinates of sampling locations are given in Table 1.

2.2 DNA Extraction, PCR and Sequencing

Salting out procedure was adopted to extract the DNA from the samples (Ajmal Khan et al. 2010; Akbar John et al. 2010; Prasanna Kumar et al. 2011). Approximately 570 bp of COI gene from mitochondrial DNA was amplified using Forward (Fish F2_t1: TGTAAAACGACGGCCAGTCGACTAATCATAAAGATATCGGCAC) and Reverse primer (FishR2_t1: CAGGAAACAGCTATGACACTTCAGGGTGACCGAAGAATCAGAA) under the PCR condition of an initial step of 2 min at 95 °C followed by 35 cycles of 0.5 min at 94 °C, 0.5 min at 54 °C and 1 min at 72 °C, followed in turn by 10 min at 72 °C and then held at 4 °C.

Table 1 Sample details and coordinates of each sampling locations

Scientific name	Type of sample	Sex	Sampling area	Latitude and longitude	
Carcinoscorpius rotundicauda	Immatured egg	F	Sitieu forest	3°36.157′ N	103°23.952′ E
Carcinoscorpius rotundicauda	Flesh	F	Sitieu forest	3°36.157′ N	103°23.952′ E
Tachypleus gigas	Flesh	M	Balok	3°56.218′ N	103°22.623′ E
Tachypleus gigas	Flesh	F	Balok	3°56.201′ N	103°22.615′ E
Tachypleus gigas	Flesh	F	Pulau Gaya	6.016666° N	116.0333333° E
Tachypleus gigas	Fertilized egg	F	Pekan	3°36.157′ N	103°23.952′ E
Tachypleus gigas	Immatured egg	F	Pekan	3°36.157′ N	103°23.952′ E
Tachypleus gigas	Pretrilobite	–	Pekan	3°36.157′ N	103°23.952′ E
Tachypleus gigas	Trilobite	–	Pekan	3°36.157′ N	103°23.952′ E
Tachypleus gigas	Immatured egg	F	Pulau Gaya	6.0166667° N	116.0333333° E

PCR products were visualized on 2 % China Agarose gel and the photographed using Gel imager Under UV light. Products were labeled using Qiagen sequencing kit and sequenced unidirectionally using a MegaBace capillary sequencer at Bioserve biotechnologies pvt. Ltd. Hyderabad, India. Generated sequences were edited using Chromas Pro 2.33v. Sequences were deposited in NCBI under the Genbank ID JF896105-JF896114 (Table 2).

2.3 *Software Analysis*

DNA sequences generated in this study together with the horseshoe crab and predicted sister taxa sequences (retrieved from public DNA data banks were run in Clustal X 2.0.6v for multiple sequence alignment under default setting (Larkin et al. 2007). Nucleotide composition was determined using BioEdit 7.0.9v (Hall 1999). Molecular Evolutionary Genetics Analysis (MEGA) beta 4.1v was used to generate phylogram using distance matrix methods such as Neighbor Joining (NJ) method and Un-weighted Pair Group *Method* with Arithmetic Mean (UPGMA) (Tamura et al. 2007). Kimura 2 Parameter (K2P) was used as a distance model to generate phylogenetic tree in both the methods (Kimura 1980). Genetic Distance (GD) data were also retrieved at each codon position using the same K2P distance model.

Table 2 Details of sequences and corresponding developmental stages submitted in National Centre for Biotechnological Information (NCBI)

Scientific name	Type of sample	Sex	Genbank accession no	Protein ID	Sequence length in bp
Carcinoscorpius rotundicauda	Immatured egg	F	JF896105	AEG75796	635
Carcinoscorpius rotundicauda	Flesh	F	JF896106	AEG75797	647
Tachypleus gigas	Flesh	M	JF896107	AEG75798	565
Tachypleus gigas	Flesh	F	JF896108	AEG75799	486
Tachypleus gigas	Flesh	F	JF896109	AEG75800	537
Tachypleus gigas	Fertilized egg	F	JF896110	AEG75801	534
Tachypleus gigas	Immatured egg	F	JF896111	AEG75802	506
Tachypleus gigas	Pretrilobite	–	JF896112	AEG75803	537
Tachypleus gigas	Trilobite	–	JF896113	AEG75804	538
Tachypleus gigas	Immatured egg	F	JF896114	AEG75805	691

Note Sequences available online from June 5th 2011 on NCBI portal

3 Results

3.1 Neighbor Joining (NJ Method)

Phylogenetic tree was constructed using Neighbor Joining method to verify the efficiency of *cox*1 gene in delineating closely related and morphologically cryptic species of horseshoe crabs (Fig. 1). Phylogram was constructed with 36 sequences (horseshoe crabs = 21; Insects = 5; Scorpion = 8 and out groups = 2). The out groups used were *Portunus pelagicus* (Class: Malacostraca) and *Artemia franciscana* (Class: Branchiopoda) distinctly segregated in separate branch in phylogram proving its reliability besides the higher bootstrap values in the internal branch nods. Four distinct clads were found in the phylogram having horseshoe crabs in Clad 1, Scorpions in Clad 2, Insects in Clad 3 and out group organisms in Clad 4. The phylogeographical signals were apparent in Clad 1, segregating *T. gigas* collected from Pulau Gaya (Borneo) from the very species sampled in East coast of Malaysia (Balok and Pekan). The efficiency of cox1 gene in species level identification of various developmental stage of *T. gigas* was noted but it was inefficient in sex determination and precise identification of various developmental stages. The constructed phylogram proved the monophyletic nature of *T. gigas* with *T. tridentatus*. Atlantic horseshoe crab (*L. polyphemus*) was genetically distinct from other species of horseshoe crabs. Insects were clumped together in Clad 3 which shared a common branch node with horseshoe crabs proving their closer genetic relatedness with horseshoe crabs. Terrestrial scorpion species used in this analysis

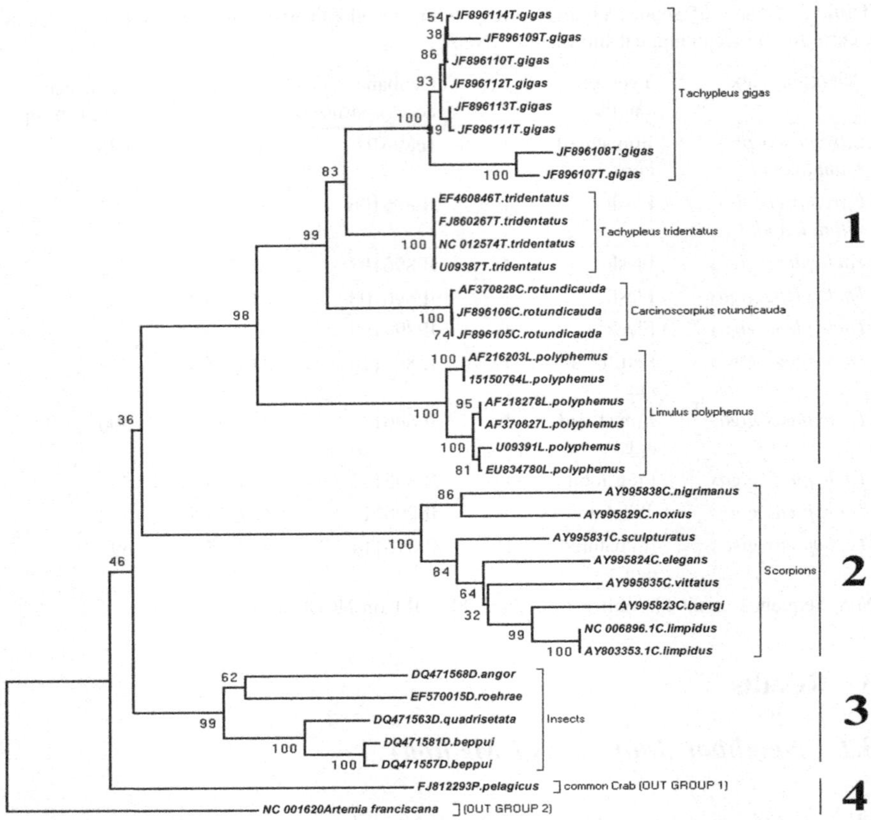

Fig. 1 Neighbor-Joining (NJ) phylogenetic tree was constructed to determine the evolutionary history of horseshoe crabs with related sister taxa (Saitou and Nei 1987)

were clumped together in Clad 2 indicating their distant genetic relatedness with horseshoe crab species. Among the horseshoe crabs, *L. polyphemus* had comparatively higher GC content than the other species of horseshoe crabs. Average GC content in *L. polyphemus* was 38.02 % followed by *C. rotundicauda* 36.49 %, *T. gigas* (33.55 %) and *T. tridentatus* (32.78 %).

3.2 Unweighted Pair Group Method with Arithmetic Mean (UPGMA) Method

UPGMA method was adopted to construct hierarchical clustering Phylogram to infer genetic relatedness of selected group of species (Fig. 2). This method strictly follows the molecular clock hypothesis (constant rate of evolution) and hence useful in verifying the information obtained through NJ-method. Same sequence

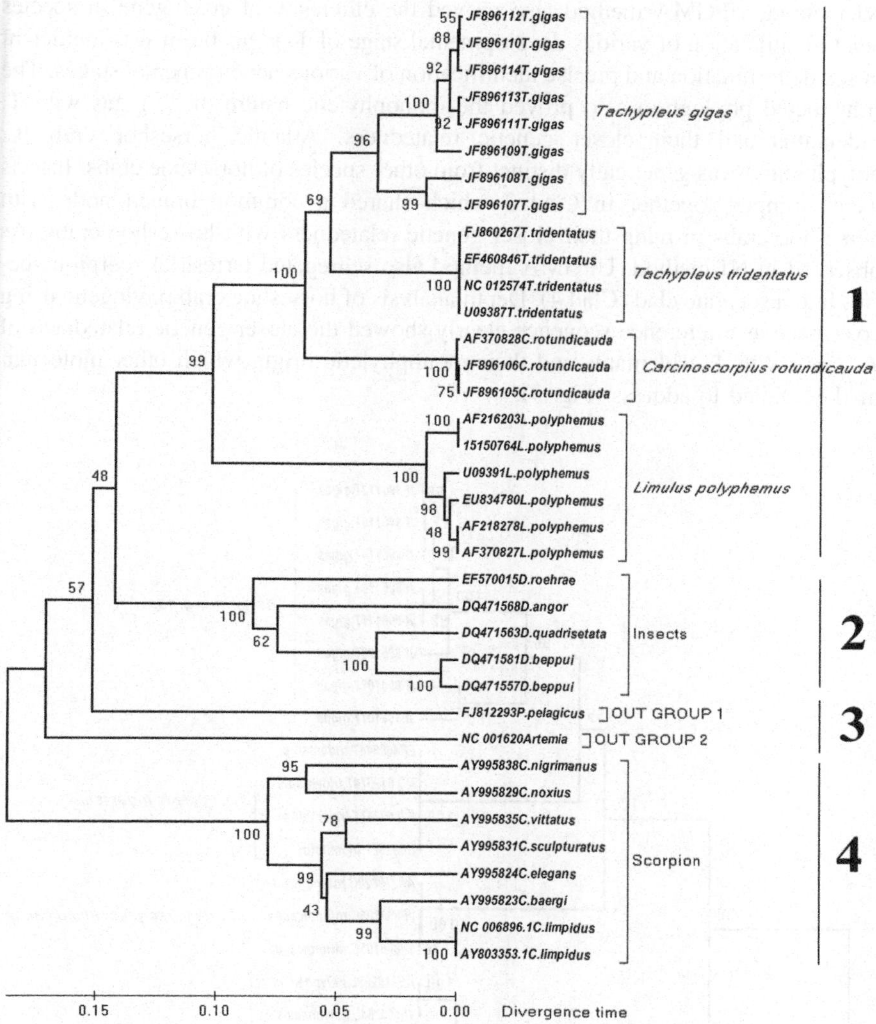

Fig. 2 Un-weighted Pair Group Method with Arithmetic Mean (UPGMA) phylogenetic tree was constructed to determine the evolutionary history of horseshoe crabs with related sister taxa (Saitou and Nei 1987)

file used for NJ method was also used for phylogram construction. As observed in NJ method, the out groups used were clearly segregated in separate branch in phylogram proving its reliability besides the higher bootstrap values in the internal branch nods. Four distinct clads were noted in the phylogram having horseshoe crabs in Clad 1, Insects in Clad 2, out group organisms in Clad 3 and Scorpions in Clad 4. Unlike the NJ method, UPGMA method could not show more apparently the phylogeographical signals in the constructed phylogram. However, similar to

NJ method, UPGMA method also proved the efficiency of cox1 gene in species level identification of various developmental stage of T. gigas but it was inefficient in sex determination and precise identification of various developmental stages. The constructed phylogram also proved the monophyletic nature of T. gigas with T. tridentatus and their closer genetic relatedness. Atlantic horseshoe crab (L. polyphemus) was genetically distinct from other species of horseshoe crabs. Insects were clumped together in Clad 2 which shared a common branch node with horseshoe crabs proving their closer genetic relatedness with horseshoe crabs. As observed in NJ method, UPGMA method also segregated terrestrial scorpion species into a separate clad (Clad 4). Depth analysis of horseshoe crab phylogene using cox1 gene as a reference sequence clearly showed the closer genetic relatedness of *T. gigas* with T. tridentatus and their monophyletic origin which other molecular markers failed to address (Fig. 3).

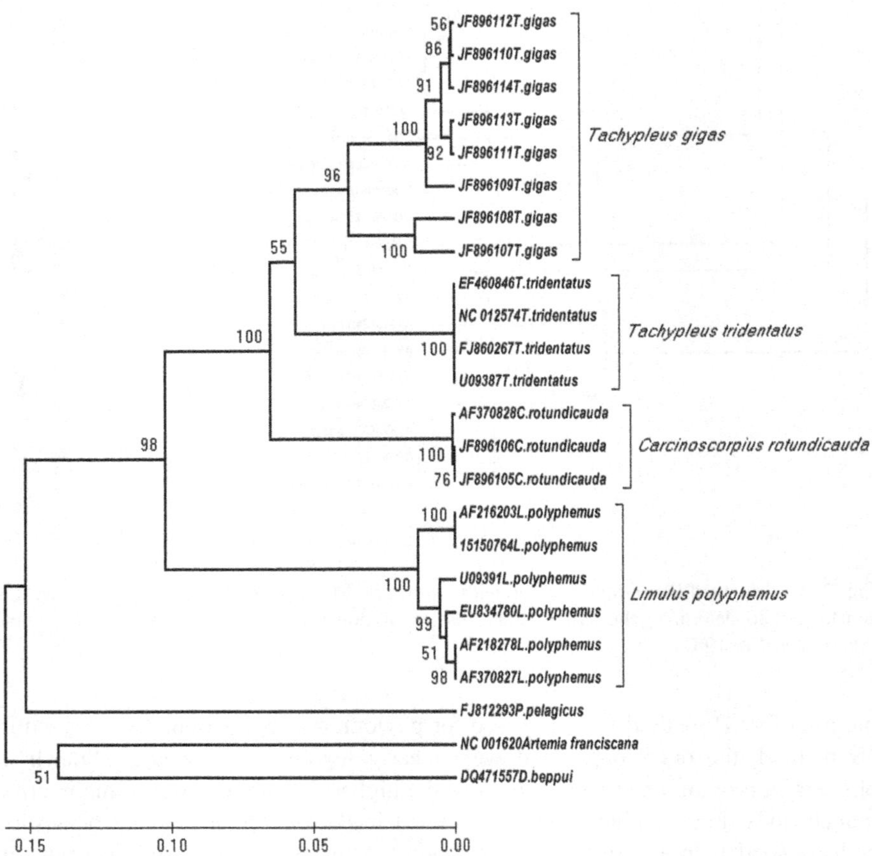

Fig. 3 Monophyletic origin of *Tachypleus gigas* with *T. tridentatus* was determined using UPGMA phylogenetic tree

Genetic Distance Data Analysis

Mean genetic distance within *L. polyphemus* at 1st, 2nd and 3rd codon position was 0.006, 0 and 0.048 respectively. Intra-species genetic distance among *T. gigas* and *C. rotundicauda* at 1st, 2nd and 3rd codon position were 0.016, 0.03 and 0.064 (*T. gigas*) and 0.003, 0 and 0.05 (*C. rotundicauda*) respectively. *T. tridentatus* showed no variation in GD value in all the codon positions. Among the Asian horseshoe crab species *T. gigas* showed closer genetic relatedness (lower GD value) with *T. tridentatus* with GD values of 0.037, 0.017 and 0.307 at 1st, 2nd and 3rd codon position respectively. This observation clearly revealed their monophyletic origin. The genetic distance values between *C. rotundicauda* and *T. gigas* were 0.053, 0.017 and 0.395 at 1st, 2nd and 3rd codon position respectively. Calculated genetic distance data showed higher genetic distance value in third codon position than its corresponding first and second codon positions. The GD data also proved the distance of genetic relatedness of *L. polyphemus* with Asian horseshoe crabs (Table 3).

Mean GD value within the four species of horseshoe crabs was 0.132, 0.053, 0.019 and 0.405 at 1st, 2nd and 3rd codon position respectively (at 1st + 2nd + 3rd + Noncoding gene). Average GD value between representative Insects and horseshoe crabs at 1st, 2nd and 3rd codon position were 0.207, 0.077 and 0.718 respectively. On the other hand mean GD values between representative

Table 3 Average genetic distance (GD) between four available species of horseshoe crabs observed in only 1st, 2nd and 3rd codon positions

		L. polyphemus			*T. gigas*			*T. tridentatus*			*C. rotundicauda*		
	CP	1st	2nd	3rd	1st	2nd	3rd	1st	2nd	3rd	1st	2nd	3rd
LP	1st	0.006			0.092			0.077			0.080		
	2nd		0			0.032			0.015			0.015	
	3rd			0.048			0.659			0.66			0.629
TG	1st				0.016			0.037			0.053		
	2nd					0.030			0.017			0.017	
	3rd						0.064			0.307			0.395
TT	1st							0			0.037		
	2nd								0			0	
	3rd									0			0.363
CR	1st										0.003		
	2nd											0	
	3rd												0.05

Note CP—Codon positions; LP—*L. polyphemus*; TG—*T. gigas*; TT—*T. tridentatus*; CR—*C. rotundicauda*

Scorpion and horseshoe crabs at 1st, 2nd and 3rd codon position were 0.217, 0.063 and 1.264 respectively (Table 4). These observations proved the closer genetic relatedness of horseshoe crabs with insects than with scorpions.

Nucleotide diversity between horseshoe crab species was comparatively smaller ($\pi = 0.150633$) than between related sister group taxa and horseshoe crabs. Positive Tajima test statistic values of overall and within horseshoe crabs population signified low levels of both low and high frequency polymorphisms in the sequences (Table 5).

Nucleotide substitution pattern observed among the test organisms (including horseshoe crabs, insects, scorpions and common crabs) clearly showed that transitional substitutions are very common in the gene sequence than transversional substitutions with transition/transversion bias value $R = 0.988$. The transition/transversion rate ratios were $k_1 = 2.243$ (for purines) and $k_2 = 2.414$ (for pyrimidines) (Table 6).

Nucleotide substitution pattern observed within the horseshoe crabs also showed that transitional substitutions are very common in the gene sequence than transversional substitutions with transition/transversion bias value $R = 1.711$. The transition/transversion rate ratios were $k_1 = 4.143$ (for purines) and $k_2 = 4.375$ (for pyrimidines) (Table 7).

Table 4 Mean genetic distance (GD) values of different groups of organisms with reference to horseshoe crab at all the possible codon position indicating 3rd codon position shows higher GD value

	1st + 2nd + 3rd codon position	1st codon position	2nd codon position	3rd codon position
	Horseshoe crabs	Horseshoe crabs	Horseshoe crabs	Horseshoe crabs
Horseshoe crabs	0.132	0.053	0.019	0.405
Insect	0.284	0.207	0.077	0.718
Scorpions	0.377	0.217	0.063	1.264

Table 5 Results from Tajima's Neutrality test calculated for 36 sequences (including horseshoe crabs, insects and scorpions) and 24 sequences (only horseshoe crabs)

	m	S	p_s	π	D
Overall	36	332	0.536349	0.210329	2.372310
Between horseshoe crabs	24	290	0.461783	0.150633	0.878823

Note The Tajima test statistic (Tajima 1989) was estimated using MEGA4 (Tamura et al. 2007). All positions containing gaps and missing data were eliminated from the dataset (Complete deletion option). The abbreviations used are as follows: m = number of sites, S = Number of segregating sites, $p_s = S/m$, $\Theta = p_s/a_1$, and π = nucleotide diversity. D is the Tajima test statistic

Table 6 Pattern of Nucleotide substitution observed among the test organisms (including horseshoe crabs, insects, scorpions and common crabs) calculated using Maximum Composite Likelihood method

	A	T	C	G
A	–	*8.43*	*4.35*	**9.79**
T	*5.91*	–	**10.49**	*4.37*
C	*5.91*	**20.36**	–	*4.37*
G	**13.25**	*8.43*	*4.35*	–

Note Each entry shows the probability of substitution from one base (row) to another base (column) instantaneously. Only entries within a row should be compared. Rates of different transitional substitutions are shown in bold and those of transversional substitutions are shown in italics. The nucleotide frequencies are 0.256 (A), 0.366 (T), 0.189 (C), and 0.189 (G). The overall transition/transversion bias is R = 0.988, where $R = \frac{A \times G \times k1 + T \times C \times k}{(A+G) \times (T+C)}$. Codon positions included were 1st + 2nd + 3rd + Noncoding. All positions containing gaps and missing data were eliminated from the dataset (Complete-deletion option)

Table 7 Pattern of Nucleotide substitution observed within horseshoe crabs, calculated using Maximum Composite Likelihood method

	A	T	C	G
A	–	*5.47*	*3.33*	**11.05**
T	*4.49*	–	**14.56**	*2.67*
C	*4.49*	**23.91**	–	*2.67*
G	**18.59**	*5.47*	*3.33*	–

Note Each entry shows the probability of substitution from one base (row) to another base (column) instantaneously. Only entries within a row should be compared. Rates of different transitional substitutions are shown in bold and those of transversional substitutions are shown in italics. The nucleotide frequencies are 0.281 (A), 0.343 (T), 0.209 (C), and 0.167 (G). The overall transition/transversion bias is $R = 1.711$, where $R = \frac{A \times G \times k1 + T \times C \times k}{(A+G) \times (T+C)}$. Codon positions included were 1st + 2nd + 3rd + Noncoding. All positions containing gaps and missing data were eliminated from the dataset (Complete-deletion option)

4 Discussion

4.1 Phylogenetic Study on Horseshoe Crab

Earlier studies on the serology of Atlantic species, *(L. polyphemus)* and Indo-Pacific conspecifics (Shuster 1962), phylogenetic analyses of amino acid sequences of coagulogen and the fibrinopeptide-like peptide C (Shishikura et al. 1982; Srimal et al. 1985; Sugita and Shishikura 1995), immunological comparisons of hemocyanins (Sugita 1988), two-dimensional electrophoresis of general proteins (Miyazaki et al. 1987), interspecific hybridization experiments (Sekiguchi and Sugita 1980), cladistic appraisals of morphological characters (Fisher 1984), and mtDNA genes (Avise 1994) had clearly proved the sister taxon status of *L. polyphemus* to Indo-Pacific conspecifics. However, these molecular tools failed to address the monophyletic origin of *Tachypleus gigas* with *T. tridentatus*. Instead, all these molecular tools clumped *T. gigas* with *C. rotundicauda* and showed lower genetic distance between *T. gigas* and *C. rotundicauda* than between *T. gigas* and *T. tridentatus* (Kamaruzzaman et al. 2011). It was strongly believed that these three species constitute a phylogenetically irresolvable trichotomy (Xia 2000). However, all the previous studies failed to address the sampling size used in the phylogenetic tree construction. Ward et al. (2005) proved the importance of sampling size in determining and identifying the species using inter and intra species genetic distance data analysis. Recent studies on horseshoe crab phylogeny (Xia 2000; Kamaruzzaman et al. 2011) have a serious drawback due to the sample size they used to evaluate the topology of the phylogenetic tree. In this study, considerable number of sample (only horseshoe crabs N = 21) size gave substantial amount of information on the monophyletic origin of *T. gigas* and *T. tridentatus* which in turn was verified using UPGMA phylogram and genetic distance data analysis. Present study also showed the phylogeographical cues in cox1 gene.

4.2 *Distance Matrix Method*

4.2.1 Neighbor Joining (NJ) Tree

The phylogeographical signals were apparent in Clad 1, segregating *T. gigas* collected from Pulau Gaya (East Malaysia: Borneo) from the very species sampled in East coast of Peninsular Malaysia (Balok and Pekan). But the phylogram failed to segregate the samples collected from Balok and Pekan. This might probably due to 1. The short geographical distance between these sampled area (~80 km), 2. Constant gene flow between the horseshoe crab populations from these two sampling stations. Similar observation was reported by various researchers on fishes (Akbar John et al. 2010), ticks (Song et al. 2011) spiders (Zhang et al. 2005) and others (Hickerson and Cunningham 2000; MuÑOz et al. 2008). As observed in previous studies, the distant relatedness of American horseshoe crab (*L. polyphemus*) to the Indo Pacific conspecifics was apparent in the phylograms. More interestingly, the monophyletic origin of *T. gigas* with *T. tridentatus* was evident in Neighbor Joining (NJ) phylogram and this observation was cross examined using genetic distance (GD) data analysis which showed the GD value of (0.037, 0.017 and 0.307 at 1st, 2nd and 3rd codon position respectively) between *T. gigas* and *T. tridentatus*, whereas the GD values between *T. gigas* and *C. rotundicauda* was 0.053, 0.017 and 0.395 at 1st, 2nd and 3rd codon position respectively. Though this observation is in contrast to the previous studies, it is virtually true because of the large number of individual species of horseshoe crab samples used in this study. The monophyletic origin of *T. gigas* and *T. tridentatus* and the reliability of the phylogram were also checked by constructing sub phylogram constituting only representative horseshoe crab species and the selected out groups from possible sister taxa which also showed the similar results.

The phylogram clearly showed the genetic relatedness of horseshoe crabs and insects indicating that horseshoe crabs might have probably evolved from the ancient aquatic insects. This observation is in agreement with the conclusions of recent studies on horseshoe crab phylogeny (Xia 2000; Kamaruzzaman et al. 2011). However, Eurypterids (e.g., sea scorpions) have traditionally been regarded as close relatives of horseshoe crabs. Subsequent studies placed eurypterids closer to the arachnids (e.g., spiders, terrestrial scorpions, mites and ticks) in a group called Metastomata (Pavlicek et al. 2008). There has also been a belief that eurypterids are closely related to terrestrial scorpions (Raz et al. 2009). Recent study on the genetic relationships between arachnids and their relatives recognized Eurypterida, Xiphosura and Arachnida as three major groups, and their genetic relatedness at present cannot be resolved using available molecular markers (Shultz 2007). Similar conclusion was evident in the phylogram which was cross checked with genetic distance data that terrestrial scorpions which are closely related to sea scorpions are distantly related to horseshoe crabs.

4.2.2 PGMA Method

Unweighted Pair Group Method with Arithmetic Mean (UPGMA) was used to construct un-rooted phylogenetic tree to 1. Infer the genetic relationship between horseshoe crabs, 2. To check the reliability of information on the monophyletic origin of *T. gigas* with *T. tridentatus,* and 3. To verify the observation obtained in NJ tree method. As noted in NJ tree, UPGMA tree had 4 distinct clads segregating representative species to their respective clads. Unlike NJ tree, the basic principle and hypothesis in UPGMA method (constant rate of evolution) restricted its efficiency in showing the distinct phylogeographical cues in the phylogram. However, higher bootstrap value in internal nods and distinct segregation of representative species showed its efficiency in phylogenetic tree construction. The phylogram also proved the monophyletic origin of *T. gigas* with *T. tridentatus* and their closer genetic relatedness. Atlantic horseshoe crab (*L. polyphemus*) was genetically distinct from other species of horseshoe crabs. To verify this observation, a sub tree was constructed consisting of only 4 species of horseshoe crab and 3 out group species which also showed similar facts. Insects were clumped together in Clad 2 which shared a common branch node with horseshoe crabs proving their closer genetic relatedness with horseshoe crabs. Phylogram also showed the distant genetic relatedness of terrestrial scorpions with horseshoe crabs similar to NJ tree. Both NJ and UPGMA phylogram clearly proved the efficacy of COI gene in delineating the members of evolutionarily cryptic groups of organisms, besides revealing the monophyletic origin of south East Asian horseshoe crabs (*T. gigas* and *T. tridentatus*).

4.3 Genetic Distance (GD) Data Analysis

The basic principle behind the DNA barcoding technology is there should be low rates of DNA sequence divergence among individuals of the same species than between the species. In other words, intra species genetic distance should be low compared to inter species genetic distance. These were quite evident in GD data analysis (Hebert et al. 2003, 2004). In this study we mainly concentrated on the calculation of GD within horseshoe crabs because of the representation of all the four extant species in the constructed phylogram. Mean genetic distance within the horseshoe crabs at all the codon positions were 0.053, 0.019 and 0.405 at 1st, 2nd and 3rd Codon position respectively. Genetic Distance between the selected animal groups showed that the GD value was lower between horseshoe crabs and insects with 0.207, 0.077 and 0.718 at 1st, 2nd and 3rd codon position respectively compared to horseshoe crabs and scorpions with 0.217, 0.063 and 1.264 codon position respectively. This observation clearly indicated the closer genetic relatedness of insects with horseshoe crabs. Similar observation was made by Kamaruzzaman et al. (2011). It was also observed that the horseshoe crabs are genetically related to common crabs than the scorpion. This might be the reason why out group (*Portunus pelagicus*) used in this study clustered with horseshoe crabs closely than scorpions.

Another interesting observation made from the genetic distance data was higher genetic distance observed in third codon position than its corresponding first and second codon positions. Similar observation was made by Ward et al. (2005) while barcoding fishes from Australian waters. Simmons et al. (2006) also observed that greater phylogenetic signal is often found in parsimony-based analyses of third codon positions of protein-coding genes relative to their corresponding first and second codon positions, even for early-derived basal clades (Siemion and Przemyslaw 1994; Ajmal Khan et al. 2010). Average genetic distance among the different groups of test organisms used in this study showed higher GD value at 3rd codon position indicating that detailed study on 3rd codon position might reveal possible evolutionary information among the closely related groups of organisms.

4.4 Tajima's Neutrality Test

Nucleotide diversity between horseshoe crab species was comparatively smaller ($\pi = 0.150633$) than between related sister group taxa and horseshoe crabs ($\pi = 0.210329$). Positive Tajima test statistic values of overall and within horseshoe crabs population signified that there were low levels of both low and high frequency polymorphisms in the horseshoe crab sequences which ultimately helped them in retaining the genetic makeup virtually unchanged over millions of years.

4.5 Nucleotide Substitution Analysis

It is a well-known fact that during DNA sequence evolution the rate of transitional changes differs from the rate of transversional changes, with transitions generally occurring more frequently than transversions. This difference is often referred to as transition bias, and estimation of the extent of transition bias may be of interest, since it may vary for different organisms and for different genes within a collection of organisms. In general, there are twice as many possible transversions as transitions due to the relatively high rate of mutation of methylated cytosines to thymine (Brown et al. 1982; Gojobori et al. 1982; Curtis and Clegg 1984; Graur and Li 2000). Proper estimation is also important because the ratio of the rates of transitional to transversional changes (often called the Ti:Tv ratio) play a role in evolutionary distance correction methods and is used in several common evolutionary models (e.g., the F84 model) (Wakeley 1996). However, it was observed that Ti:Tv ratio is strongly influenced by sampling size. Hence, we also calculated ti/tv bias which is more realistic and widely applied reliable estimate. Transition/Transversion (ti/tv) bias is known to be a general property of DNA sequence evolution, it is more pronounced in animal mitochondrial DNAs (mtDNAs) than in nuclear or chloroplast DNAs (Wakeley 1996). Estimation of the ti/tv bias is important not only to our understanding of the patterns of DNA sequence evolution, but also to reliable estimation of sequence distance and phylogeny reconstruction (Rosenberg et al. 2003).

The calculated ti/tv bias among the test organisms (including horseshoe crabs, insects, scorpions and common crabs) was $R = 0.988$ whereas ti/tv bias within horseshoe crabs was 1.711. This observation clearly proved that horseshoe crab *cox1* undergoes more transition mutations compared to transversion mutations. This observation was also proved by calculating Ti:Tv ratio where this ratio was $k_1 = 4.143$ (for purines) and $k_2 = 4.375$ (for pyrimidines) within the horseshoe crabs.

5 Conclusion

Molecular taxonomic study clearly segregated American conspecific (*Limulus polyphemus*) from Indo Pacific horseshoe crabs and thereby proving their distant intra species genetic relatedness. More interestingly, the efficiency of *cox1* gene in species level delineation of the cryptic taxonomy of horseshoe crabs proved the monophyletic origin of *Tachypleus gigas* and *T. tridentatus* which other molecular markers failed to address. Close genetic relatedness of horseshoe crabs with insects showed that they might have evolved from ancient aquatic insects. The efficiency of cytochrome oxydase C subunit 1 gene in species level identification evolutionarily conserved horseshoe crab genome was apparent in the constructed phylogram together with the precise identification of their differential developmental stages to the species level. It was also evident from the phylogram that *cox1* gene has sound phylogeographical signals. Higher Genetic Distance (GD) value obtained from 3rd codon position than its corresponding 1st and 2nd codon positions proved the presence of grater genetic cues in 3rd codon position that could be used to study the genetic relatedness of evolutionarily conserved species in future.

References

Ajmal Khan S, Lyla PS, Akbar John B, Prasanna Kumar C, Murugan S, Jalal KCA (2010) DNA barcoding of Stolephorus indicus, Stolephorus commersonnii and Terapon jarbua of Parangipettai coastal waters. Biotechnology 9:373–377

Akbar John B, Kamaruzzaman BY, Jalal KCA, Zaleha K (2011) Hydrology of the horseshoe crab nesting grounds at Pahang coast, Malaysia. Orient J Chem 27(4):1475–1483

Akbar John B, Kamaruzzaman BY, Jalal KCA, Zaleha K (2012) Sediment profiling of the nesting grounds of horseshoe crabs at east peninsular Malaysia. Int J Biol 4(2):159–165

Akbar John B, Prasanna kumar C, Lyla PS, Ajmal Khan S, Jalal KCA (2010) DNA barcoding of lates calcarifer (Bloch 1970). Res J Biol Sci 5:414–419

Avise JC (1994) Molecular markers, natural history and evolution. Chapman & Hall, New York

Brown WM, Prager EM, Wang A, Wilson AC (1982) Mitochondrial DNA sequences of primates: tempo and mode of evolution. J Mol Evol 18(4):225–239

Chiu HMC, Morton B (2003) The morphological differentiation of two horseshoe crab species, Tachypleus tridentatus and Carcinoscorpius rotundicauda (Xiphosura), in Hong Kong with a regional Asian comparison. J Nat Hist 37(19):2369–2382

Curtis SE, Clegg MT (1984) Molecular evolution of chloroplast DNA sequences. Mol Biol Evol 1 (4):291–301

Fisher DC (1984) The Xiphosurida: archetypes of bradytely. In: Eldredge N, Stanley SM (eds) Living fossils. Springer, New York, pp 196–213

Gojobori T, Li WH, Graur D (1982) Patterns of nucleotide substitution in pseudogenes and functional genes. J Mol Evol 18(5):360–369

Graur D, Li WH (2000) Fundamentals of molecular evolution. Sinauer, Sunderland

Hall TA (1999) BioEdit: a user-friendly biological sequence alignment editor and analysis program for Windows 95/98/NT. Nucleic Acids Symp Ser 41:95–98

Hebert PDN, Cywinska A, Ball SL, deWaard JR (2003) Biological identifications through DNA barcodes. Proc R Soc Lond Ser B Biol Sci 270(1512):313–321

Hebert PDN, Penton EH, Burns JM, Janzen DH, Hallwachs W (2004) Ten species in one: DNA barcoding reveals cryptic species in the neotropical skipper butterfly Astraptes fulgerator. Proc Natl Acad Sci USA 101(41):14812–14817

Hickerson MJ, Cunningham CW (2000) Dramatic mitochondrial gene rearrangements in the hermit crab Pagurus longicarpus (Crustacea, Anomura). Mol Biol Evol 17(4):639–644

Kamaruzzaman BY, John BA, Zaleha K, Jalal KCA (2011) Molecular phylogeny of horseshoe crab. Asian J Biotechnol 3(3):302–309

Kimura M (1980) A simple method for estimating evolutionary rate of base substitutions through comparative studies of nucleotide sequences. J Mol Evol 16:111–120

Larkin MA, Blackshields G, Brown NP, Chenna R, McGettigan PA, McWilliam H, Higgins DG (2007) Clustal W and Clustal X version 2.0. Bioinformatics 23(21):2947–2948

Mikkelsen T (1988) The secret in the blue blood. Science Press, Beijing

Miyazaki JI, Sekiguchi K, Hirabayashi T (1987) Application of an improved method of two-dimensional electrophoresis to the systematic study of horseshoe crabs. Biol Bull 172 (2):212–224

MuÑOz J, GÓMez A, Green AJ, Figuerola J, Amat F, Rico C (2008) Phylogeography and local endemism of the native Mediterranean brine shrimp Artemia salina (Branchiopoda: Anostraca). Mol Ecol 17(13):3160–3177

Obst M, Faurby S, Bussarawit S, Funch P (2012) Molecular phylogeny of extant horseshoe crabs (Xiphosura, Limulidae) indicates Paleogene diversification of Asian species. Mol Phylogeny Evol 62(1):21–26

Pavlicek T, Mienis HK, Raz S, Hassid V, Rubenyan A, Nevo E (2008) Gastropod biodiversity at the evolution canyon microsite, lower Nahal Oren, Mount Carmel, Israel. Biol J Linn Soc 93:147–155

Prasanna Kumar C, Akbar John B, Ajmal Khan S, Lyla PS, Murugan S, Rozihan M, Jalal KCA (2011) Efficiency of universal barcode gene (Coxi) on morphologically cryptic mugilidae fishes delineation. Trends Appl Sci Res 6:1028–1036

Raz S, Retzkin S, Pavlíček T, Hoffman A, Kimchi H, Zehavi D, Nevo E (2009) Scorpion biodiversity and interslope divergence at "evolution canyon", lower Nahal Oren microsite, Mt. Carmel, Israel. PLoS ONE 4(4):e5214

Rosenberg MS, Subramanian S, Kumar S (2003) Patterns of transitional mutation biases within and among mammalian genomes. Mol Biol Evol 20(6):988–993

Rudkin DM, Young GA (2009) Horseshoe crabs—an ancient ancestry revealed. In: Tanacredi JT, Botton ML, Smith DR (eds) Biology and conservation of horseshoe crabs. Springer, US, pp 25–44

Saitou N, Nei M (1987) The neighbor-joining method: a new method for reconstructing phylogenetic trees. Mol Biol Evo 4:406–425

Sekiguchi K, Nakamura K (1979) Ecology of the extant horseshoe crabs. In: Cohen E (ed) Biomedical applications of the horseshoe crab (Limulidae). Alan R. Liss, New York, pp 37–45

Sekiguchi K, Sugita H (1980) Systematics and hybridization in the four living species of horseshoe crabs. Evolution 34(4):712–718

Shishikura F, Nakamura S, Takahashi K, Sekiguchi K (1982) Horseshoe crab phylogeny based on amino acid sequences of the fibrino-peptide-like peptide C. J Exp Zool 223(1):89–91

Shultz JW (2007) A phylogenetic analysis of the arachnid orders based on morphological characters. Zoolog J Linn Soc 150:221–265

Shuster CN (1962) Serological correspondence among horseshoe crabs (Limulidae). Zoologica 47:1–8

Siemion IZ, Przemyslaw JS (1994) The informational context of the third base in amino acid codons. Biosystems 33:139–148

Simmons MP, Zhang L-B, Webb CT, Reeves A (2006) How can third codon positions outperform first and second codon positions in phylogenetic inference? An empirical example from the seed plants. Syst Biol 55(2):245–258

Song S, Shao R, Atwell R, Barker S, Vankan D (2011) Phylogenetic and phylogeographic relationships in Ixodes holocyclus and Ixodes cornuatus (Acari: Ixodidae) inferred from COX1 and ITS2 sequences. Int J Parasitol 41(8):871–880

Srimal S, Miyata T, Kawabata S-I, Miyata T, Iwanaga S (1985) The complete amino acid sequence of coagulogen isolated from Southeast Asian horseshoe crab, Carcinoscorpius rotundicauda. J Biochem 98(2):305–318

Sugita H (1988) Immunological comparisons of hemocyanins and their phylogenetic implications. In: Sekiguchi K (ed) Biology of horseshoe crabs. Science House, Tokyo, pp 315–334

Sugita H, Shishikura F (1995) A case of orthologous sequences of hemocyanin subunits for an evolutionary study of horseshoe crabs: amino acid sequence comparison of immunologically identical subunits of Carcinoscorpius rotundicauda and Tachypleus tridentatus. Zoolog Sci 12 (5):661–667

Tamura K, Dudley J, Nei M, Kumar S (2007) MEGA4: molecular evolutionary genetics analysis (MEGA) software version 4.0. Mol Biol Evol 24(8):1596–1599

Tajima F (1989) Statistical methods to test for nucleotide mutation hypothesis by DNA polymorphism. Genetics 123:585–595

Wakeley J (1996) The excess of transitions among nucleotide substitutions: new methods of estimating transition bias underscore its significance. Trends Ecol Evol 11:158–163
Ward RD, Zemlak TS, Innes BH, Last PR, Hebert PDN (2005) DNA barcoding Australia's fish species. Philos Trans R Soc B Biol Sci 360(1462):1847–1857
Xia X (2000) Phylogenetic relationship among horseshoe crab species: effect of substitution models on phylogenetic analyses. Syst Biol 49(1):87–100
Yamasaki T, Makioka T, Saito J (1988) Morphology. In: Sekiguchi K (ed) Bioligy of horseshoe crabs. Tokyo, Science House Co. Ltd., pp 69–132
Zaleha K, Kamaruzzaman BY, Akbar John B, Ong MC (2010) Cd, Cu and Pb concentration levels in horseshoe crab nesting grounds of Pahang coast, Malaysia. J Biol Sci 10(8):790–794
Zhang AB, Kubota K, Takami Y, Kim JL, Kim JK, Sota T (2005) Species status and phylogeography of two closely related Coptolabrus species (Coleoptera: Carabidae) in South Korea inferred from mitochondrial and nuclear gene sequences. Mol Ecol 14(12):3823–3841

DNA Barcoding in Marine Nematodes: Successes and Pitfalls

Punyasloke Bhadury

Abstract The phylum Nematoda is one of the most diverse and abundant in every ecological niche including terrestrial and aquatic environments. It has been estimated that the abundance of nematodes can reach up to 10^8 individuals per square metre. Like any other environment, sedimentary layer of marine environments are rich in benthic fauna and free-living marine nematodes constitute an important component of benthic domain. These organisms are present in all types of marine habitat including from shallow coastal environment to deep-sea. The structural and functional organization of nematode assemblages can provide precious information on the 'health' of ecosystem in which they live, and they are widely recognized as excellent bioindicators of natural and anthropogenic disturbances. The long term success of DNA barcoding as a tool for speeding up identification of marine nematodes during biodiversity and ecosystem monitoring studies will also depend on large scale international efforts such as along the line of Census of Marine Life (CoML) programs. To conclude, morpho-taxonomy coupled with DNA taxonomy using DNA barcoding as a technique could be implemented for undertaking studies on marine nematodes.

Keywords Bioindicators · Coastal · Census of Marine Life · DNA barcodes · Nematoda

1 Introduction

The phylum Nematoda is one of the most diverse and abundant in every ecological niche including terrestrial and aquatic environments (Lambshead 2004). It has been estimated that the abundance of nematodes can reach up to 10^8 individuals per square metre (Lambshead 2004). Like any other environment, sedimentary layer of

P. Bhadury (✉)
Integrative Taxonomy and Microbial Ecology Research Group,
Department of Biological Sciences, Indian Institute of Science Education and Research
Kolkata, Mohanpur, Nadia 741246, West Bengal, India
e-mail: pbhadury@iiserkol.ac.in

S. Trivedi et al. (eds.), *DNA Barcoding in Marine Perspectives*,
DOI 10.1007/978-3-319-41840-7_8

marine environments are rich in benthic fauna and free-living marine nematodes constitute an important component of benthic layer. The structural and functional organization of nematode assemblages can provide precious information on the 'health' of ecosystem in which they live, and they are widely recognized as excellent bioindicators of natural and anthropogenic disturbances (e.g. Balsamo et al. 2010; Boufahja et al. 2011a, b; Semprucci et al. 2015).

The number of valid free-living marine nematode species based on morpho-taxonomy, even after two centuries of research stands at 6900, which is approximately 19 % of the total number of species that has been estimated to be present in marine environment (Appeltans et al. 2012). Identification of free-living marine nematodes is mainly carried out based on careful observation of morphological characters under a microscope. However, such observation is extremely time-consuming and problematic, mainly because of the nature of superficial taxonomy, high phenotypic plasticity among populations and absence of clear taxonomic diagnostic characters for identification of cryptic species (Avise and Walker 1999; Derycke et al. 2005; Bhadury et al. 2008; Fonseca et al. 2008; Rodrigues Da Silva et al. 2010). Adding to these, during biodiversity surveys many of the encountered marine nematodes are found in juvenile forms, thereby rendering identification extremely difficult as some of the superficial characters are found only in adult specimens (see Fig. 1).

The application of molecular techniques and the development of DNA sequence-based approaches have revolutionized biological research. Techniques including polymerase chain reaction dependent denaturing gradient gel

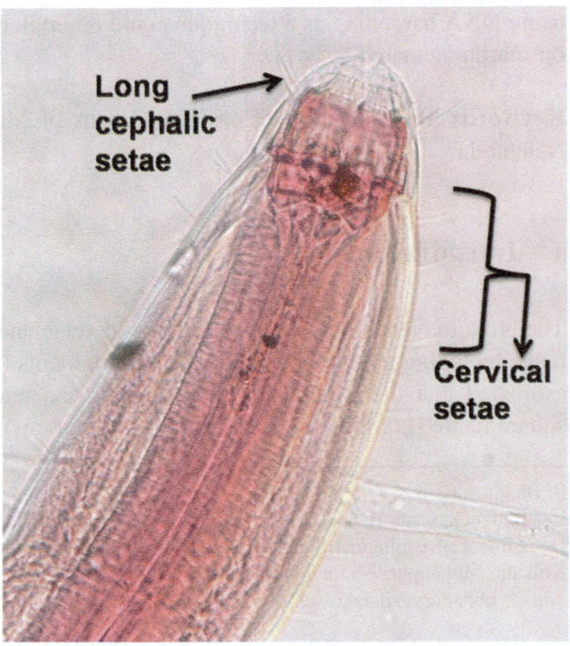

Fig. 1 Morphological characters such as long cephalic setae and cervical setae used for marine nematode identification are encountered in adult specimens

electrophoresis (PCR-DGGE), clone library and sequencing approach, terminal-restriction fragment length polymorphism (T-RFLP) and in recent times the application of next generation sequencing (NGS) are increasingly used in nematology research (e.g. Waite et al. 2003; Cook et al. 2005; Donn et al. 2008; Chen et al. 2010; Fonseca et al. 2010). Amongst the sequence-based approaches, DNA barcoding has gained importance with increased focus on application of the same towards elucidating biodiversity of nematode communities across different types of habitats and scientists have argued this tool can offer promising prospects in order to overcome taxonomic impediments associated with marine nematode systematic (e.g. Bhadury and Annapurna 2011a).

In this chapter, the successes and constraints of DNA barcoding as a tool for unraveling marine nematode biodiversity has been mainly discussed. Additionally, a perspective has been provided on how NGS techniques coupled with DNA barcoding can ultimately accelerate marine nematode biodiversity related research on a global scale.

2 DNA Barcoding

In 2003, a paper was published in the Proceedings of the Royal Society by Paul Hebert et al. in which the authors stated that mitochondrial COI gene could serve as a genetic code for all animal life and proposed the idea of DNA barcoding (Hebert et al. 2003). DNA barcoding relies on the use of a standardized DNA region as a tag for rapid and accurate species characterization (Hebert et al. 2003). In case of animals, the locus of choice as proposed for barcoding is a 658 base pair region (also known as Folmer region) of the cytochrome *c* oxidase I (COI), a gene which is involved in the respiratory chain of mitochondria (Valentini et al. 2008). This gene shows interesting level of variability and differences are low among individuals of the same species and high between individuals of different species. The primary goal of DNA barcoding is to identify all eukaryotic species (Miller 2007). In addition, the broader objectives of DNA barcoding as laid down by the International Barcode of Life Initiative (iBOL) include discovery of new species and facilitate identification, particularly in cryptic microscopic and other organisms with complex or inaccessible morphology as well as increase massively the speed of processing larger data sets (Hebert and Gregory 2005). To date, DNA barcoding has been tested across range of phyla and habitats with promising results (e.g. Hebert et al. 2004; Radulovici et al. 2010; Bucklin et al. 2011). It has been also applied as a tool to investigate evolutionary relationships and in cladistics (Savolainen et al. 2005).

In some taxa such as Porifera and Anthozoa issues pertaining to the low resolution level of targeted COI region have been also found in addition to the presence of nuclear copies of COI gene which may lead to over-estimation of species diversity during biodiversity surveys (Erpenbeck et al. 2006; Huang et al. 2008; Song et al. 2008). In case of marine nematodes, the COI amplification has proved to be relatively difficult because of rampant gene rearrangement, hypervariation among haplotypes, multipartionining and frequent recombination in their mitochondrial genomes (Bhadury et al. 2006; Lunt and Hyman 1997; Armstrong et al. 2000; Rodrigues Da Silva et al. 2010; Hyman et al. 2011). Therefore, several research groups have argued for the use of other regions of nematode genome such as the small and large subunits of nuclear ribosomal RNA cistron (18S rRNA and 28S rRNA) for DNA barcoding (De Ley et al. 2005; Bhadury et al. 2006).

The nuclear subunit ribosomal RNA cistron is a promising candidate due to its greater abundance (multiple copies) in the metazoan genome which makes it easier to target using PCR based approaches and its relatively conserved flanking regions that can provide classifications into molecular taxonomic units. Additionally, there is intense selection because of the vital role of rRNA in the assembly of proteins in ribosome. Thus, parts of them are strongly conserved as well as there are variable regions which can be effectively used for species delineation and resolving phylogenetic relationships (Floyd et al. 2002; Holterman et al. 2008).

2.1 DNA Barcoding Studies for Marine Nematodes

The study by De Ley et al. (2005) is one of the foremost where the applicability of DNA barcoding was evaluated based on the amplification and sequencing of large subunit ribosomal RNA (28S rRNA). In their study 37 nematode specimens were barcoded based on the D2D3 expansion segments of 28S rRNA and these represent at least 32 species, none of which matched with nematode 28S rRNA sequences available in sequence databases. They highlighted that data-rich surveys and phylogenetic tools for analysis of barcode sequences are an essential component for biodiversity discovery. The study by De Ley et al. (2005) also stressed the important of photo vouchering of specimens that are used for DNA barcoding so that digital and sequence data could be optimally used in large scale surveys. The applicability of DNA barcoding approach for rapid identification of marine nematodes during biodiversity survey from estuarine and coastal sedimentary environments of British Isles was evaluated for the first time based on the amplification, sequencing and phylogeny of nuclear small subunit ribosomal RNA (18S rRNA) marker (Bhadury et al. 2006). The authors showed that the accuracy of DNA barcoding approach for identification of marine nematodes up to species level can be more than 97 % and unidentified specimens can be correctly assigned to genus or species level based on molecular phylogeny, thereby confirming the reliability of this technique with potential application during large scale biodiversity surveys. Additionally, the authors suggested a 345 bp stretch of the 18S rRNA marker as a potential DNA barcode region for undertaking high throughput studies on marine nematodes during coastal ecosystem surveys (Bhadury et al. 2006).

Subsequently, a number of studies attempted to use DNA barcoding as a tool for identification of marine nematodes in other ecoregions. For example, Pereira et al. (2010) combined morpho-taxonomy and DNA barcoding to evaluate the diversity of free-living marine nematodes (Order Enoplida) from four coastal sites located in the Gulf of California and three on the Pacific coast of Baja California, Mexico. Their results indicated that 28S rRNA sequences (D2D3 domain) may be better suited for DNA barcoding of marine nematodes belonging to the Order Enoplida as compared to 18S rRNA, particularly for differentiating closely related or cryptic species. A study undertaken by Floyd et al. (2005) on marine nematodes from deep-sea sediment from the Equatorial Pacific Ocean highlighted the development of more robust nematode specific 18S rRNA primers during DNA barcoding studies. In another study focused on New Jersey coast of United States of America, Bhadury and Austen (2010) showed that DNA barcoding can be also used to understand the functional significance of marine nematodes in coastal ecosystems. The authors recovered several Monhysteridae-like 18S rRNA sequences which was then linked to the presence of oxic-anoxic boundary of studied intertidal sediments. In the same study, the authors reported the additional development of more stringent nematode specific 18S rRNA primers that could be used for biodiversity studies to overcome co-amplification of other eukaryotic phyla (Bhadury and Austen 2010). Armenteros et al. (2014a) compared the nuclear 18S rRNA and mitochondrial COI barcode sequences in tropical marine Desmodorid species and showed that COI barcode sequences can be more effective to disentangle relationship among closely related species provided reference sequence databases are substantially enriched.

3 DNA Barcoding of Marine Nematodes Using Mitochondrial COI Approach

In the last five years there has been significant development with respect to the application of mitochondrial COI for barcoding of marine nematodes. In a recent study, Derycke et al. (2010a) successfully evaluated the effectiveness of COI in 41 species of marine nematodes belonging to 33 genera and representing almost all orders from European coasts. The authors identified a region of COI (I3-M11 partition) that could be used for the purpose of DNA barcoding. At the same time, the authors highlighted the need for a strict scrutiny of the obtained COI sequences since co-maplification from contamination, nuclear pseudogenes and endosymbionts may lead to wrong identification of nematodes. Overall, they concluded that the variability pattern observed in mitochondrial COI could be used to aid in identification of nematode species provided a good reference sequence database is equally available. To date, the list of available 18S rRNA, 28S rRNA and mitochondrial COI primers have been detailed in Table 1.

Table 1 List of commonly used primers for undertaking DNA barcoding in marine nematodes

Molecular marker of interest for DNA barcoding	Primer combinations	Size of amplicon	Reference
18S rRNA	G18S4[5′-GCTTGTCTCAAAGATTAAGCC-3′] 26R [5′-CATTCTTGGCAAATGCTTTCG-3′] 22F [5′-TCCAAGGAAGGCAGCAGGC-3′] 13R [5′-GGGCATCACAGACCTGTTA-3′] 24F1 [5′-AGAGGTGAAATTCTTGGATC-3′] 18P [5′-TGATCCWKCYGCAGGTTCAC-3′]	1600 (bp) approx.	Blaxter et al. (1998), Meldal et al. (2007), Bik et al. (2010b)
	MN18F [5′-CGCGAATRGCTCATTACAACAGC-3′] NEM18S_R[5′-GGGCGGTATCTGATCGCC-3′]	925 (bp)	Bhadury et al. (2006)
	MN18F [5′-CGCGAATRGCTCATTACAACAGC-3′] 22R [5′-GCCTGCTGCCTTCCTTGGA-3′]	345 (bp)	Bhadury et al. (2006)
	M18F [5′-AGRGGTGAAATYCGTGGAC-3′] M18R [5′-TCTCGCTCGTTATCGGAAT-3′]	427 (bp)	Bhadury and Austen (2010)
28S rRNA	D2A [5′-ACAAGTACCGTGAGGGAAAGTTG-3′] D3b [5′-TCGGAAGGAACCAGCTACTA-3′]	583 (bp)	De Ley et al. (2005)
COI	JB2F [5′-ATGTTTTGATTTTACCWGCWTTYGGTGT-3′] JB5GEDR [5′-AGCACCTAAACTTAAAACATARTGRAA RTG-3′]	396 (bp)	Derycke et al. (2007) Derycke et al. (2010b)
	JB3F [5′-TTTTTTGGGCATCCTGAGGTTTAT-3′] JB5R [5′-AGCACC TAAACTTAAAACATAATGAAAATG-3′]	418 (bp)	Derycke et al. (2005) Derycke et al. (2010a, b)

4 Application of DNA Barcoding in Marine Nematode Population Genetics and Taxonomy

The approach of DNA barcoding for marine nematodes has extended beyond biodiversity assessment. It has been applied to test pertinent questions in areas of marine nematode population genetics as well as in resolving their taxonomy. The mitochondrial COI sequences have been used to investigate population genetic structure in the cryptic marine nematode species, *Pellioditis marina* (Derycke et al. 2005). The authors found strong genetic differentiation among populations of *P. marina* and concluded that autecological characteristics, including short generation time, high colonization potential and local adaptation play important roles in observed trend. In another study, Derycke et al. (2007) investigated the population genetic structure of a marine nematode, *Geomonhystera disjuncta* based on COI and nuclear ITS (inter transcribed spacer) regions and found the evidence of presence of five cryptic taxa within this species. There are other reports of the limited use of COI in population genetic studies undertaken in selective species of marine nematodes (Derycke et al. 2006, 2008). The 18S rRNA has been also used as a molecular marker to study the population structure of marine nematode *Terschellingia longicaudata* based on specimens collected from several locations globally (Bhadury et al. 2008). The authors showed the presence of cryptic species complexes in *T. longicaudata* population based on 18S rRNA molecular phylogeny and they also highlighted that currently available morpho-taxonomy may not be adequate to discriminate these complexes (Bhadury et al. 2008). Besides these two molecular markers, the D2/D3 region of 28S rRNA has been also used to investigate cryptic species complexes in the marine nematode *Thoracostoma trachygaster* (Derycke et al. 2010b).

DNA barcoding has been also tested as a tool to resolve taxonomy of marine nematodes and also for describing new species. For example, Neres et al. (2010) described a new species, *Daptonema matrona* based on morphological characters and 18S rRNA sequences using integrative taxonomy approaches. Other studies have used molecular phylogeny (based on 18S rDNA and mt COI barcode regions) as a tool for describing new species of marine nematodes (e.g. Rho et al. 2011; Cunha et al. 2013; Armenteros et al. 2014b; Tchesunov et al. 2015). However, the rate of description of new species of marine nematodes has not accelerated possibly due to the lack of integration of morpho-taxonomy with molecular methods such as DNA barcoding. Indeed, a comprehensive review undertaken by Radulovici et al. (2010) showed that only 21 new species of marine nematodes are reported every year as compared to other marine faunal groups (e.g. Mollusca: 354 new species described every year).

5 Limited Coverage of DNA Barcode Sequences

One of the major factors that have restricted the wider applicability of DNA barcoding as a tool in marine nematode diversity and taxonomy studies is the paucity of reference sequences available from different geographic locations globally. As of today (10th September, 2015) more than 850 marine nematode nucleotide sequences representing molecular markers such as nuclear small subunit ribosomal DNA (18S rRNA), large subunit ribosomal DNA (28S rRNA) and mitochondrial cytochrome c oxidase I (COI) are available in sequence databases such as GenBank, EMBL, DDBJ and PDB. However, majority of these sequences represent coastal and marine ecosystems from Western Europe and North America. Indeed, representative marine nematode sequences from biodiversity rich continents such as Asia, South America and Africa remain largely unrepresented in public sequence databases. For example, a recent study undertaken by Kumar et al. (2015) based on clone library and sequencing approach of a region of 18S rRNA, which is also used for DNA barcoding, showed the presence of several novel species of marine nematodes in the North Eastern Indian Ocean region (West coast of India) and that many of the sequences could be identified only up to genus level highlighting the limitation of existing sequence databases. Therefore it is extremely important to broaden the coverage of marine biodiversity rich geographical locations for wider use of DNA barcoding as a tool during biodiversity surveys.

6 Next Generation Sequencing and Incorporation of DNA Barcodes

High-throughput DNA sequencing technology is a new revolution that can increase our understanding of marine nematode diversity and their role in ecosystem processes across different environments. Next generation sequencing (NGS) platforms such as Roche 454, Illumina (HiSeq and MiSeq) and ABI SOLiD, with capacities to process millions of sequences in parallel, have revolutionized biological research in recent years including applications towards understanding marine ecosystem (e.g. Lecroq et al. 2011; Fonseca et al. 2010, 2014; Bik et al. 2012). Specification of some of the NGS platforms has been detailed in Table 2. Individual read lengths in majority of next-generation platforms are usually limited, but the depth of coverage per base pair and advanced sequence assembly software allow sequencing of giga base pairs (Gbp) in a single run. Concurrently, there has been significant development in bioinformatic pipelines, which can process these large set of raw data generating out of NGS platforms (Stoltzfus et al. 2013; Darling et al. 2014). The NGS methodology has been predominantly tested using molecular markers such as 18S rRNA and the regions that have been targeted represent the same regions tested previously in DNA barcoding studies of marine nematodes.

Table 2 Specification of some of the NGS platforms

Sequencing platform	454	Illumina	ABI SOLID	Ion torrent
Year of availability	2005	2006	2006	2010
Length of sequenced fragment	200–700 bp	150 bp approx.	35–50 bp	200 bp approx.
Run time	23 h	27 h to 11 days	7–8 days	2 h
Technology used for sequencing	emPCR, pyrosequencing	Polonies, cleavable dye terminators	emPCR, ligation with cleavable dye terminators	emPCR, H+ detection

NGS has been tested towards assessment of marine nematode diversity from coastal areas of United Kingdom. Creer et al. (2010) analyzed sediment samples collected from intertidal zone at Littlehampton on the south coast of England, UK as well as soil, litter and understory habitats sampled at La Selva Biological Station, Costa Rica. Community PCR and 454 sequencing based on 18S rDNA approaches yielded a total of 29,756 high-quality sequences over 200 bases from the marine samples while for the tropical rain forest samples yielded a total of 40,334 high-quality sequences of at least 200 bases. In the terrestrial data set, the soil habitat had fewer nematode operational cluster taxonomic units (OCTUs) (35) than either the marine littoral habitat (149) or canopy (97). Plant-parasitic nematodes were more diverse and abundant in soil environment, with bacterial and fungal-feeding nematodes predominating in the litter and canopy. This study evaluated the effectiveness of 454 sequencing in nematode diversity studies from varied environments including marine realms.

Fonseca et al. (2010) applied a metagenetic approach using next generation sequencing of the 18S nuclear small subunit ribosomal RNA (rRNA) marker to assess simultaneously the relative levels of richness and patterns of diversity of multiple metazoan phyla from eight benthic samples collected from low-tide zone of an estuarine beach near Prestwick on the West coast of Scotland, and from one sample from a beach in Littlehampton in the South of England. The amplicons were processed for sequencing on a Roche 454 FLX platform generating a total of 353,896 sequences which were subsequently filtered to 305,702 for downstream analysis. Based on the annotation 374 OCTUs were assigned to phylum level. Of the metazoan OCTUs, 182 were from Nematoda, at least three times more than from any other individual meiofaunal taxon. Platyhelminthes (61 OCTUs) was the second richest phylum, followed by Arthropoda (29 OCTUs including Copepoda, Ostracoda and Malacostraca), Mollusca (22 OCTUs), Gastrotricha (7 OCTUs), Annelida (6 OCTUs) and five less-rich phyla (for example, Bryozoa, Echinodermata, Cercozoa, Rotifera and Alveolata with between 1 and 3 OCTUs each). Based on the comparisons of the OCTU sequences with the NCBI databases, authors found that majority (95 %) of Nematoda OCTUs have never been sequenced before. For other phyla only a small fraction of the OCTUs showed 100 % identity with previously sequenced specimens. Only fifteen of the OCTUs with <90 % identity (300 sequences) were robustly placed within identified phyla (mainly Nematoda). For Prestwick samples the authors detected 182 Nematoda OCTUs, compared with 450 species of free-living marine nematodes that have been described from around the entire British Isles. Geographically, these datasets represent the discovery of 40 % of the previously known phylum richness from transect that represents 0.004 % of the length of the British coastline (~ 17,820 km, Ordnance Survey). In all, 70 % of Nematoda OCTUs were unique to Prestwick and 58–100 % of the OCTUs for the other phyla were only present in Prestwick. Based on 454 sequencing approach, a study was undertaken to assess microbial eukaryotic communities across depth (shallow water to abyssal) and ocean basins (deep-sea Pacific and Atlantic) (Bik et al. 2012). Within the 12 sites examined, the authors found that some taxa can maintain eurybathic ranges and cosmopolitan deep-sea distributions, but majority of species appear to be regionally restricted. Contrary to previous observation of nematodes being the most abundant members in sediment, the authors found equal or more dominant role for other taxonomic groups in some of the deep-sea sites (e.g. unicellular eukaryotes in the Pacific) based on 454 sequencing. In addition, low genetic divergence between geographically disparate deep-sea sites suggested either a shorter coalescence time between deep-sea regions or slower rates of evolution across this vast oceanic ecosystem (Bik et al. 2012). In another study involving deep-sea sites in the Southern Ocean, the authors investigated association between deep-sea nematodes and marine fungi and potential implications including food preference of nematodes based on capillary and 454 sequencing approaches (Bhadury et al. 2011b).

In a recent study undertaken by Lallias et al. (2015) based on 454 Roche sequencing of 18S rRNA barcodes, the authors showed that marine nematode species richness accounted for almost 22 and 31 % in Thames and Mersey estuaries of United Kingdom respectively and more than 55 % of the generated sequence reads (out of 957,216 reads) represented nematode signatures. NGS approaches can provide information of marine nematode diversity from less explored sites or regions; however at the same time some of these studies as detailed above also reflect that a significant proportion of the sequences generated show limited identity with sequences that are available in published databases such as GenBank/EMBL/DDBJ. This is due to the restricted number of deposited sequences for valid marine nematode species including DNA barcode sequences from geographically less represented regions. Thus for high throughput sequencing tools to work it is important that coordinated effort should be taken in DNA barcoding of marine nematodes with special emphasis on continents such as South America, Africa and Asia.

7 Way Forward

The flow diagram provides an overview of approaches to be undertaken to facilitate and speed up marine nematode identification and elucidation of their community structure by integration of morpho-taxonomy, DNA barcoding and NGS methodologies during biodiversity surveys (see Fig. 2). Such approaches could result in enriched reference sequence databases and ultimately help towards resolving marine nematode species identification during large scale biodiversity surveys. In addition, potential new DNA barcode regions need to identified, in particular the need to further modify and develop the applicability of mitochondrial COI DNA barcode with the designing of more robust set of primers. For example, COI primers

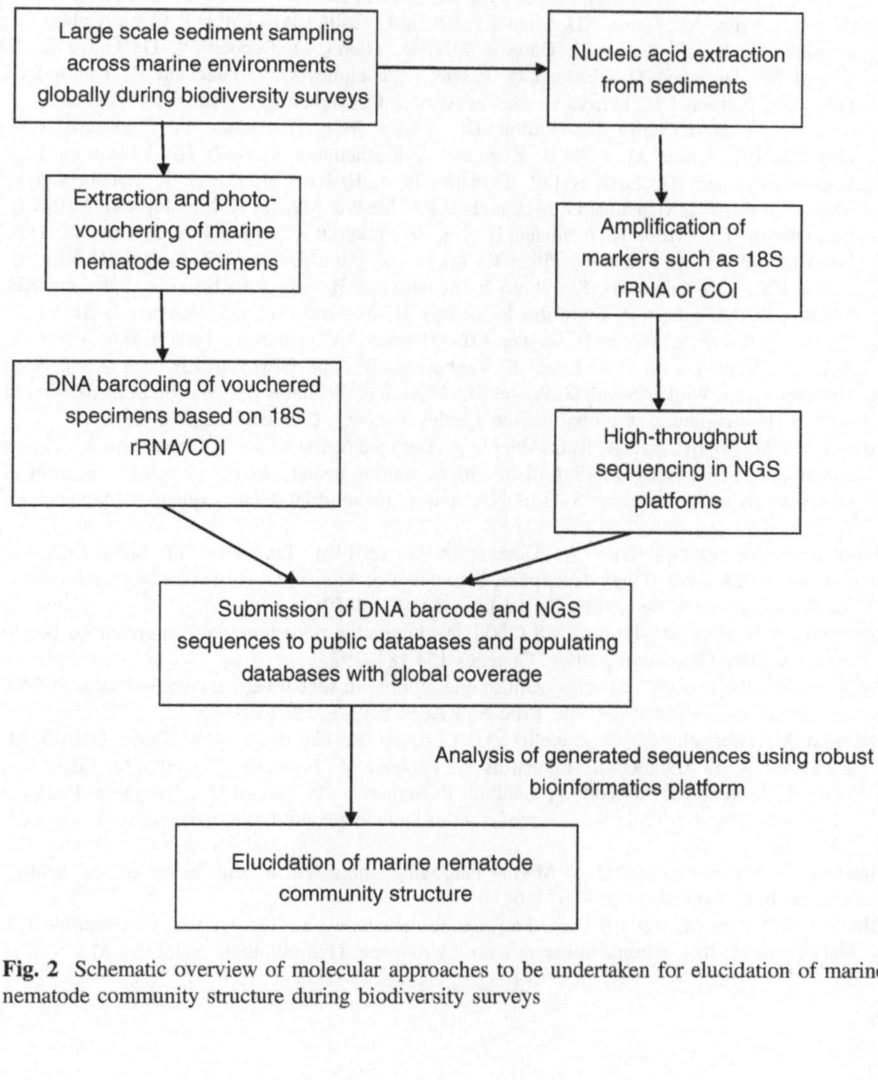

Fig. 2 Schematic overview of molecular approaches to be undertaken for elucidation of marine nematode community structure during biodiversity surveys

could be developed at order and family levels to broaden the taxonomic coverage of marine nematodes in future studies. The long term success of DNA barcoding as a tool for speeding up identification of marine nematodes will also depend on coordinated international efforts such as along the line of Census of Marine Life (CoML) programs. To conclude, morpho-taxonomy coupled with DNA taxonomy involving DNA barcoding as a tool could be potentially made as a pre-requisite for undertaking studies on marine nematodes.

References

Appeltans W, Ahyong ST, Anderson G, Angel MV, Artois T, Bailly N, Bamber R, Barber A, Bartsch I, Berta A, Błażewicz-Paszkowycz M, Bock P, Boxshall G, Boyko CB, Brandão SN, Bray RA, Bruce NL, Cairns SD, Chan TY, Cheng L, Collins AG, Cribb T, Curini-Galletti M, Dahdouh-Guebas F, Davie PJ, Dawson MN, De Clerck O, Decock W, De Grave S, de Voogd NJ, Domning DP, Emig CC, Erséus C, Eschmeyer W, Fauchald K, Fautin DG, Feist SW, Fransen CH, Furuya H, Garcia-Alvarez O, Gerken S, Gibson D, Gittenberger A, Gofas S, Gómez-Daglio L, Gordon DP, Guiry MD, Hernandez F, Hoeksema BW, Hopcroft RR, Jaume D, Kirk P, Koedam N, Koenemann S, Kolb JB, Kristensen RM, Kroh A, Lambert G, Lazarus DB, Lemaitre R, Longshaw M, Lowry J, Macpherson E, Madin LP, Mah C, Mapstone G, McLaughlin PA, Mees J, Meland K, Messing CG, Mills CE, Molodtsova TN, Mooi R, Neuhaus B, Ng PK, Nielsen C, Norenburg J, Opresko DM, Osawa M, Paulay G, Perrin W, Pilger JF, Poore GC, Pugh P, Read GB, Reimer JD, Rius M, Rocha RM, Saiz-Salinas JI, Scarabino V, Schierwater B, Schmidt-Rhaesa A, Schnabel KE, Schotte M, Schuchert P, Schwabe E, Segers H, Self-Sullivan C, Shenkar N, Siegel V, Sterrer W, Stöhr S, Swalla B, Tasker ML, Thuesen EV, Timm T, Todaro MA, Turon X, Tyler S, Uetz P, van der Land J, Vanhoorne B, van Ofwegen LP, van Soest RW, Vanaverbeke J, Walker-Smith G, Walter TC, Warren A, Williams GC, Wilson SP, Costello MJ (2012) The magnitude of global marine species diversity. Curr Biol 22:2189–2202

Armenteros M, Rojas-Corzo A, Ruiz-Abierno A, Derycke S, Backeljau T, Decraemer W (2014a) Systematics and DNA barcoding of free-living marine nematodes with emphasis on tropical desmodorids using nuclear SSU rDNA and mitochondrial COI sequences. Nematology 16:979–989

Armenteros M, Ruiz-Abierno A, Decraemer W (2014b) Taxonomy of Stilbonematinae (Nematoda: Desmodoridae): description of two new and three known species and phylogenetic relationships within the family. Zool J Linn Soc 171:1–21

Armstrong MR, Blok VC, Phillips MS (2000) A multipartite mitochondrial genome in the potato cyst nematode *Globodera pallida*. Genetics 154:181–192

Avise JC, Walker D (1999) Species realities and number in sexual vertebrates: perspectives from an asexually transmitted genome. Proc Natl Acad Sci USA 96:992–995

Balsamo M, Albertelli G, Ceccherelli VU, Coccioni R, Colangelo MA, Curini-Galletti M, Danovaro R, D'Addabbo R, Leonardis C, Fabiano M, Frontalini F, Gallo M, Gambi C, Guidi L, Moreno M, Pusceddu A, Sandulli R, Semprucci F, Todaro MA, Tongiorgi P (2010) Meiofauna of the Adriatic Sea: current state of knowledge and future perspective. Chem Ecol 26:45–63

Bhadury P, Annapurna C (2011a) Marine barcoding—how will it help Indian marine benthic studies. Ind J Geo-Mar Sci 40:645–647

Bhadury P, Austen MC (2010) Barcoding marine nematodes: an improved set of nematode 18S rRNA primers to overcome eukaryotic co-interference. Hydrobiologia 641:245–251

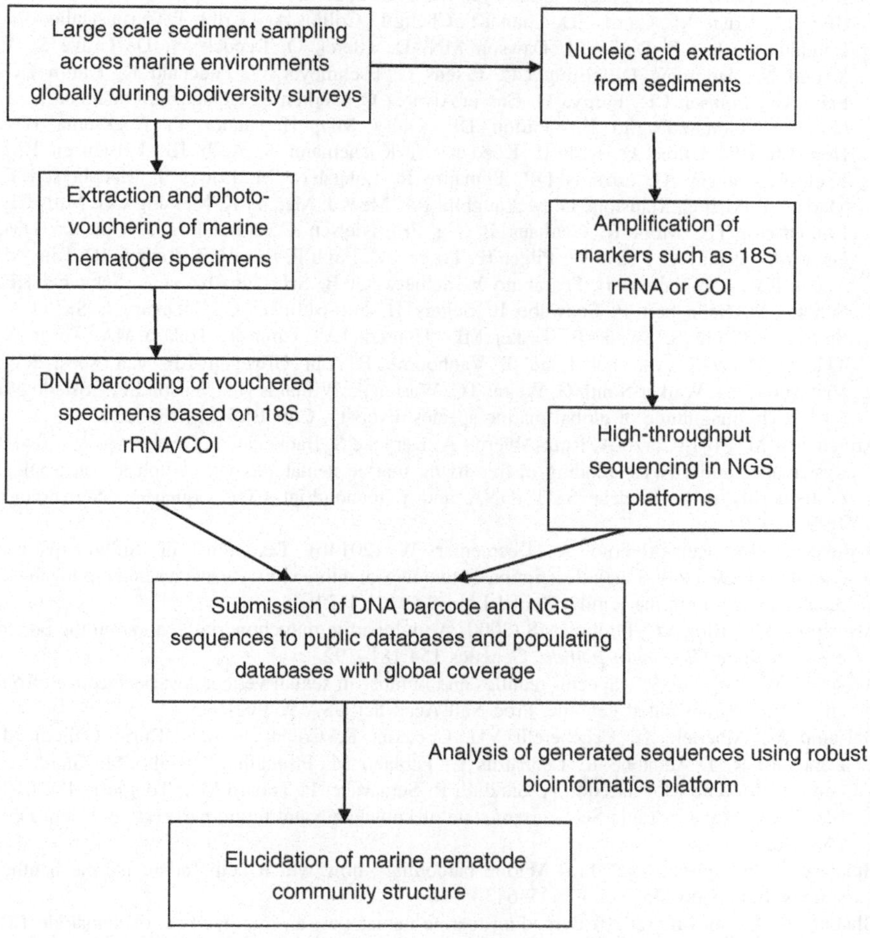

Bhadury P, Austen MC, Bilton DT, Lambshead PJD, Rogers AD, Smerdon GR (2006) Development and evaluation of a DNA barcoding approach for the rapid identification of nematodes. Mar Ecol Prog Ser 320:1–9

Bhadury P, Austen MC, Bilton DT, Lambshead PJD, Rogers AD, Smerdon GR (2008) Evaluation of combined morphological and molecular techniques for marine nematode (*Terschellingia* spp.) identification. Mar Biol 154:509–518

Bhadury P, Bik H, Lambshead JD, Austen MC, Smerdon GR, Rogers AD (2011b) Molecular diversity of fungal phylotypes co-amplified alongside nematodes from coastal and deep-sea marine environments. PLoS ONE 6:e26445

Bik HM, Porazinska DL, Creer S, Caporaso JG, Knight R, Thomas WK (2012) Sequencing our way towards understanding global eukaryotic biodiversity. Trends Ecol Evol 27:233–243

Bik HM, Lunt DH, Thomas WKT, Lambshead PJD (2010b) Low endemism, continued deepshallow interchanges, and evidence for cosmopolitan distributions in free-living marine nematodes (order Enoplida). BMC Evol Biol 10:389

Blaxter ML, De Ley P, Garey JR, Liu LX, Scheldeman P, Vierstraete A, Vanfleteren JR, Mackey LY, Dorris M, Frisse LM, Vida JT, Thomas WK (1998) A molecular evolutionary framework for the phylum Nematoda. Nature 392:71–75

Boufahja F, Sellami B, Dellali M, Aïssa P, Mahmoudi E, Beyrem H (2011a) A microcosm experiment on the effects of permethrin on a free-living nematode assemblage. Nematology 13:901–909

Boufahja F, Hedfi A, Amorri J, Aïssa P, Beyrem H, Mahmoudi E (2011b) Examination of the bioindicator potential of *Oncholaimus campylocercoides* (Nematoda, Oncholaimidae) from Bizerte bay (Tunisia). Ecol Ind 11:1139–1148

Bucklin A, Steinke D, Blanco-Bercial L (2011) DNA barcoding of marine metazoa. Ann Rev Mar Sci 3:471–508

Chen XY, Daniell TJ, Neilson R, O'Flaherty V, Griffiths BS (2010) A comparison of molecular methods for monitoring soil nematodes and their use as biological indicators. Eur J Soil Biol 46:319–324

Cook AA, Bhadury P, Debenham NJ, Meldal BHM, Blaxter ML, Smerdon GR, Austen MC, Lambshead PJD, Rogers AD (2005) Denaturing gradient gel electrophoresis as a tool for the identification of marine nematodes. Mar Ecol Prog Ser 291:103–113

Creer S, Fonseca VG, Porazinska DL, Giblin-Davis RM, Sung W, Power DM, Packer M, Carvalho GR, Blaxter ML, Lambshead PJD, Thomas WK (2010) Ultrasequencing of the meiofaunal biosphere: practice, pitfalls and promises. Mol Ecol 19:4–20

Cunha BP, Brito S, Fonseca G (2013) Zygonemella: the forgotten genus of the family Xyalidae (Nematoda). Zootaxa 3669:179–183

Darling A, Bik HM, Jospin G, Lowe E, Eisen JA (2014) PhyloSift: a pipeline for phylogenetic taxonomy assignments from environmental metagenome data. Peer J 2:e243

De Ley P, De Ley IM, Morris K, Abebe E, Mundo Ocampo M, Yoder M, Heras J, Waumann D, Rocha Olivares A, Burr AHJ (2005) An integrated approach to fast and informative morphological vouchering of nematodes for applications in molecular barcoding. Phil Trans R Soc Lond B 360:1945–1958

Derycke S, Remerie T, Vierstraete A, Backeljau T, Vanfleteren J, Vincx M, Moens T (2005) Mitochondrial DNA variation and cryptic speciation within the free-living marine nematode *Pellioditis marina*. Mar Ecol Prog Ser 300:91–103

Derycke S, Backeljau T, Vlaeminck C, Vierstraete A, Vanfleteren J, Vincx M, Moens T (2006) Seasonal dynamics of population genetic structure in cryptic taxa of the *Pellioditis marina* complex (Nematoda: Rhabditida). Genetica 128:307–321

Derycke S, Backeljau T, Vlaeminck C, Vierstraete A, Vanfleteren J, Vincx M, Moens T (2007) Spatiotemporal analysis of population genetic structure in *Geomonhystera disjuncta* (Nematoda, Monhysteridae) reveals high levels of molecular diversity. Mar Biol 15:1799–1812

Derycke S, Fonseca G, Vierstraete A, Vanfleteren J, Vincx M, Moens T (2008) Disentangling taxonomy within the *Rhabditis* (Pellioditis) *marina* (Nematoda, Rhabditidae) species complex using molecular and morphological tools. Zool J Linn Soc 152:1–15

Derycke S, Vanaverbeke J, Rigaux A, Backeljau T, Moens T (2010a) Exploring the use of Cytochrome Oxidase c subunit 1 (COI) for DNA barcoding of free-living marine nematodes. PLoS ONE 5:e13716

Derycke S, De Ley P, De Ley IT, Holovachov O, Rigaux A, Moens T (2010b) Linking DNA sequences to morphology: cryptic diversity and population genetic structure in the marine nematode *Thoracostoma trachygaster* (Nematoda, Leptosomatidae). Zool Scripta 39:276–289

Donn S, Griffiths BS, Neilson R, Daniell TJ (2008) DNA extraction from soil nematodes for multi-sample community studies. Appl Soil Ecol 38:20–26

Erpenbeck D, Hooper JNA, Worheide G (2006) CO1 phylogenies in diploblasts and the 'Barcoding of Life'—are we sequencing a suboptimal partition? Mol Ecol Notes 6:50–553

Floyd R, Abebe E, Papert A, Blaxter ML (2002) Molecular barcodes for soil nematode identification. Mol Ecol 11:839–850

Floyd R, Rogers AD, Lambshead PJD, Smith CR (2005) Nematode-specific PCR primers for the 18S small subunit rRNA gene. Mol Ecol Notes 5:611–612

Fonseca G, Derycke S, Moens T (2008) Integrative taxonomy in two free-living nematode species complexes. Biol J Linn 94:737–753

Fonseca VG, Carvalho GR, Sung W, Johnson HF, Power DM, Neill SP, Packer M, Blaxter ML, Lambshead PJD, Thomas WK, Creer S (2010) Second-generation environmental sequencing unmasks marine metazoan biodiversity. Nat Comm. doi:10.1038/ncomms1095

Fonseca VG, Carvalho GR, Nichols B, Quince C, Johnson HF, Neill SP, Lambshead JD, Thomas WK, Power DM, Creer S (2014) Metagenetic analysis of patterns of distribution and diversity of marine meiobenthic eukaryotes. Glob Ecol Biogeogr 23:1293–1302

Hebert PDN, Gregory TR (2005) The promise of DNA barcoding for taxonomy. Syst Biol 54: 852–859

Hebert PDN, Cywinska A, Ball SL, Dewaard JR (2003) Biological identifications through DNA barcodes. Proc R Soc Lond B 270:313–321

Hebert PDN, Stoeckle MY, Zemlak TS, Francis CM (2004) Identification of birds through DNA barcodes. PLoS Biol 2:e312

Holterman M, Rybarczyk K, Van Den Elsen S, Van Megen H, Mooyman P, Peña Santiago R, Bongers T, Bakker J, Helder J (2008) A ribosomal DNA-based framework for the detection and quantification of stress-sensitive nematode families in terrestrial habitats. Mol Ecol Res 8:23–34

Huang DW, Meier R, Todd PA, Chou LM (2008) Slow mitochondrial COI sequence evolution at the base of the metazoan tree and its implications for DNA barcoding. J Mol Evol 66:167–174

Hyman BC, Lewis SC, Tang S, Wu Z (2011) Rampant gene rearrangement and haplotype hypervariation among nematode mitochondrial genomes. Genetica 139:611–615

Kumar A, Sen D, Bhadury P (2015) Unraveling free-living marine nematode community structure from a biodiversity rich tropical coastal setting based on molecular approaches. Mar Biodiversi 45:537–547

Lallias D, Hiddink JG, Fonseca VG, Gaspar JM, Sung W, Neill SP, Barnes N, Ferrero T, Hall N, Lambshead PJD, Packer M, Kelley Thomas W, Creer S (2015) Environmental metabarcoding reveals heterogeneous drivers of microbial eukaryote diversity in contrasting estuarine ecosystems. ISME J 9:1208–1221

Lambshead PJD (2004) Marine nematode biodiversity. In: Chen ZX, Chen SY, Dickson DW (eds) Nematology: advances and perspectives, vol 1. CABI Publishing, Wallingford, pp 439–468

Lecroq B, Lejzerowicz F, Bachar D, Christen R, Esling P, Baerlocher L, Østerås M, Farinelli L, Pawlowski J (2011) Ultra-deep sequencing of foraminiferal microbarcodes unveils hidden richness of early monothalamous lineages in deep-sea sediments. Proc Natl Acad Sci USA 108:13177–13182

Lunt DH, Hyman BC (1997) Animal mitochondrial DNA recombination. Nature 387:247

Meldal BHM, Debenham NJ, De Ley P, De Ley IT, Vanfleteren JR, Vierstraete AR, Bert W, Borgonie G, Moens T, Tyler PA, Austen MC, Blaxter ML, Rogers AD, Lambshead PJ (2007) An improved molecular phylogeny of the Nematoda with special emphasis on marine nematode taxa. Mol Phylogenet Evol 42:622–636

Miller SE (2007) DNA barcoding and the renaissance of taxonomy. Proc Natl Acad Sci USA 104:4775–4784

Neres PF, Da Fonseca-Genevois VG, Torres RA, Da Fonseca Cavalcanti M, De Castro FJV, Da Silva NRR, Rieger TT, Decraemer W (2010) Morphological and molecular taxonomy of a new *Daptonema* (Nematoda, Xyalidae) with comments on the systematics of some related taxa. Zool J Linn Soc 158:1–15

Pereira TJ, Fonseca G, Mundo-Ocampo M, Guilherme BC, Rocha-Olivares A (2010) Diversity of free-living marine nematodes (Enoplida) from Baja California assessed by integrative taxonomy. Mar Biol 157:1665–1678

Radulovici AE, Archambault P, Dufresne F (2010) DNA barcodes for marine biodiversity: moving fast forward? Diversity 2:450–472

Rho HS, Decraemer W, Sørensen MV, Min WG, Jung J, Kim W (2011) Megadraconema cornutum, a new genus and species from Korea, with a discussion of its classification and relationships within the family Draconematidae (Nematoda, Desmodorida) based on morphological and molecular characters. Zool Sci 28:68–84

Rodrigues Da Silva NR, Da Silva MC, Fonseca Genevois V, De Esteves AM, De Ley P, Decraemer W, Rieger TT, Dos Santos Correia MT (2010) Marine nematode taxonomy in the age of DNA: the present and future of molecular tools to assess their biodiversity. Nematology 12:661–672

Savolainen V, Cowan RS, Vogler AP, Roderick GK, Lane R (2005) Towards writing the encyclopedia of life: an introduction to DNA barcoding. Phil Trans R Soc Lond B 360:1805–1811

Semprucci F, Losi V, Moreno M (2015) A review of Italian research on free-living marine nematodes and the future perspectives in their use as Ecological Indicators (EcoInd). Mediterr Mar Sci 16:352–365

Song H, Buhay JE, Whiting MF, Crandall KA (2008) Many species in one: DNA barcoding overestimates the number of species when nuclear mitochondrial pseudogenes are coamplified. Proc Natl Acad Sci USA 105:13486–13491

Stoltzfus A, Lapp H, Matasci N, Deus H, Sidlauskas B, Zmasek CM, Vaidya G, Pontelli E, Cranston K, Vos R, Webb CO, Harmon LJ, Pirrung M, O'Meara B, Pennell MW, Mirarab S, Rosenberg MS, Balhoff JP, Bik HM, Heath T, Midford P, Brown JW, McTavish EJ, Sukumaran J, Westneat M, Alfaro ME, Steele A (2013) Phylotastic! making tree-of-life knowledge accessible, re-usable and convenient. BMC Bioinform 14:158

Tchesunov AV, Portnova DA, Campenhout JV (2015) Description of two free-living nematode species of Halomonhystera disjuncta complex (Nematoda: Monhysterida) from two peculiar habitats in the sea. Helgoland Mar Res 69:57–85

Valentini A, Pompanon F, Taberlet P (2008) DNA barcoding for ecologists. Trends Ecol Evol 24:110–117

Waite IS, O'Donnell AG, Harrison A, Davies JT, Colvan SR, Ekschmitt K, Dogan H, Wolters V, Bongers T, Bongers M, Bakonyi G, Nagy P, Papatheodorou EM, Stamou GP, Boström S (2003) Design and evaluation of nematode 18S rDNA primers for PCR and denaturing gradient gel electrophoresis (DGGE) of soil community DNA. Soil Biol Biochem 35:1165–1173

DNA Barcoding of Calanoid Copepods from the Gulf of California

Juan Ramon Beltrán-Castro and Sergio Hernández-Trujillo

Abstract The diversity of pelagic copepods was studied in different sites around the Gulf of California, to confirm their morphological identification with of CO1 gen analysis. This is the first study of its kind for the Gulf of California and we report the results of 101 barcode sequences for 27 species of copepods. The separation of species based on morphological characters was very clear for most species and consistent with the formation of genetic groups obtained with the CO1 gene and its corresponding barcode without overlap between the sequences, thus becoming the initial records a database of genetic sequences for the area.

Keywords Gulf of California · Calanoid copepods · Barcoding · Cytochrome Oxidase 1 · Marine zooplankton

1 Introduction

The crustaceans are one of the most diverse marine groups, and have a significant role in the food web and be represented in all ecosystems; These and others features make the group present ambiguous morphological characters from the earliest stages of development to adulthood, so the DNA barcode becomes a powerful tool for reliable identification of species (Radulovici et al. 2010).

Pelagic copepods are maxillopod crustaceans that its abundance and frequency of occurrence in the epi and mesopelagic zone are one of the most important taxa in the marine food chain. More than 2500 species of marine planktonic copepods have been described, with distributions ranging from shallow waters to abyss to hadopelagic depths (Razouls et al. 2011). Its great diversity needs accuracy on species identification because, in some cases, share morphometric and meristic characteristics that tend to confuse the identity of the species. To help solve cryptic species

J.R. Beltrán-Castro · S. Hernández-Trujillo (✉)
Instituto Politécnico Nacional., CICIMAR,
Av. IPN s/n, Playa Palo Santa Rita, 23096 La Paz, BCS, Mexico
e-mail: strujil@ipn.mx

S. Trivedi et al. (eds.), *DNA Barcoding in Marine Perspectives*,
DOI 10.1007/978-3-319-41840-7_9

occurrence and other taxonomic issues, molecular tools are used since the late 20th century, specifically through the use of a DNA fragment known as cytochrome C Oxidase 1 (Hebert et al. 2003).

Recent research of DNA barcode on marine copepods have been focused on the exploration of species diversity mostly in ocean regions, the problem of taxa, biogeographic analysis and finding cryptic species (Blanco-Bercial et al. 2014).

In Central and Southern Gulf of California, have been identified between 53 (Hernández-Trujillo and Esquivel-Herrera 1989) and 84 (Brinton et al. 1986) copepods pelagic species; at the Bay and Ensenada of La Paz have been recorded 129 species (Palomares 1996; González-Navarro and Saldierna-Martínez 1997; Lavaniegos-Espejo and González-Navarro 1999; Aceves-Medina et al. 2007); these figures, compared with 197 species identified at national level (Hernández-Trujillo and Esqueda-Escárcega 2002) indicate the importance of the Gulf of California from the standpoint of not only the holoplankton biodiversity, but also the meroplankton.

The Cytochrome c oxidase 1 (CO1), such as has been previously mentioned, it has been widely used in studies of diversity of fungi (Kurtzman 1985), bacteria (Wilson et al. 1995), on marine copepods (Bucklin et al. 1998), lepidoptera (Brown et al. 1999), onychophora (Trewick 2000), annelidae, arthropoda, chordata, cnidarians, echinoderms and flatworms (Hebert et al. 2003), birds (Hebert et al. 2004), fishes (Trivedi et al. 2014) euphausiids (Bucklin et al. 2007) Branchiopoda resistance eggs, Copepoda, Rotifera, Bryozoa and Ascidia in ballast water (Briski et al. 2011), among other metazoarians.

Bucklin et al. (2010a, b) estimated at 230,000 the number of marine animals of known species and one million which have not yet been described, so consider that the process of description of new species will be continued in the coming decades.

For copepods, studies of DNA barcode has focused to reviewing the copepod morphological descriptions which by their level of uncertainty may result in sibling or cryptic species, and CO1 is a gene that has shown very useful for separating copepods species of genus *Calanus* (Hill et al. 2001; Unal et al. 2006), *Clausocalanus* (Bucklin et al. 2003), *Neocalanus* (Machida et al. 2009), *Pseudocalanus* (Bucklin et al. 2003), *Oncaea* and *Triconia* (Böttger-Schnack and and Machida 2011).

In order to document the biodiversity of pelagic copepods from Gulf of California, we identified by gene cytochrome C Oxidase 1 (CO1) several of the species that inhabit the study area, and contribute to mainstreaming a barcode library.

2 Materials and Methods

Specimens were collected with a plankton net (300 μm mesh size) from the Gulf of California mostly. The sampling polygon extended from 24 to 29° N and 110 to 112° W (Fig. 1) while specific geographic coordinates and collection dates for all localities are recorded in the project files Copepods of La Paz Bay, BCS, and Copepods of the NW of Mexico in the Barcode of Life Data System (Ratnasingham and Hebert 2007).

Whenever possible, we barcoded at least five adults of each species. Individuals were photographed and are kept as vouchers in Copepoda Collection at Centro Interdisciplinario de Ciencias Marinas of Instituto Politécnico Nacional. All identifications were based on specialized literature and direct comparison with previously deposited material in the same collection, as well as a validation by the curator of the collection.

Extraction and amplification of CO1 gene was carried out according to protocol proposed by Hajibabaei et al. (2007) and modified by Elías-Gutiérrez and Valdez-Moreno (2008a, b) and sequencing as described in Hajibabaei et al. (2006).

Having obtained the sequences were aligned using the MEGA 5 software, afterwards manually alignment was made to 550 bp with GeneDoc software; MEGA 5 software 5 similarity tree was constructed using statistical Maximun Likelihood method, with bootstrap 1000, genetic differences were calculated using the distance model of two parameters or K2P Kimura (Tamura et al. 2011). The sequences are deposited in www.boldsystems.org.

3 Results

483 individuals belonging to 101 species of copepods were morphologically identified, of which 27 were subjected to the process of DNA extraction and amplification of mitochondrial CO1 gene (Table 1); 93 sequences, of which 89 % had more than 500 base pairs (Fig. 2), were obtained. The largest number of specimens sequenced came from the area of the Islas Marias, whereas the rest come from the central Gulf of California (Fig. 1).

The tree of similarity shows the clustering of specimens of Calanoida and Poecilostomatoida orders (Fig. 3). The average distance (K2P) among the species was 0.36 % and the divergence between genus 28.6 % (Table 2); the highest frequency of interspecific divergence was between 0 and 0.5 %, and in the case of the genera all had more than 20 %.

In the Calanoida order 4 families were pooled: Pontellidae, Paracalanidae, Eucalanidae and Candaciidae with bootstrap values of 20.2, 17.5, 23.7 and 54.9 % respectively (Fig. 2); for Poecilostomatoida there were only representatives of one family which was grouped (52.7 % bootstrap).

In order Calanoida, for example, Pontellidae family had a similarity value of 77 %; *Pontellina plumata*, *Pontellopsis armata*, *Calanopia elliptica*, *Labidocera acuta* and *Labidocera johnsoni* were the most divergent with 22 %. The outside group was genus *Candacia* and *Candacia catula, Candacia curta* and *Candacia simplex* were similar to each other by approximately 41 %.

Table 1 Copepod species identified

Order	Family	Species
Calanoida	Acartidae	*Acartia danae* (Giesbrecht 1889)
	Aetideidae	*Aetideus armatus* (Boeck 1872)
	Calanidae	*Undinula vulgaris* (Dana 1852)
		Canthocalanus pauper (Giesbrecht 1888)
	Paracalanidae	*Acrocalanus gibber* (Giesbrecht 1888)
		Paracalanus parvus (Claus 1863)
	Calocalanidae	*Calocalanus pavo* (Dana 1849)
	Candaciidae	*Candacia curta* (Dana 1849)
		Candacia simplex (Giesbrecht 1889)
	Centropagidae	*Centropages furcatus* (Dana 1849)
	Eucalanidae	*Pareucalanus sewelli* (Fleminger 1973)
		Rhincalanus nasutus (Giesbrecht 1888)
		Subeucalanus mucronatus (Giesbrecht 1888)
		Subeucalanus subcrassus (Giesbrecht 1888)
	Clausocalanidae	*Clausocalanus furcatus* (Brady 1883)
	Pontellidae	*Calanopia elliptica* (Dana 1849)
		Labidocera johnsoni (Fleminger 1964)
		Labidocera acutifrons (Dana 1849)
		Labidocera acuta (Dana 1849)
		Pontellopsis armata (Giesbrecht 1889)
		Pontellopsis occidentalis (Esterly 1906)
		Pontellina plumata (Dana 1849)
	Euchaetidae	*Euchaeta indica* (Wolfenden 1905)
	Temoridae	*Temora discaudata* (Giesbrecht 1889)
Poecilostomatoida	Sapphirinidae	*Copilia mirabilis* (Dana 1849)
		Sapphirina intestinata (Giesbrecht 1891)
		Sapphirina scarlata (Giesbrecht 1892)

4 Discussion

The separation of species based on morphological characters was clear for most species, although in the case of *Labidocera diandra* male the existence of two morphotypes hampered identification (Beltrán-Castro 2014). In the other species studied so far, no cryptic genotypes were found; on the other hand, for the morphological identification of the species in this study proved characteristics of sufficient quality for proper taxonomic identification, coinciding with findings in the Mediterranean with Oncaeidae (Böttger-Schnack and Machida 2011) family, hydrothermal vents with Dirivultidae (Gollner et al. 2011) family and various areas of the world with families Clausocalanidae and Calanidae (Bucklin et al. 2003).

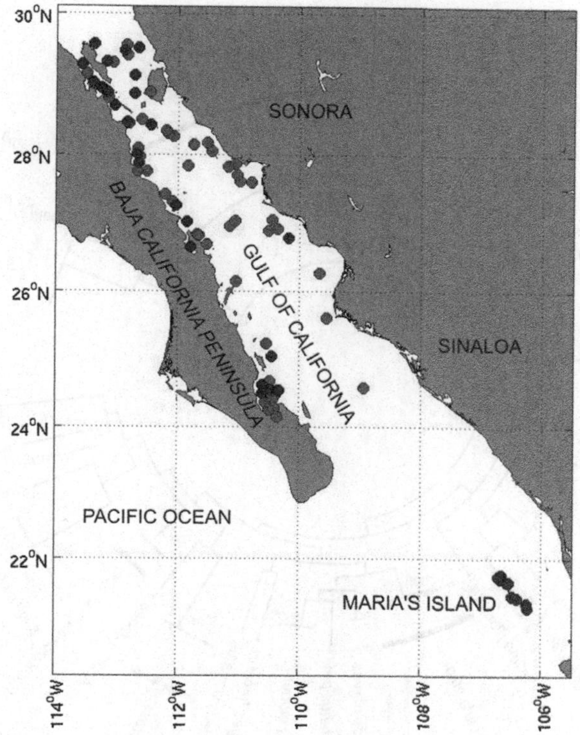

Fig. 1 Sampling stations in the study area

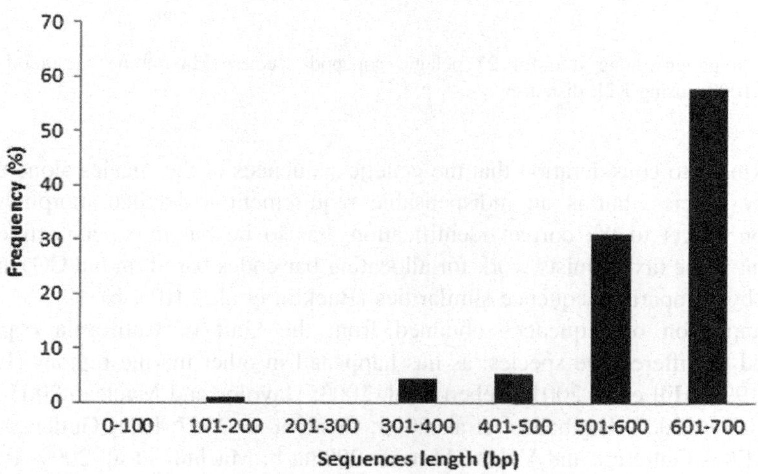

Fig. 2 Sequence length distribution of pelagic copepods from Gulf of California

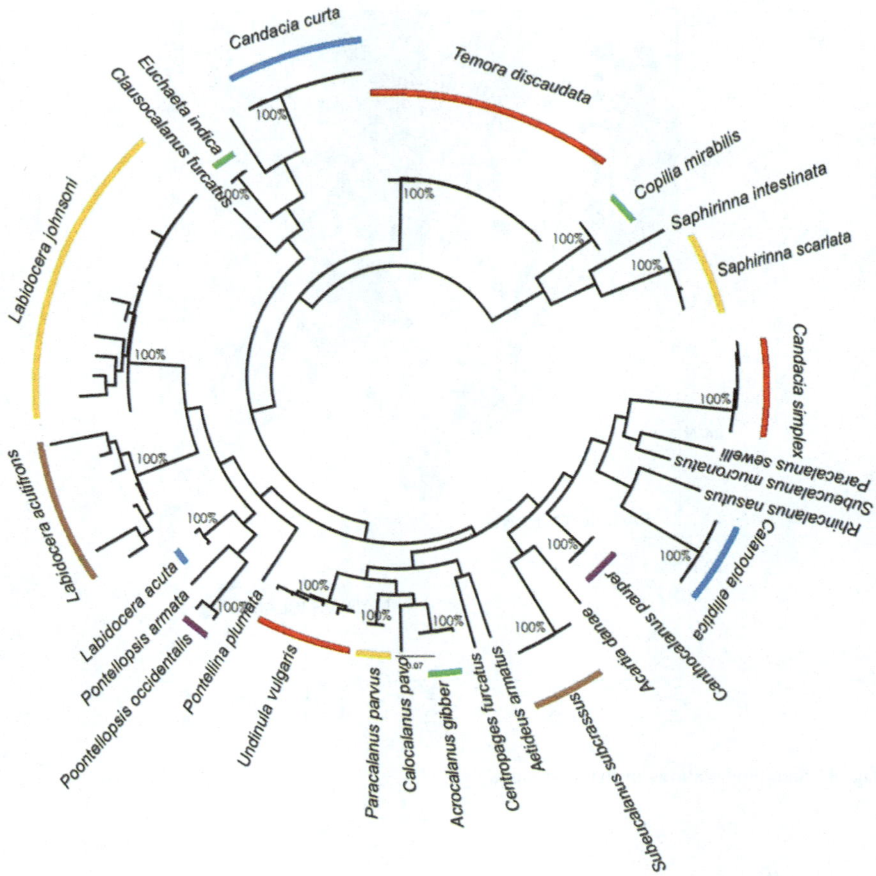

Fig. 3 Neighbor-joining tree for 27 pelagic copepod species (Maximum Likelihood, bootstrap = 1000), using K2P distances

Taking into consideration that the genetic sequences of the species alone do not identify species, but is an indispensable requirement a detailed morphological revision to get to the correct identification, has to be put into perspective how relevant is the taxonomists work for allocating bar codes based on the CO1 or 12S genes by comparing sequence similarities (Bucklin et al. 2010a, b).

Comparison of sequences obtained from the Gulf of California copepods allowed to differentiate species, as has happened in other marine regions (Folmer et al. 1994; Hill et al. 2001; Hebert et al. 2003; Davolos and Maclean 2005; Matz and Nielsen 2005; Hajibabaei et al. 2006; Costa et al. 2007; Elías-Gutiérrez et al. 2008; Elías-Gutiérrez and Valdez-Moreno 2008a, b; Machida et al. 2009; Bucklin et al. 2010a, b; Camacho et al. 2011; Raupach et al. 2010; Blanco-Bercial et al. 2011); in most species variation lower than 2 % of CO1, confirms as a gene with systematic feature (Unal et al. 2006) for reliable identification, although there are

Table 2 Genetic divergences (K2P) at different taxonomic levels

Comparisons within	Taxa	Number of comparisons	Minimum (%)	Distance mean (%)	Maximum (%)	SE dist (%)
Species	15	216	0	0.364	2.823	0.033
Genus	13	61	21.276	28.593	35.257	0.458
Family	8	104	13.159	25.204	33.896	0.454
Order	2	2132	13.393	28.508	38.565	0.092
Class	1	568	31.745	37.354	43.468	0.106

families like Oncaeidae in which the gene 12S better performance (Böttger-Schnack and Machida 2011).

Morphological identification of species was consistent with the formation of genetic groups obtained with CO1 gene and its corresponding barcode without overlap between the sequences. This morphological-genetic match is the start of a library of barcodes copepods in the Gulf of California.

Because of the relatively rapid evolution of CO1 gene, its variation may help solve taxonomic problems associated with geographic variation in distribution of some species of copepods; for instance *Calanus pacificus* have three subspecies which differ geographically as well as their CO1 sequences almost 3 % (Hill et al. 2001). In this study, were found haplotype of *Undinula vulgaris, Calanus pacificus, Canthocalanus pauper, Euchaeta indica, Candacia simplex, Temora discaudata, Paracalanus parvus, Clausocalanus furcatus, Acartia danae, Subeucalanus subcrassus, Rhincalanus nasutus, Aetideus armatus* and *Centropages furcatus*, which are located at various sampling sites and are representing different habitat types, as has been observed with family Dirivultidae (Gollner et al. 2011).

This tool opens an option for the study of biodiversity in the study area since it is distribution area of numerous planktonic species, not just of copepods, so a larger sampling effort has to be carried out to reveal the presence of more species, to correct misidentification that lead to a reclassification and/or description of the "problem" species (Elías-Gutiérrez and Valdez-Moreno 2008a, b; Camacho et al. 2011), revitalize the biological collections, accelerate inventory biodiversity, performing phylogenetic and biogeographic analysis and learn more about the process of speciation.

Molecular analysis of copepods in the Gulf of California will surely help you discover the taxonomic significance of intraspecific genetic separation discovering cryptic species (Bucklin et al. 1998), especially in the calanoid copepods in which numerous examples of sibling species have been discriminated with few morphological characters. The ability to understand the dynamics of plankton community depends on the ability to accurately measure the diversity of species and to accurately identify individuals' species morphologically similar.

This work confirms that the molecular and morphological methods can be considered complementary and when applied in combination, constitute a powerful tool for identification with minimal errors not only of copepods in the Gulf of California, but in the adjoining marine areas; the results are the first step in building databases of sequences and update morphological identification keys.

Acknowledgments To the Instituto Politécnico Nacional and Centro Interdisciplinario de Ciencias Marinas by funding SIP20110224, 20120950 and 20130752 projects. Thank to Dr. Jaime Gómez by providing access to zooplankton samples from Islas Marias and Gulf of California. Thank the node Chetumal of Laboratorio Nacional de Código de Barras, especially Dr. Manuel Elias Gutierrez and MC Areli M. Arce for support in the processing of DNA extraction. To mexican Barcode iniciative by financially supporting the authors in obtaining results and their disclosure. Authors are thankful to BEIFI Programme, EDI and COFAA-IPN by grants received.

References

Aceves-Medina G, Esqueda-Escárcega GM, Pacheco-Chávez R, Zárate-Villafranco A, Hernández-Alonso JR, Hernández-Trujillo S (2007) Cambios diarios en la composición y abundancia de copépodos planctónicos al sur de Bahía de La Paz (Octubre 2002). Hidrobiológica 17(2):185–188

Beltrán-Castro R (2014) Diversidad de copépodos pontélidos (Copepoda: Calanoida: Pontellidae) en el Golfo de California y el valor taxonómico de un estudio con Citocromo c Oxidasa 1 (CO1). Master thesis, CICIMAR-IPN, La Paz, BCS, Mexico, p 80

Blanco-Bercial L, Cornils A, Copley N, Bucklin A (2014) DNA barcoding of marine copepods: assessment of analytical approaches to species identification. PLOS Curr Tree Life. doi:10.1371/currents.tol.cdf8b74881f87e3b01d56b43791626d2

Blanco-Bercial L, Álvarez-Marqués F, Bucklin A (2011) Comparative phylogeography and connectivity of sibling species of the marine copepod Clausocalanus (Calanoida). J Exp Mar Biol Ecol 404:108–115

Brinton E, Fleminger A, Siegel-Causey D (1986) The temperate and tropical planktonic biotas of the Gulf of California. CalCOFI Rep 27:228–266

Briski E, Cristescu ME, Bailey SA, MacIsaac HJ (2011) Use of DNA barcoding to detect invertebrate invasive species from diapausing eggs. Biol Invasions 13:1325–1340

Brown B, Emberson RM, Paterson AM (1999) Mitochondrial COI and II provide useful markers for *Wiseana* (Lepidoptera: Hepialidae) species identification. Bull Entomol Res 89:287–293

Bucklin A, Bentley AM, Franzen SP (1998) Distribution and relative abundance of the copepods, *Pseudocalanus moultoni* and *P. newmani*, on Georges Bank based on molecular identification of sibling species. Mar Biol 132:97–106

Bucklin A, Frost BW, Bradford-Grieve J, Allen LD, Copley NJ (2003) Molecular systematic and phylogenetic assessment of 34 calanoid copepod species of the Calanidae and Clausocalanidae. Mar Biol 142:333–343

Bucklin A, Hopcroft RR, Kosobokova KN, Nigro LM, Ortman BD, Jennings RM, Sweetman CJ (2010a) DNA barcoding of Arctic Ocean holozooplankton for species identification and recognition. Deep-Sea Res II 57:40–48. doi:10.1016/j.dsr2.2009.08.005

Bucklin A, Ortman BD, Jennings RM, Nigro LM, Sweetman CJ, Copley NJ, Sutton T, Wiebe PH (2010b) A "Rosetta Stone" for metazoan zooplankton: DNA barcode analysis of species diversity of the Sargasso Sea (Northwest Atlantic Ocean). Deep-Sea Res II 57:2234–2247. doi:10.1016/j.dsr2.2010.09.025

Bucklin A, Wibe PH, Smolenack SB, Copley NJ, Beaudet JG, Bonner KG, Farber-Lorda J, Pierson JJ (2007) DNA Barcodes for species identification of euphausiids (Euphausiacea, Crustacea). J Plankton Res 29:483–493

Böttger-Schnack R, Machida RJ (2011) Comparison of morphological and molecular traits for species identification and taxonomic grouping of oncaeid copepods. Hidrobiologia 666:111–125

Camacho AI, Rey I, Dorda BA (2011) Identifying cryptic speciation across ground wáter populations: first COI sequences of Bathynellidae (Crustacea, Syncarida). Graellsia. doi:10.3989/graellsia.2011.v67.031

Costa FO, deWaard JR, Boutillier J, Ratnasingham S, Dooh RT, Hajibabaei M, Hebert PDN (2007) Biological identifications through DNA barcodes: the case of the Crustacea. Can J Fish Aquat Sci 64:272–295

Davolos D, Maclean N (2005) Mitochondrial COI-NC-COII sequences in talitrid amphipods (Crustacea). Nature 94:81–86

Elías-Gutiérrez M, Martínez FJ, Ivanova NV, Valdez-Moreno M, Hebert PDN (2008) DNA barcodes for Cladocera and Copepoda from Mexico and Guatemala, highlights and new discoveries. Zootaxa 1839:1–42

Elías–Gutiérrez M, Valdez-Moreno M (2008a) A new cryptic species of *Leberis* Smirnov, 1989 (Crustacea, Cladocera, Chydoridae) from the Mexican semi-desert region, highlighted by DNA barcoding. Hidrobiológica 18(1):63–74

Elías-Gutiérrez M, Valdez-Moreno M (2008b) Código de barras, una alternativa para estudiar la biodiversidad. Ciencia y Desarrollo 34(218):60–67

Folmer O, Black M, Hoeh W, Lutz R, Vrijenhoek R (1994) DNA primers for amplification of mitochondrial cytochrome c oxidase subunit I from diverse metazoan invertebrates. Mol Mar Biol Biotechnol 3(5):294–299

Gollner S, Fontaneto D, Martínez-Arbizu P (2011) Molecular taxonomy confirms morphological classification of deep-sea hydrothermal vent copepods (Dirivultidae) and suggests broad physiological tolerance of species and frequent dispersal along ridges. Mar Biol 158:221–231

González-Navarro E, Saldierna-Martínez R (1997) Zooplancton de la Bahía de La Paz. In: Ramírez JU, Rodríguez MR (eds) La Bahía de La Paz, investigación y conservación. UABCS-CICIMAR-SCRIPPS, La Paz, pp 43–57

Hajibabaei M, Smith MA, Janzen DH, Rodriguez JJ, Whitfield JB, Hebert PDN (2006) A minimalist barcode can identify a specimen whose DNA is degraded. Mol Ecol Notes 6(4):959–964

Hajibabaei M, Singer GAC, Clare EL, Hebert PDN (2007) Design and applicability of DNA arrays and DNA barcodes in biodiversity monitoring. BMC Biology 5:24. doi:10.1186/1741-7007-5-24

Hebert PA, Cywinska SLB, de Waard JR (2003) Biological identifications through DNA barcodes. Proc R Soc Lond B 270:313–322

Hebert PDN, Stoeckle MY, Zemlak TS, Francis CM (2004) Identification of birds through DNA barcodes. PLoS Biol 2(10):e312. doi:10.1371/journal.pbio.0020312

Hernández-Trujillo S, Esqueda-Escárcega GM (2002) La diversidad de copépodos marinos en México. Oceánides 17(1):57–68

Hernández-Trujillo S, Esquivel-Herrera A (1989) Los copépodos y quetognatos del Golfo de California en la primavera de 1985. Inv Mar CICIMAR 4(2):151–164

Hill RS, Allen LD, Bucklin A (2001) Multiplexed species-specific PCR protocol to discriminate four N. Atlantic *Calanus* species, with an mtCOI gene tree for ten *Calanus* species. Mar Biol 139:279–287

Kurtzman CP (1985) Molecular taxonomy on the fungi. In: Bennett W, Lasure LL (eds) Gene manipulation in fungi, pp 35–63

Lavaniegos-Espejo B, González-Navarro E (1999) Cambios en la comunidad de copépodos durante el ENSO 1992–93 en el Canal de San Lorenzo, Golfo de California. Ciencias Marinas 25(2):239–265

Machida R, Hashiguchi Y, Nishida M, Nishida S (2009) Zooplankton diversity analysis through single-gene sequencing of a community sample. BMC Genom 10:438

Matz MV, Nielsen R (2005) A likelihood ratio test for species membership based on DNA sequence data. Philos Trans R Soc Lond B Biol Sci 360:1969–1974. doi:10.1098/rstb.2005.1728

Palomares R (1996) Estructura espacial y variación estacional de los copépodos en la Ensenada de La Paz. Oceánides 11(1):29–43

Radulovici AE, Archambault P, Dufresne F (2010) DNA barcodes for marine biodiversity: moving fast forward? Diversity 2:450–472. doi:10.3390/d2040450

Ratnasingham S, Hebert PDN (2007) Barcoding bold: the barcode of life data system. Mol Ecol Notes. www.barcodinglife.org

Raupach M, Astrin JJJ, Hannig K, Peters MK, Stoeckle MY, Wägele J-W (2010) Molecular species identification of Central European ground beetles (Coleoptera: Carabidae) using nuclear rDNA expansion segments and DNA barcodes. Front Zool 7:26

Razouls C, de Bovée F, Kouwenberg J, Desreumaux N (2011) 2005–2011 Diversity and geographic distribution of marine planktonic copepods. http://copepodes.obs-banyuls.fr/en. Accessed 7 Sept 2011

Tamura K, Peterson D, Peterson N, Stecher G, Nei M, Kumar S (2011) MEGA5: molecular evolutionary genetics analysis using maximum likelihood, evolutionary distance, and maximum parsimony methods. Mol Biol Evol 28:2731–2739

Trewick SA (2000) Mitochondrial DNA sequences support allozyme evidence for cryptic radiation of New Zealand *Peripatoides* (Onychophora). Mol Ecol 9:269–281

Trivedi S, Affan R, Alessa AHA et al (2014) DNA barcoding of Red Sea fishes from Saudi Arabia —the first approach. DNA Barcodes 2:17–20. http://dx.doi.org/10.2478/dna-2014-0003

Unal E, Frost BW, Armbrust V, Kideys AE (2006) Phylogeography of *Calanushelgolandicus* and the Black Sea copepod *Calanuseuxinus*, with notes on *Pseudocalanuselongatus* (Copepoda, Calanoida). Deep-Sea Res II 53:1961–1975

Wilson K, Thorndyke M, Nilsen F, Rogers A, Martínez P (1995) Marine systems: moving into the genomics era. Mar Ecol 26:3–16. doi:10.1111/j.1439-0485.2005.00041.x

DNA Barcoding of Primitive Species-Nemertine from Sundarbans Marine Bio-resource

Bishal Dhar, Apurba Ghose, Sharbadeb Kundu, Amalesh Choudhury, Sudipta Ghorai, Subrata Trivedi, Joyobrato Nath and Sankar Kumar Ghosh

Abstract The phylum Nemertea is known as "ribbon worms" or "proboscis worms" or nemertine. Most are very slim, usually, only a few millimetres wide, although a few have relatively short but wide bodies. They are believed to be an ancient order with their origins in the Cambrian period over 500 million years ago. These nemerteans are distributed globally mostly in the temperate tropical region. Nevertheless, this group is mostly neglected and its taxonomy is jumbled with some blurry and incomplete descriptions and thus created perplexity in the identification process. In a case study, the nemertine samples were collected from the Sundarbans delta (one of the largest biodiversity hotspots) and were subjected to DNA Barcoding approach for species level identification. The sequence analysis was done in comparison with the previously characterized species, calculating its genetic distance (Kimura-2-parameter) as well as similarity match with the published sequence. It was found that the samples from the Sundarbans clustered distinctly as a separate clade with respect to other species, which was congruent with the genetic distance. From this study, it was confirmed that the species from this region was novel as compared to other distinguished species, which was nearly impossible with the conventional morphology due to lack of valid diagnostic keys. In this context, DNA taxonomy has proved itself to be a more powerful tool to systematizing taxa in the definite clades, recognition of possible bio-geographic patterns of these species or to uncover possible hidden species.

B. Dhar · A. Ghose · S. Kundu · J. Nath · S.K. Ghosh (✉)
Department of Biotechnology, Assam University, Silchar, Assam, India
e-mail: drsankarghosh@gmail.com

A. Choudhury
Department of Marine Science, S.D. Marine Biological Research Institute, Sunderban, West Bengal, India

S. Ghorai
Department of Zoology, Egra SSB College, Vidyasagar University, Midnapore, West Bengal, India

S. Trivedi
Department of Biology, University of Tabuk, Tabuk 71491, Saudi Arabia

S. Trivedi et al. (eds.), *DNA Barcoding in Marine Perspectives*,
DOI 10.1007/978-3-319-41840-7_10

Keywords Biodiversity · DNA barcode · Nemertine · Sundarbans

1 Introduction

1.1 Nemertea: A Primitive Group

Nemertea is a phylum of invertebrate animals also known as "ribbon worms" or "proboscis worms" or nemertines (Fig. 1). Alternative names for the phylum have included Nemertini, Nemertinea. Although most are less than 20 cm (7.9 in) long, one specimen has been estimated at 54 m (177 ft) (Leasi and Norenburg 2014). Most are very slim, usually only a few millimeter wide, although a few have relatively short but wide bodies. It has been suggested that three fossil species may be nemerteans, but none is confirmed. Traditional taxonomy divides the phylum in two classes, Anopla ("unarmed"—their proboscises do not have a little dagger) with two orders, and Enopla ("armed" with a dagger) with two orders. However, it is now accepted that Anopla are paraphyletic (have given rise to another group), as one order is more closely related to Enopla than to the other order of Anopla. The phylum Nemertea is monophyletic, whose synapomorphies include the rhynchocoel and eversible proboscis. Traditional taxonomy says that nemerteans are closely related to flatworms and that both are relatively "primitive" acoelomates. Now both phyla are regarded as members of the Lophotrochozoa, a very large "super-phylum" that also includes molluscs, annelids, brachiopods, bryozoa and many other protostomes (Sundberg et al. 2010).

The synapomorphies (trait shared by an ancestor and all its descendants, but not by other groups) include the eversible proboscis located in the rhynchocoel (Thollesson and Norenburg 2003). Nemerteans' affinities with Annelida (including Echiura, Pogonophora, Vestimentifera and perhaps Sipuncula) and Mollusca make the ribbon-worms members of Lophotrochozoa, which include about half of the extant animal phyla (Giribet 2008). Lophotrochozoa groups: those animals that feed using a lophophore (Brachiopoda, Bryozoa, Phoronida, Entoprocta); phyla in which most members' embryos develop into trochophore larvae (for example Annelida and Mollusca); and some other phyla (such as Platyhelminthes, Sipuncula, Gastrotricha, Gnathostomulida, Micrognathozoa, Nemertea, Phoronida, Platyhelminthes and Rotifera) (Giribet 2008; Bik et al. 2010). These groupings are based on molecular phylogeny, which compares sections of organisms' DNA and RNA. While analyses by molecular phylogeny are confident that members of Lophotrochozoa are more closely related to each other than of non-members, the relationships between members are mostly unclear (Bik et al. 2010).

1.2 Sundarbans: A Natural Biodiversity Hotspot

The Sundarbans is a natural expanse in the Bengal region comprising Bangladesh and Eastern India. It is the largest single block of tidal halophytic mangrove forest in the world. The Sundarbans covers approximately 10,000 km^2 (3,900 sq mi) of which 60 % is in Bangladesh with the remainder in India (Manna et al. 2010). The Indian Sundarbans Delta (ISD) is the part of the delta of the Ganga-Brahmaputra-Meghna (GBM) basin in Asia. The Sundarbans, shared between India and Bangladesh is home to one of the largest mangrove forest in the world. The ISD, dispersed over between 21° 40′ 04″ N and 22° 09′ 21″ N latitude, and 88° 01′ 56″ E and 89° 06′ 01″ E longitude, is the smaller and also the western part of the complete Sundarbans delta (Fig. 2).

The Sundarbans flora is characterized by the abundance of sundari (*Heritiera fomes*), gewa (*Excoecaria agallocha*), goran (*Ceriops decandra*) and keora (*Sonneratia apetala*), all of which occur prominently throughout the area. The characteristic tree of the forest is the sundari (*Heritiera littoralis*), from which the

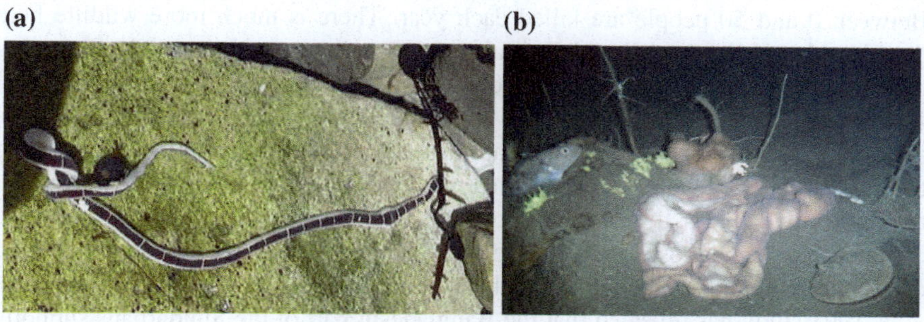

Fig. 1 Nemertean species: **a** *Basiodiscus mexicanus* (http://www.ucmp.berkeley.edu/nemertini/nemertini.html). **b** *Parborlasia corrugatus*, or the proboscis worm (http://www.usap.gov/)

Fig. 2 The Ganges begins its delta (GD) in northwest Bengal. This is at the apex of the *red triangle* (the third side is the sea). The Brahmaputra delta (BD) is marked in *brown*. The two rivers have continually shifted courses, and continue to do so. Their confluence changes and this change is reflected in the merging of the two deltas. The Sundarbans region is shown in *pink ellipse*, the western portion of which is termed as the Indian Sundarbans Delta

name of the forest had probably been derived. It yields a hard wood, used for building houses and making boats, furniture and other things. New forest accretions are often conspicuously dominated by keora (*Sonneratia apetala*) and tidal forests. It is an indicator species for newly accreted mudbanks and is an important species for wildlife, especially spotted deer (*Axis axis*). There is abundance of dhundul or passur (*Xylocarpus granatum)* and kankra (*Bruguiera gymnorrhiza*) though distribution is discontinuous. Among palms, *Poresia coaractata*, *Myriostachya wightiana* and golpata (*Nypa fruticans*), and among grasses spear grass (*Imperata cylindrica*) and khagra (*Phragmites karka*) are well distributed.

The Sundarbans provides a unique ecosystem and a rich wildlife habitat. According to the 2011 tiger census, the Sundarbans have about 270 tigers. Although previous rough estimates had suggested much higher figures close to 300, the 2011 census provided the first ever scientific estimate of tigers from the area (Thollesson and Norenburg 2003). Tiger attacks are frequent in the Sundarbans. Between 0 and 50 people are killed each year. There is much more wildlife here than just the endangered Royal Bengal tiger (*Panthera tigris tigris*). Most importantly, mangroves are a transition from the marine to freshwater and terrestrial systems, and provide critical habitat for numerous species of small fish, crabs, shrimps and other crustaceans that adapt to feed and shelter, and reproduce among the tangled mass of roots, known as pneumatophores, which grow upward from the anaerobic mud to get the supply of oxygen. Fishing cats, macaques, wild boars, common grey mongooses, foxes, jungle cats, flying foxes, pangolins, and spotted deer are also found in abundance in the Sundarbans.

A 1991 study has revealed that the Bangladeshi part of the Sundarbans supports diverse biological resources including at least 150 species of commercially important fish, 270 species of birds, 42 species of mammals, 35 reptiles and 8 amphibian species. This represents a significant proportion of the species present in Bangladesh (i.e. about 30 % of the reptiles, 37 % the birds and 34 % of the mammals) and includes a large number of species which are now extinct elsewhere in the country (Thollesson and Norenburg 2003). Two amphibians, 14 reptiles, 25 aves and five mammals are endangered. The Sundarbans is an important wintering area for migrant water birds and is an area suitable for watching and studying avifauna (Thollesson and Norenburg 2003). Crustaceans include Crabs, Squilla, Prawn and Shrimps. Varieties of crabs like the most commonly sighted Fiddler crab, Red crab, Hermit crab, Tree crab, Mud crab are displayed in specimen jars. Above all, one gets the rare opportunity to discover the two varieties of the endangered living fossil Horse-shoe Crab, which are found in Sundarban.

The Sundarban National Park is a National Park, Tiger Reserve, and a Biosphere Reserve in West Bengal, India. It is part of the Sundarbans on the Ganges Delta, and adjacent to the Sundarbans Reserve Forest in Bangladesh. The delta is densely covered by mangrove forests, and is one of the largest reserves for the Bengal tiger. It is also home to a variety of bird, reptile and invertebrate species, including the salt-water crocodile. The present Sundarbans National Park was declared as the core area of Sundarbans Tiger Reserve in 1973 and a wildlife sanctuary in 1977. On 4 May 1984 it was declared a National Park.

2 Molecular Identification of Species: DNA Barcoding Approach

DNA barcoding is a new taxonomic tool for identifying biological specimens and managing species diversity (Bhattacharjee et al. 2012; Chakraborty and Ghosh 2014; Trivedi et al. 2016a). Intraspecific variation in this sequence is an order of magnitude less than that observed inter-specifically and this provides the means by which species are differentiated. It is being used as a research tool for refining our understanding of biological diversity, and as a system for assigning biological samples to their species of origin. It is a really generalist identification method and could be pivotal for the identification certain life stages (e.g. eggs, larvae, nymphs or pupae), which are often impossible to identify otherwise.

There are 3 aspects of barcoding: taxon identification, species delimitation, and phylogenetic placement. DNA barcoding is a technique in which species identification is performed by using DNA sequences from a small fragment of the genome, with the aim of contributing to a wide range of ecological and conservation studies in which traditional taxonomic identification is not practical (Laskar et al. 2013; Mahadani and Ghosh 2013; Dhar and Ghosh 2015). DNA barcoding is well established in animals, but there is not yet any universally accepted barcode for plants. DNA barcoding is based on the premise that a short standardized sequence can distinguish individuals of a species because genetic variation between species exceeds that within species (Hebert et al. 2003). The barcode sequence from each unknown specimen is then compared with a library of reference barcode sequences derived from individuals of known identity. A specimen is identified if its sequence closely matches one in the barcode library. Otherwise, the new record can lead to a novel barcode sequence for a given species (i.e. a new haplotype or geographical variant), or it can suggest the existence of a newly encountered species.

3 DNA Barcoding of Nemertines

Despite of being the least controversial application, an increasing number of studies question whether it is at all achievable and useful (Sundberg et al. 2010; Cameron et al. 2006), many studies have already proven the power of DNA barcoding being instrumental in identifying species (Bhattacharjee et al. 2012; Trivedi et al. 2014, 2015, 2016b, c; Chakraborty and Ghosh 2014; Laskar et al. 2013; Mahadani and Ghosh 2013; Dhar and Ghosh 2015; Hebert et al. 2003). The Barcoding of Life Database has a remarkable number of data entries and the content and is growing daily. Nevertheless, data is mainly entered for taxonomically known groups, i.e. the taxa where species are already well and reliably described and named. Organisms having few understandable morphological characters are many times difficult both to identify and to classify. Barcoding on the other hand be very useful and beneficial for such groups (Chakraborty and Ghosh 2014; Laskar et al. 2013; Hebert et al. 2004; Hajibabaei et al. 2006), but they may, at the same time, be the most difficult to study taxonomically.

In 2009 by a group of Scientist Sundeberg et al. sampled 62 different specimens (Table 1) near shore on the Swedish west coast during 2006 and 2007, with DNA vouchers deposited at The Swedish Museum of Natural History (Table 1). Specimens were assigned to morphospecies based on original descriptions and additional information (Sundberg et al. 2010).

The all-inclusive species identification was made through standard protocol of DNA barcoding along with morphological analysis. The study confirms the status of the *Cerebratulus spp.* Which is difficult to distinguish based on conventional taxonomy due to morphological similarities or due to some uniqueness in color or shape. The study confirmed that although there exist some difference in morphology (Color or shape) but these may not corresponds to genetic variation. Thus they disprove the hypothesis that external coloration and pigment patterns indicate valid species and gave emphasis on the valid identification keys along with genetic data for proper identification of species.

The meiofauna community in marine benthic ecosystems often has been and remains overlooked because of taxonomic identification difficulties, and because these species were thought by some to be cosmopolitan. However, meiofauna constitute among the most diverse, species-rich, and abundant communities of marine biocenoses; having completely different evolutionary histories in the same habitat and in a relatively small population (Giere 2009). This can serve as an invaluable model to identify generalities in macroecology and biogeography that transcend phylogenetic constraints (Leasi and Norenburg 2014; Curini-Galletti et al. 2012).

Precise estimation of species level diversity, in case of microscopic meiofaunal (minute interstitial animals living in soil and aquatic sediments) nemerteans, is always challenging. The soft body forms of these organisms often lack clear diagnostic morphological traits. Moreover, morphological taxonomy of Nemertea is pretty much complicated, as there is a scarcity of unambiguous character states. So, this in turn compromise the identification process for most of the species (and higher clades) because of their inadequacy and also for having some ambiguous characters and character states. In this context, DNA taxonomy is an efficient approach to uncover possible hidden cryptic species, to systematize taxa in definite clades, and also to reveal possible biogeographic patterns of these species (Leasi and Norenburg 2014).

A study to uncover actual diversity of meiofaunal nemerteans across different sites in Central America examined some worms belonging to the genera *Cephalothrix*, *Ototyphlonemertes*, and *Tetrastemma* in their sampling. This DNA barcoding based faunistic and taxonomic investigation then revealed (i) the presence of several hidden cryptic species and (ii) numerous potential misidentifications due to traditional taxonomy. The misidentification generally happened because of (i) possible human mistakes, (ii) incorrect use of morphological traits, or (iii) unpredictable presence of cryptic species. In addition to these, they also found some more discrepancies with the traditional taxonomy suggesting for the recognized

Table 1 List of specimens included in analysis

Morphospecies	Specimen ID	Locality	Position	Accession nr	Voucher	Coll. date	Collector	Identifier
C. *aerugatus*	1	Humlesäcken	N58° 16,075′/E011° 24,849′	FJ811457	SMNH106090	Sept 3, 2006	M Strand	E Vodoti
	8	Humlesäcken	N58° 16,075′/E011° 24,849′	FJ811458	SMNH106091	Sept 3, 2006	M Strand	E Vodoti
	23	Tjärnö	N58° 53,304′/E011° 5,241′	FJ811459	SMNH106092	Oct 6, 2006	M Strand	E Vodoti
C. *fuscus*	2	Humlesäcken	N58° 16,075′/E011° 24,849′	FJ811460	SMNH106093	Sept 3, 2006	M Strand	E Vodoti
	3	Humlesäcken	N58° 16,110′/E011° 24,761′	FJ811461	SMNH106094	Sept 3, 2006	M Strand	E Vodoti
	7	Flatholmen	N58° 15,704′/E011° 24,105′	FJ811462	SMNH106095	Sept 4, 2006	M Strand	E Vodoti
	9	Flatholmen	N58° 15,704′/E011° 24,105′	FJ811463	SMNH106096	Sept 4, 2006	M Strand	E Vodoti
	10	Kristineberg	N58° 15,400′/E011° 26,200′	FJ811464	SMNH106097	Sept 4, 2006	M Strand	E Vodoti
	11	Kristineberg	N58° 15,400′/E011° 26,200′	FJ811465	SMNH106098	Sept 4, 2006	M Strand	E Vodoti
	12	Grundsund	N58° 13,173′/E011° 24,589′	FJ811466	SMNH106099	Aug 28, 2006	M Strand	E Vodoti
	13	Grundsund	N58° 13,173′/E011° 24,589′	FJ811467	SMNH106100	Aug 28, 2006	M Strand	E Vodoti
	18	Kristineberg	N58° 16,094′/E011° 24,864′	FJ811468	SMNH106101	Sept 5, 2006	M Strand	E Vodoti
	19	Kristineberg	N58° 16,094′/E011° 24,864′	FJ811469	SMNH106102	Sept 5, 2006	M Strand	E Vodoti
	20	Kristineberg	N58° 16,094/E011° 24,864′	FJ811470	SMNH106103	Sept 5, 2006	M Strand	E Vodoti
	21	Kristineberg	N58° 16,075′/E011° 24,849′	FJ811471	SMNH106104	Sept 5, 2006	M Strand	E Vodoti
	22	Kristineberg	N58° 16,075′/E011° 24,849′	FJ811472	SMNH106105	Sept 5, 2006	M Strand	E Vodoti
	24	Flatholmen	N58° 15,503′/E011° 24,115′	FJ811473	SMNH106106	Aug 12, 2006	M Strand	E Vodoti
	25	Flatholmen	N58° 15,503′/E011° 24,115′	FJ811474	SMNH106107	Aug 12, 2006	M Strand	E Vodoti
	26	Tjärnö	N58° 52,522′/E010, 59,818′	FJ811475	5MNH106108	Oct 7, 2006	M Strand	E Vodoti
	27	Tjärnö	N58° 50,963′/E011° 2,832′	FJ811476	SMNH106109	Oct 7, 2006	M Strand	E Vodoti
	115	Bonden	N58° 13,087′/E011° 19,533′	FJ811477	SMNH106110	Apr, 2006	T Kånneby	T Kånneby
	135	Gflaz2	N58° 23,936′/E011° 37,618′	FJ811478	SMNH106111	Apr, 2006	T Kånneby	T Kånneby
	152	Stångholmen	N58° 15,905′/E011° 25,240′	FJ811479	SMNH106112	Apr 24, 2006	T Kånneby	T Kånneby
	167	Humlesäcken	N58° 16,137′/E011° 24,818′	FJ811480	SMNH106113	Apr 20, 2006	T Kånneby	T Kånneby
	211	Skåreskären	N58° 17,552′/E011° 31,918′	FJ811481	SMNH106114	May 2, 2006	T Kånneby	T Kånneby

(continued)

Table 1 (continued)

Morphospecies	Specimen ID	Locality	Position	Accession nr	Voucher	Coll. date	Collector	Identifier
	212	Grundsund	N58° 13,173′/E011° 24,589′	FJ811482	5MNH106115	Apr 28, 2006	T Kånneby	T Kånneby
	232	Flatholmsrännan	N58° 15,388′/E011° 24,620′	FJ811483	SMNH106116	May 5, 2006	T Kånneby	T Kånneby
	263	Humlesäcksränna	N58° 16,110′/E011° 24,761′	FJ811484	SMNH106117	May 10, 2006	T Kånneby	T Kånneby
	269	Humlesäcksränna	N58° 16,058′/E011° 24,761′	FJ811485	SMNH106118	May 10, 2006	T Kånneby	T Kånneby
	300	Humlesäcken	N58° 16,103′/E011° 24,845′	FJ811486	SMNH106119	May 16, 2006	T Kånneby	T Kånneby
	367	Humlesäcken	N58° 16,049′/E011° 24,910′	FJ811487	SMNH106120	May 16, 2006	T Kånneby	T Kånneby
	507	Skagerack	N57° 58,996′/E011° 10,944′	FJ811488	SMNH106121	Aug 23, 2006	T Kånneby	T Kånneby
	517	Skagerack	N57° 58,996′/E011° 10,944′	FJ811489	5MNH106122	Aug 23, 2006	T Kånneby	T Kånneby
	518	Kattegatt	N57° 13,555′/E011° 47,850′	FJ811490	SMNH106123	Aug 29, 2006	T Kånneby	T Kånneby
	519	Kattegatt	N57° 13,555′/E011° 47,850′	FJ811491	SMNH106124	Aug 29, 2006	T Kånneby	T Kånneby
C. *marginatus*	14	Kristineberg	N58° 16,027′/E011° 24,923′	FJ811492	SMNH106125	Aug 28, 2006	T Kånneby	E Vodoti
	36	Tjärnö	N58° 51,459′/E011° 8,563′	FJ811493	SMNH106126	March 23, 2007	M Strand	M Strand
	38	WA, USA	–	AJ436931	–	–	–	–
	124	Väderöarna	N58° 27,046′/E010° 45,384′	FJ811495	SMNH106127	Apr 25, 2006	T Kånneby	T Kånneby
	125	Bonden	N58° 13,087′/E011° 19,533′	FJ811496	SMNH106128	Apr 26, 2006	T Kånneby	T Kånneby
	270	Kristineberg	N58° 16,075′/E011° 24,849′	FJ811497	SMNH106129	May 5, 2006	T Kånneby	T Kånneby
	273	Kristineberg	N58° 16,110′/E011° 24,845′	FJ811498	SMNH106130	May 5, 2006	T Kånneby	T Kånneby
	353	Kristineberg	N58° 16,110′/E011° 24,761′	FJ811499	SMNH106131	May 5,2006	T Kånneby	T Kånneby
C. *pantherinus*	28	Tjärnö	N58° 52,522′/E010, 59,818′	FJ811500	SMNH106133	Oct 7, 2006	M Strand	E Vodoti
	29	Tjärnö	N58° 52,522′/E010, 59,818′	FJ811501	SMNH106134	Oct 7, 2006	M Strand	E Vodoti
C. *roseus*	5	Tjärnö	N58° 57,738′/E011° 5,045′	FJ811502	SMNH106135	Aug 29, 2006	T Kånneby	E Vodoti
C. sp	4	Tjärnö	N58° 57,738′/E011° 5,045′	FJ811503	SMNH106136	Aug 29, 2006	T Kånneby	E Vodoti
	15	Kristineberg	N58° 16,027′/E011° 24,923′	FJ811504	SMNTH106137	Aug 28, 2006	T Kånneby	E Vodoti
	16	Kristineberg	N58° 16,075′/E011° 24,849′	FJ811505	SMNH106138	Sept 5, 2006	M Strand	E Vodoti

(continued)

Table 1 (continued)

Morphospecies	Specimen ID	Locality	Position	Accession nr	Voucher	Coll. date	Collector	Identifier
	17	Kristineberg	N58° 16,075′/E011° 24,849′	FJ811506	SMNH106139	Sept 5, 2006	M Strand	E Vodoti
	30	Tjärnö	N58° 52,687′/E011° 06,864′	FJ811507	SMNH106140	Dec 3, 2006	M Strand	M Strand
	31	Tjärnö	N58° 52,687′/E011° 06,864′	FJ811508	SMNH106141	Dec 3, 2006	M Strand	M Strand
	32	Tjärnö	N58° 52,650′/E011° 06,907′	FJ811509	SMNH106142	Dec 3, 2006	M Strand	M Strand
	33	Tjärnö	N58° 52,650′/E011° 06,907′	FJ811510	SMNH106143	Dec 3, 2006	M Strand	M Strand
	34	Tjärnö	N58° 50,919′/E011° 07,505′	FJ811511	SMNH106144	Dec 3, 2006	M Strand	M Strand
	35	Tjärnö	N58° 50,919′/E011° 07,505′	FJ811512	SMNH106145	Dec 3, 2006	M Strand	M Strand
	114	Bonden	N58° 13,087′/E011° 19,533′	FJ811513	SMNH106146	Apr 26, 2006	T Kånneby	T Kånneby
	166	Humlesäcken	N58° 16,137′/E011° 24,818′	FJ811514	SMNH106147	Apr 20, 2006	T Kånneby	T Kånneby

DNA vouchers deposited at Swedish Museum of Natural History, SMNH (Sundberg et al. 2010)

morphotypes. Therefore, nowadays DNA taxonomy is essential to estimate the actual diversity of meiofaunal and other nemerteans (Leasi and Norenburg 2014).

4 Status of the Nemertines from Sunderban Marine Bio-resource

The sequences obtained from the samples (Nemertines) was subjected to similarity match approach on BOLD identification engine as well as Genbank with the pre-existing database sequences and previously published records. But, the species in the case study did not find any match both in GenBank and BOLD database Table 2. Therefore those species can be considered as novel species and warrant for the detailed study.

Similarly, in the Neighbour Joining Cluster the Nemertine collected from the Sunderban (Nodes marked dotted) Clustered distinctly and separated from the reported species as illustrated in the Fig. 3. The four specimens (SGN01 i—SGN01 iv) clustered as cohesive unit with strong bootstrap support (100 %) which signifies that all those belong to same species, whereas the other two specimen SGN03 and SGN04 clustered separately as two different unit. This confirms the presence of three unreported noble species from the region.

Table 2 Confused species status of the studied Sunderban species based on similarity match with the database

Sl. no.	Sample ID	Supplied specimen	BOLD-IDS		Close match in GenBank (BLASTN similarity in %)	Remarks
			Species level barcode records process ID	Public record barcode database process ID		
1	SGN 01 i	Nemertine	No match	No match	*Cerebratulus leucopsis* (88)	Unidentified
2	SGN 01 ii	Nemertine	No match	No match	*Cerebratulus leucopsis* (88)	Unidentified
3	SGN 01 iii	Nemertine	No match	No match	*Cerebratulus leucopsis* (88)	Unidentified
4	SGN 01 iv	Nemertine	No match	No match	*Cerebratulus leucopsis* (88)	Unidentified
5	SGN 03 R	Nemertine	No match	No match	*Poseidonemertes sp.* (87)	Unidentified
6	SGN 04 R	Nemertine	No match	No match	*Lumbrineris erecta* (82)	Unidentified

The developed sequences of the specimen had 'no match' in the Public Record Barcode Database of BOLD-IDS as well as genbank

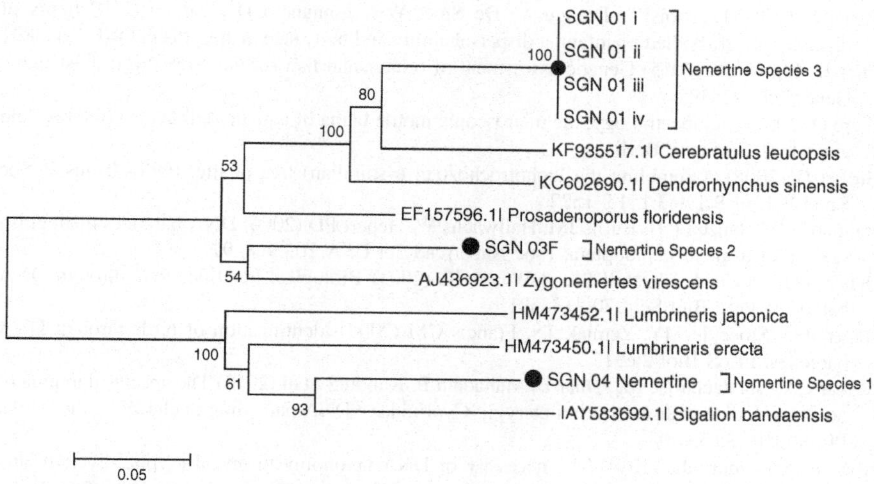

Fig. 3 Neighbour-joining tree construction of the Nemertine species of Sundarbans area included in the study. *Black dots* indicates the generated sequences from the specimens collected from Sunderbans and the rest of the representative of the available species sequences reported so far

The result of this study established two facts indicating the feasibility of DNA barcoding based on *COI* genes. It strongly proved the efficacy of the *COI* sequences in identifying the species collected from the Sundarbans belonging to the rare and primitive phylum of Nemertines. Also, it can provide some basic information for phylogenetic analyses among the studied species giving some implication for evolutionary relationship. DNA barcoding can be used to answer questions of biodiversity assessment, conservation and estimation of sampling efficiency without waiting for the slower process of taxonomy. In parallel, the process also provided a taxonomic enterprise for those taxa that have often been neglected due to their perceived taxonomic opacity. In a nutshell DNA barcoding may lead to species discovery by flagging cryptic species, although more data than *COI* sequences are necessary for describing a new species.

References

Bhattacharjee MJ, Laskar BA, Dhar B, Ghosh SK (2012) Identification and re-evaluation of freshwater catfishes through DNA barcoding. PLoS ONE 7:e49950

Bik HM, Lambshead PJ, Thomas WK, Lunt DH (2010) Moving towards a complete molecular framework of the Nematoda: a focus on the Enoplida and early-branching clades. BMC Evol Biol 10:353

Cameron S, Rubinoff D, Will K (2006) Who will actually use DNA barcoding and what will it cost? Syst Biol 55:844–847

Chakraborty M, Ghosh SK (2014) An assessment of the DNA barcodes of Indian freshwater fishes. Gene 537:20–28

Curini-Galletti M, Artois T, Delogu V, De Smet WH, Fontaneto D et al (2012) Patterns of diversity in soft-bodied meiofauna: dispersal ability and body size matter. PLoS ONE 7:e33801
Dhar B, Ghosh SK (2015) Genetic assessment of ornamental fish species from North East India. Gene 555:382–392
Giere O (2009) Meiobenthology: the microscopic motile fauna of aquatic sediments (2nd. rev. and extended). Springer, Berlin
Giribet G (2008) Assembling the lophotrochozoan (=spiralian) tree of life. Philos Trans R Soc Lond B Biol Sci 363:1513–1522
Hajibabaei M, Janzen DH, Burns JM, Hallwachs W, Hebert PD (2006) DNA barcodes distinguish species of tropical Lepidoptera. Proc Natl Acad Sci USA 103:968–971
Hebert PD, Cywinska A, Ball SL, deWaard JR (2003) Biological identifications through DNA barcodes. Proc Biol Sci 270:313–321
Hebert PD, Stoeckle MY, Zemlak TS, Francis CM (2004) Identification of birds through DNA barcodes. PLoS Biol 2:e312
Laskar BA, Bhattacharjee MJ, Dhar B, Mahadani P, Kundu S et al (2013) The species dilemma of northeast Indian mahseer (Actinopterygii: Cyprinidae): DNA barcoding in clarifying the riddle. PLoS ONE 8:e53704
Leasi F, Norenburg JL (2014) The necessity of DNA taxonomy to reveal cryptic diversity and spatial distribution of meiofauna, with a focus on Nemertea. PLoS ONE 9:e104385
Mahadani P, Ghosh SK (2013) DNA barcoding: a tool for species identification from herbal juices. DNA Barcodes 1:35–38
Manna S, Chaudhuri K, Bhattacharyya S, Bhattacharyya M (2010) Dynamics of Sundarban estuarine ecosystem: eutrophication induced threat to mangroves. Saline Syst 6:8
Sundberg P, Thuroczy Vodoti E, Strand M (2010) DNA barcoding should accompany taxonomy—the case of Cerebratulus spp (Nemertea). Mol Ecol Resour 10:274–281
Thollesson M, Norenburg JL (2003) Ribbon worm relationships: a phylogeny of the phylum Nemertea. Proc Biol Sci 270:407–415
Trivedi S, Affan R, Allesa AHA, Ansari AA, Dhar B, Mahadani P, Ghosh SK (2014) DNA barcoding of Red Sea fishes from Saudi Arabia—the first approach. DNA Barcodes 2:17–20
Trivedi S, Aloufi AA, Ansari AA, Ghosh SK (2015) Molecular phylogeny of oysters belonging to the genus *Crassostrea* through DNA barcoding. J Entomol Zool Stud 3(1):21–26
Trivedi S, Aloufi AA, Ansari AA, Ghosh SK (2016a) Role of DNA barcoding in marine biodiversity assessment and conservation: an update. Saudi J Biol Sci 23(2):161–171
Trivedi S, Aloufi AA, Rehman H, Saggu S, Ghosh SK (2016b) DNA barcoding: tool for assessing species identification in Reptilia. J Entomol Zool Stud 4(1):132–137
Trivedi S, Ansari AA, Rehman H, Saggu S, Abbas ZK, Ghosh SK (2016c) DNA barcoding as a molecular tool for the assessment of plant biodiversity. In: Ansari AA, Gill SS (eds) Plant biodiversity: monitoring assessment and conservation. CABI Publications, UK (In press)

Future Perspectives of DNA Barcoding in Marine Zooplanktons and Invertebrates

Farhina Pasha, Shalini Saggu and Maryam Fahad Albalawi

Abstract The World registry of marine species as of May, 2015 show 229,409 accepted species; of which 220,461 checked (96 %) 421,632 species names including synonyms 527,721 taxon names. Therefore there was a desperate need for developing new molecular method for clarifying obscurities in customary taxonomy. As most of the organisms in marine habitat are microscopic, sample size is large, and most of all the sample preservation is a tricky and expertise requiring job, marine taxonomic identification progress has been very slow. Therefore there was an urgent need of "fast, simple, reliable and inexpensive" method for identification of marine diversity and the answer came as a technique known as "DNA Barcoding". DNA barcoding has emerged as an ideal technique for taxonomic identification of marine taxa, as morphological characters are less reliable and often lead to cryptic species overlapping. Combining genomic study with DNA barcode can be a very effective solution. With all the advantages and limitations there is a strong implementation of DNA barcode on large scale barcode campaigns that will provide enormous amount of data for proper marine taxonomy especially in marine invertebrates.

Keywords DNA barcoding · Marine invertebrates · Biodiversity · Cytochrome oxidase gene · Species identification

1 Introduction

Determination of species is an essential part of identifying and describing biodiversity. Conventionally, identification has been based on morphological analysis provided by taxonomic studies. Nowadays interest in biodiversity has been increased in the fields of ecology, evolutionary biology, agriculture and economics, among others; it has become increasingly important to precisely identify species.

F. Pasha (✉) · S. Saggu · M.F. Albalawi
Faculty of Sciences, Department of Biology, University of Tabuk, Tabuk, Saudi Arabia
e-mail: fbasha@ut.edu.sa; farhinabpl@gmail.com

S. Trivedi et al. (eds.), *DNA Barcoding in Marine Perspectives*,
DOI 10.1007/978-3-319-41840-7_11

However, the number of taxonomists and other identification experts has drastically decreased. Therefore, for non-experts alternative and accurate identification are required. One of the most promising approaches is, the use of molecular instead of morphological data for identifying taxa, which has long been a fundamental idea of many biologists (Busse et al. 1996; Blaxter 2003). Developments in DNA-sequencing technologies have facilitated researchers studying biodiversity to conduct simple, cost–effective and rapid DNA analyses. This progress plays a large role in the creation of DNA barcoding.

1.1 Marine Diversity

More than 70 % of our planet is covered by oceans that have higher biodiversity compared to terrestrial or freshwater ecosystems. The massive marine ecosystem is the habitat for a large number of flora and fauna, both macro and micro. It is estimated around 167,817 known species (that is about 318,004 taxa, species to phyla) are known s as per World Register of Marine Species (WoRMS) (Bouchet 2006). Still more than one million species are to be revealed. As a known fact 70 % of the planet is occupied by Oceans, therefore it is a prime concern to develop new technologies for the investigation of marine life and its vast biodiversity, with especial reference to deep sea to allow unearthing of novel species (Grassle and Maciolek 1992) and also to clarify the status of obscure once (Wong and Hanner 2008). In marine ecosystem, invertebrates living in or on the sediments known as (infauna) (epifauna) respectively forma largest marine group.

The World registry of marine species as of May, 2015 show 229,409 accepted species; of which 220,461 checked (96 %) 421,632 species names including synonyms 527,721 taxon names (infraspecies to kingdoms). The marine species registry is raised by the European Register of Marine Species (EMRS) and it combines with many new species indices and is sustained at the Flanders Marine Institute (VLIZ). Instead of constructing different sections for different ventures and to be sure that the taxonomy used by all is the same a combined database called 'Aphia' is developed and Marinespecies.org is the web interface for this database.

Marine invertebrates lack backbones from 80 % of species found on Earth and often marked by additional support systems like shells, spicules and other exoskeletons. In some invertebrate Phyla there is only one species while the others such as Arthropods include approximately 83 % of all the described marine species which is about 1 million. Marine invertebrates commonly include; sponges, marine worms, mollusks, echinoderms, cnidarians, hemichordates and the arthropods. Among which mollusks, arthropods and cnidarians are some of the biggest phyla. Taxonomic data of marine invertebrates is determined by the size of taxonomic community. For example Mollusks and crustaceans form major groups because of the large population of malacologists and carcinologists, whereas polychaetes because of large population and species diversity of macro benthic taxa require through taxonomic study (Grassle and Maciolek 1992). As there are many difficulties for biodiversity evaluation and assessment of the new species on morphological bases a new way has to be developed. There are many difficulties which arise by the morphological diagnosis during identification of the species, especially for sexually dimorphic species have different life stages (e.g. eggs or larva) and also because of the large distribution of cryptic species (Knowlton 1993). Therefore there was a desperate need for developing new molecular method for clarifying obscurities in customary taxonomy. Thereafter many methods were developed, such as RAPD (random amplified polymorphic DNA), PCR (Polymerase chain reaction), DNA hybridization, RFLP (restriction fragment length polymorphism) etc. (Wong and Hanner 2008). With all these methods having their advantages and disadvantages there was a need of "fast, simple, reliable and inexpensive" method for identification of marine diversity and the answer came as a technique known as "DNA Barcoding". A technique that relies on PCR amplification of fragments of Cytochrome oxidase gene (COI). Initially it started with identification of PCR primers for related organism and protocol for relevant taxonomic groups (e.g. marine invertebrates) were searched from the available literature. After the PCR was completed DNA sequencing was performed using BOLD and NCBI database (Fig. 1).

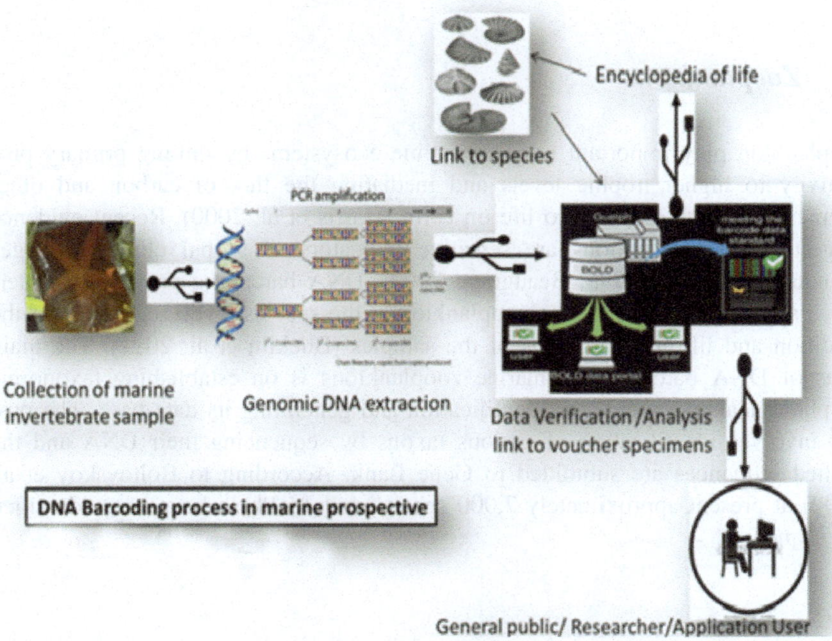

Fig. 1 The overall protocol for DNA barcoding

2 DNA Barcoding in Marine Invertebrates

DNA barcoding is an ideal technique for taxonomic identification of marine taxa, as morphological characters are less reliable and often lead to cryptic species overlapping. DNA barcoding translates expert taxonomic knowledge into a widely accessible format, DNA sequences, allowing a much broader range of scientists to identify specimens (Kerr et al. 2007). This technique of species identification is based on detecting sequence diversity in a single standardized DNA fragment, namely, mitochondrial Cytochrome c Oxidase Subunit I (COI) (Hebert et al. 2003). Examination of nucleotide sequence diversity of this gene allows the grouping of unknown specimens with a prior defined taxonomic species (Monaghan et al. 2005; Vogler and Monaghan 2006) based on the postulation that intraspecific genetic divergence is lower than the interspecific one (Hebert et al. 2003; Meyer and Paulay 2005; Waugh 2007). This method has provided a high degree of taxonomic resolution (>94 %) for most of the species examined across several animal groups (Hebert et al. 2003; Clare et al. 2007; Waugh 2007).

Marine invertebrate's represents 98 % of species on planet are organisms that lack back bone is a vast group and has a prodigious influence on the marine ecosystem and its biodiversity. The most common marine invertebrates are sponges, cnidarians, marine worms, mollusks, arthropods, echinoderms, hemichordates and a special group which is often neglected is represented by marine Diatoms and Parasites. Some of the marine invertebrate phyla are being discussed below.

2.1 *Zooplanktons*

Zooplankton play important roles in marine ecosystems by linking primary productivity to higher trophic levels and mediating the flux of carbon and other chemical elements essential to life on earth (Harris et al. 2000). Recent evidence suggests that zooplanktons are sensitive indicators of global climate changes (Planque and Taylor 1998; Beaugrand 2009). DNA barcoding provides an alternative approach for identifying zooplankton at the species level, regardless of the condition and life history stages of the samples (Bucklin et al. 2011). The main focus of DNA barcoding in marine zooplanktons is on establishing taxonomic morphological and ecological identification and generating its data base. The process involves identification of various taxons by sequencing their DNA and the verified sequences are submitted to Gene Bank. According to Boltovskoy et al. (2002) at present approximately 7,000 species and 15 Phyla have been identified and sequenced.

2.2 Diatoms

Diatoms are unicellular photoautotrophic eukaryotes which are responsible for at least 25 % of the global carbon dioxide fixation (Falkowski et al. 1998; Field et al. 1998; Mann 1999; Smetacek 1999). The diatoms are an enormous group, probably containing c. 200,000 species (Mann and Droop 1996). They are also highly significant ecologically, contributing c. 20 % of net global primary production (Mann 1999). They are an important part of benthic and planktonic biocoenoses and occur nearly ubiquitously in limnic, marine, and terrestrial ecosystems as well as in aerosols (Jahn et al. 2007). Therefore, diatoms are often used as bioindicators in water monitoring assessments and ecological studies (Smetacek 1999; Stoermer and Smol 1999). Applying the DNA barcoding concept to diatoms promises great potential to resolve the problem of inaccurate species identification and thus facilitate analyses of the biodiversity of environmental samples. In particular, the use of DNA barcodes in diatoms can serve various purposes, such as (1) DNA-based species characterization and (2) surveying the genetic diversity in an environment of interest.

Recent, 114 diatom species have been identified which accounts for 99.5 % of its taxonomic identification (Moniz and Kaczmarska 2010). Currently, the best performing barcode markers for diatoms are as follows: (i) the 3' end of the large subunit of the rbcL (rbcL-3 P), (ii) a 540 bp fragment situated 417 bp downstream of the start codon of the rbcL (540 bprbcL), (iii) the 5' end of the mitochondrial cytochrome coxidase I gene (COI-5 P), (iv) a partial sequence of the large ribosomal subunit (D1-D3 LSU, usually either D1-D2 or D2-D3), and (v) the V4 sub-region of the small ribosomal subunit (V4SSU) (Evans et al. 2007; Moniz and Kaczmarska 2009, 2010; Hamsher et al. 2011; MacGillivary and Kaczmarska 2011; Trobajo et al. 2011; Zimmermann et al. 2011; Luddington et al. 2012). The 5.8 S gene combined with the second internal transcribed spacer is also a potential barcode marker for diatoms; however, even though this marker displayed sufficient universality and good discrimination power when assessed using a dataset comprising a wide range of diatom taxa (MacGillivary and Kaczmarska 2011), it was rejected in several studies due to substantial intraclonal diversity that hampered alignment of even closely related lineages (Behnke et al. 2004; Trobajo et al. 2011). The universal plastid amplicon, the entire sequence of the small ribosomal subunit, and the entire sequence of the large subunit of rbcL are unsuitable for DNA barcoding.

2.3 Sponges

Sponges (Phylum Porifera), are diverse, sessile, benthic metazoans, occurring in marine, fresh-water and quasi-terrestrial ecosystems worldwide. In marine habitats, from coral reefs to abyssal plains, sponges play important roles. It includes ~8,500 valid species distributed world-wide in aquatic ecosystems ranging from ephemeral fresh-water bodies to coastal environments and the deep-sea. From a taxonomic and systematic point of view, Phylum Porifera is challenging because of the general paucity of characters useful for taxonomic and phylogenetic inference among sponges (Erpenbeck et al. 2006). Under the phylum Porifera approximately 9,000 to 15,000 species of sponges have been identified.

At present, this is the sole invertebrate phylum to be barcoded through a global campaign (Sponge Barcoding Project (SBP), http://www.spongebarcoding.org), Almost 8,000 taxa has been barcoded till date through SBP and World Porifera Database (WPD), which includes the current and coming future modifications in sponge taxonomic classification via DNA barcoding as reported by Morrow and Cárdenas (2015).

2.4 Cnidarians

The phylum which include, Classes like—Anthozoa (Corals and sea anemones), Hydrozoa, Scyhozoa, Staurozoa (stalked jelly fish) are the organisms only found in marine environment. They are valued for their major contribution formation of Coral Reef ecosystem. As the COI for this phylum is very slow evolving therefore they lack reliable and fast species identification and need some other COI fragment than standard 5' end (an example is <2 % interspecific divergence in scleractinian corals) (Shearer and Coffroth 2008). 16S has emerged as an important marker for the identification of Hydrozoans at species, population, genus and family level as reported by Moura et al. (2008). These authors flagged challenging issues for hydroid systematics: potential cryptic species, conspecificity (low divergence between species) or cosmopolitan species consisting of species complexes.

2.5 Mollusks

They are regarded as the largest marine invertebrate groups with more than 50,000 species are mostly classified under Gastropods and bivalvia. Another important class is Cephalopods. Some common Mollusks are represented by Oysters, snails, Octopus, Squids etc. Marine gastropods when subjected to DNA barcoding test resulted in the overlapping of intra and interspecific divergence there by leading to large omissions in species identifications especially when taxon is not properly sampled (Kemppainen et al. 2009). DNA barcode study conducted for planktonic gastropods where in samples are collected from six different oceans gave highest mean value of >3 % which represents genetic distances within the same species (Jennings et al. 2010). This serves as a vital sign for conducting taxonomic revision below species level as at the sub species level valid species can be identified.

2.6 Arthropods

It represent the biggest phylum include Insects, Crustaceans and arachnids, organisms of this phylum are characteristically having segments body. Out of the 198 species analyzed in two groups it was noted that level of nucleotide divergence was 19 to 48 times more amongst congeneric species than individual species. Therefore COI barcode has a very promising future for the taxonomic identification of Arthropods. Most of the barcoding studies have been undertaken in one of the largest phyla Crustacean which is popular for its species diversity morphologically and ecologically especially in marine perspective. A huge number of taxonomic identification of 150 species from 23 orders has been done by Costa et al. (2007). They also emphasized on taxonomic revision of the phyla. The most recent Crustaceans to be barcoded are Decapods, Somatopods and Peracarids which revels a big percentage of singletons (i.e. Species represented by one specie) (Plaisance et al. 2009). A regional approach for DNA barcoding of malacostraca crustaceans was performed by Radulovici et al. (2009) revealing the existence of hostile amphipod species *Echinogammarus ischnus*, cryptic specification reported by them was merely 5 % which is assumed to be because of incomplete taxon sampling.

2.7 Echinoderms

The organisms under this phylum are known lack of head and their radial symmetry. Usually the exoskeleton is made of calcareous material. Almost 6,000 known species are present in the marine habitat. Over 191 species from five classes were studied using DNA barcode for echinoderms, which included public data from Gen Bank (70 %) (Ward et al. 2008). Established on surface intraspecific against subterranean congeneric the variances was 97.9 % and specimens were assigned to be known species. Others were from genus *Amblypneustes*. Only a few instances of cryptic species were seen.

Species representing smaller groups, were also analyzed using DNA barcoding. 25 species were identified at morphological and molecular basis (via 18S, 16S 12S, COI) (Nielsen et al. 2009) though no new specie was revealed. Krabbe et al. (2010) reported multiple cryptic mitochondrial lineages and geographical restrictions with in small number of species. 18S gene is successfully amplified across majority of taxa with 97 % taxonomic identification success in DNA barcoding process (Bhadury et al. 2006). Although intraspecific variation is higher in other marine groups, standard COI barcoding has emerged as a successful tool for taxonomic identification for these small groups.

2.8 *Marine Parasites*

Marine parasites which are usually not neglected in marine fauna play a vital role in marine ecology. There has been a sincere attempt in recent days to undertake DNA barcoding study for parasites of intertidal marine environment (group of trematode species) (Leun et al. 2009).

3 Issues in Taxonomic Identification and Future Directions of DNA Barcoding in Marine Invertebrates

In recent scenario primers have been developed for many taxa along with markers. In case of slow mutant rate larger COI fragments being used for e.g. in sponges and cnidarians (*G. Singer, Pers. Comm.*). Although our technological capabilities can barcode the entre life present on the planet there are many hurdles to be overcome especially in case of marine invertebrates as compared to terrestrial taxa. Reasons are many but primarily, first are the use of formalin to preserve marine samples that prevents DNA amplification. In contrast fresh samples should be stored in ethanol which although is expensive and focused for some specific marine taxa. Secondly there is a drastic decrease in the number of trained marine taxonomist. Collins and Cruickshank (2013) addressed many problems like insufficient former identification of the specimens, incorrect use of joining trees, erroneously inferring the barcoding gap. Thereby consideration all the above issues a sincere effort is required by trained personal in the field to make DNA barcoding as universal barcoding for all the life forms on earth.

There is a broad range of DNA barcode studies being undertaken in near future as only 48 % of marine species have been classified till date (Table 1). Special emphasis is required for developing new methods of sampling from deep sea and their preservation methods. Making laboratories on board will be an added advantage for an early sampling and to prevent any DNA leakage there by resulting in proper DNA amplification. There is certainly a possibility to develop a DNA microarray (chips) for many marine species (Kochzius et al. 2008). Combining genomic study with DNA barcode can be a very effective solution to many problems occurring during DNA barcoding. Screening of cDNA can effectively reduce errors due to misparing. With all the advantages and limitations there is a strong implementation of DNA barcode on large scale barcode campaigns that will provide enormous amount of data for proper identifications and analysis in the near future.

Table 1 Published research articles on DNA barcoding in various marine invertebrates using PUBMED search http://www.ncbi.nlm.nih.gov/pubmed

Marine invertebrates species	Number of published researches from 2000–2015	Search criteria
Zooplanktons	3	DNA barcoding in zooplanktons
Diatoms	20	DNA barcoding in diatoms
Sponges	20	DNA barcoding in sponges
Cnidarians	24	DNA barcoding in cnidarians
Mollusks	62	DNA barcoding in mollusks
Arthropods	429	DNA barcoding in arthropods
Echinoderms	3	DNA barcoding in echinoderms
Marine parasites	9	DNA barcoding in marine parasites

Acknowledgments The authors would like to acknowledge, University of Tabuk, Tabuk, Saudi Arabia. The author would also like to thanks Department of Biology, Faculty of Sciences, Saudi Digital Library and University Library providing the facility for literature survey and collection.

References

Beaugrand G (2009) Decadal changes in climate and ecosystems in the North Atlantic Ocean and adjacent seas. Deep-Sea Res Part II 56(8–10):656–673

Behnke A, Friedl T, Chepurnov VA, Mann DG (2004) Reproductive compatibility and rDNA sequence analyses in the Sellaphorapupula species complex (Bacillariophyta). J Phycol 40:193–208

Bhadury P, Austen MC, Bilton DT, Lambshead PJD, Rogers AD, Smerdon GR (2006) Development and evaluation of a DNA-barcoding approach for the rapid identification of nematodes. Mar Ecol Prog Ser 320:1–9

Blaxter M (2003) Counting angels with DNA. Nature 421:122–124

Boltovskoy D, Correa N, Boltovskoy A (2002) Marine zooplanktonic diversity: a view from the South Atlantic. Oceanol Acta 25:271–278

Bouchet P (2006) The magnitude of marine biodiversity. In: Duarte CM (ed) The exploration of marine biodiversity: scientific and technological challenges. Fundacion BBVA, Bilbao, pp 31–64

Bucklin A, Steinke D, Blanco-Bercial L (2011) DNA barcoding of marine metazoa. Annu Rev Mar Sci 3:471–508

Busse HJ, Denner EBM, Lubitz W (1996) Classification and identification of bacteria: current approaches to an old problem. Overview of methods used in bacterial systematics. J Biotechnol 47:3–38

Clare EL, Lim BK, Engstrom MD, Eger JL, Hebert PDN (2007) DNA barcoding of neotropical bats: species identification and discovery within Guyana. Mol Ecol Notes 7:184–190

Collins RA, Cruickshank RH (2013) The seven deadly sins of DNA barcoding. Mol Ecol Resour 13:969–975

Costa FO, deWaard JR, Boutillier J, Ratnasingham S, Dooh RT, Hajibabaei M, Hebert PDN (2007) Biological identifications through DNA barcodes: the case of the Crustacea. Can J Fish Aquat Sci 64:272–295

Erpenbeck D, Hooper JNA, Wörheide G (2006) CO1 phylogenies in diploblasts and the 'Barcoding of Life'—are we sequencing a suboptimal partition? Mol Ecol Notes 6:550–553

Evans KM, Wortley AH, Mann DG (2007) An assessment of potential diatom "barcode" genes (cox1, rbcL, 18S and ITS rDNA) and their effectiveness in determining relationships in Sellaphora (Bacillariophyta). Protist 158:349–364

Falkowski PG, Barber RT, Smetacek V (1998) Biogeochemical controls and feedbacks on ocean primary production. Science 281:200–207

Field CB, Behrenfeld MJ, Randerson JT, Falkowski PG (1998) Primary production of the biosphere: integrating terrestrial and oceanic components. Science 281:237–240

Grassle JF, Maciolek NJ (1992) Deep-sea species richness—regional and local diversity estimates from quantitative bottom samples. Am Nat 139:313–341

Hamsher SE, Evans KM, Mann DG, Poulíčková A, Saunders GW (2011) Barcoding diatoms: exploring alternativesto COI-5P. Protist 162:405–422

Harris R, Wiebe P, Lenz J, Skjoldal H, Huntley M (2000) ICES zooplankton methodology manual. Academic Press, San Diego

Hebert PDN, Cywinska A, Ball SL, Ward JR (2003) Biological identifications through DNA barcodes. Proc R Soc Lond B 270:313–321

Jahn R, Zetzsche H, Reinhardt R, Gemeinholzer B (2007) Diatoms and DNA barcoding: a pilot study on an environmental sample. In: Kusber WH, Jahn R (eds) Proceedings of the 1st central European diatom meeting Berlin, Botanic Garden and Botanical Museum Berlin-Dahlem. Freie Universitat Berlin, pp 63–68

Jennings RM, Bucklin A, Ossenbrügger H, Hopcroft RR (2010) Species diversity of planktonic gastropods (Pteropoda and Heteropoda) from six ocean basins based on DNA barcode analysis. Deep-Sea Res II 57:2199–2210

Kemppainen P, Panova M, Hollander J, Johannesson K (2009) Complete lack of mitochondrial divergence between two species of NE Atlantic marine intertidal gastropods. J Evol Biol 22:2000–2011

Kerr KC, Stoeckle MY, Dove CJ, Weigt LA, Francis CM, Hebert PDN (2007) Comprehensive DNA barcode coverage of North American birds. Mol Ecol Notes 7:535–543

Knowlton N (1993) Sibling species in the sea. Annu Rev Ecol Evol Syst 24:189–216

Kochzius M, Nölte M, Weber H, Silkenbeumer N, Hjörleifsdottir S, Hreggvidsson GO, Marteinsson V, Kappel K, Planes S, Tinti F, Magoulas A, Vazquez EG, Turan C, Hervet C, Falgueras DC, Antoniou A, Land M, Blohm D (2008) DNA microarrays for identifying fishes. Mar Biotechnol 10:207–217

Krabbe K, Leese F, Mayer C, Tollrian R, Held C (2010) Cryptic mitochondrial lineages in the widespread pycnogonid Colossende ismegalonyx Hoek, 1881 from Antarctic and Subantarctic waters. Polar Biol 33:281–292

Leun TLF, Donald KM, Keeney DB, Koehle AV, Peoples RC, Poulin R (2009) Trematode parasites of Otago Harbour (New Zealand) soft-sediment intertidal ecosystems: life cycles, ecological roles and DNA barcodes. NZ J Mar Freshwater Res 43:857–865

Luddington IA, Kaczmarska I, Lovejoy C (2012) Distance and character-based evaluation of the V4 region of the 18 S rRNA gene for the identification of diatoms (Bacillariophyceae). PLoS ONE 7:e45664

MacGillivary ML, Kaczmarska I (2011) Survey of the efficacy of a short fragment of the rbcL gene as a supplemental DNA barcode for diatoms. J Eukaryot Microbiol 58:529–536

Mann DG, Droop SJM (1996) Biodiversity, biogeography and conservation of diatoms. Hydrobiologia 336:19–32

Mann DG (1999) The species concept in diatoms. Phycologia 38:437–495

Meyer CP, Paulay G (2005) DNA barcoding: error rates based on comprehensive sampling. PLoS Biol 3:2229–2238

Monaghan MT, Balke M, Ryan-Gregory T, Vogler AP (2005) DNA-based species delineation in tropical beetles using mitochondrial and nuclear markers. Philos Trans R Soc B 360: 1925–1933

Moniz MBJ, Kaczmarska I (2010) Barcoding of diatoms: nuclear encoded ITS revisited. Protist 161:7–34

Moniz MBJ, Kaczmarska I (2009) Barcoding diatoms: is there a good marker? Mol Ecol Resour 9 (Suppl. 1):65–74

Morrow C, Cárdenas P (2015) Proposal for a revised classification of the Demospongiae (Porifera). Front Zool 12:7

Moura CJ, Harris DJ, Cunha MR, Rogers AD (2008) DNA barcoding reveals cryptic diversity in marine hydroids (Cnidaria, Hydrozoa) from coastal and deep-sea environments. Zool Scr 37:93–108

Nielsen JF, Lavery S, Lorz AN (2009) Synopsis of a new collection of sea spiders (Arthropoda: Pycnogonida) from the Ross Sea, Antarctica. Polar Biol 32:1147–1155

Plaisance L, Knowlton N, Paulay G, Meyer C (2009) Reef-associated crustacean fauna: biodiversity estimates using semi-quantitative sampling and DNA barcoding. Coral Reefs 28:977–986

Planque B, Taylor AH (1998) Long-term changes in zooplankton and the climate of the North Atlantic. ICES J Mar Sci 55(4):644–654

Radulovici AE, Sainte-Marie B, Dufresne F (2009) DNA barcoding of marine crustaceans from the Estuary and Gulf of St Lawrence: a regional-scale approach. Mol Ecol Resour 9:181–187

Shearer TL, Coffroth MA (2008) Barcoding corals: limited by interspecific divergence, not intraspecific variation. Mol Ecol Resour 8:247–255

Smetacek V (1999) Diatoms and the carbon ocean cycle. Protist 150:25–32

Stoermer EP, Smol JP (1999) The diatoms: applications to the environmental and earth sciences. Cambridge University Press, Cambridge

Trobajo R, Clavero E, Evans KM, Vanormelinge P, McGregor RC, Mann DG (2011) The use of partial cox1 rbcL and LSU rDNA sequences for phylogenetics and species identification within the *Nitzschiapalea*complex (Bacillariophyceae). Eur J Phycol 45:413–425

Vogler AP, Monaghan MT (2006) Recent advances in DNA taxonomy. J Zool Syst Evol Res 45:1–10

Ward RD, Holmes BH, O'Hara TD (2008) DNA barcoding discriminates echinoderm species. Mol Ecol Resour 8:1202–1211

Waugh J (2007) DNA barcoding in animal species: progress, potential and pitfalls. BioEssays 29:188–197

Wong EHK, Hanner RH (2008) DNA barcoding detects market substitution in North American seafood. Food Res Int 41:828–837

Zimmermann J, Jahn R, Gemeinholzer B (2011) Barcoding diatoms: evaluation of the V4 subregion on the 18S rRNA gene including new primers and protocols. Organ Divers Evol 11:173–192

DNA Barcoding: Molecular Positioning of Living Fossils (Horseshoe Crab)

Bishal Dhar, Apurba Ghose, Sharbadeb Kundu, Sorokhaibam Malvika, Ningthoujam Neelima Devi, Amalesh Choudhury, Sudipta Ghorai, Subrata Trivedi and Sankar Kumar Ghosh

Abstract Living fossils are the important components of biodiversity. They represent the connection between the extinct and extant species. A living fossil is a living species that appears to be similar to a species otherwise known only from fossils, typically with no close living relatives and the extant species. The study of primitive species gives an idea about the ancestors from which they diverged and bring out many surprising facts which are unknown to the world. In a case study, species belonging to the phylum Arthropoda, Brachiopoda and Molluscs were collected from Sundarbans where Horshshoe Crab was the living fossils (*Carcinoscorpius rotundicauda). Lingula sp* and some crab species were included as the outgroup. To position this living fossil, DNA barcoding approach was employed as per standard protocol. *COI* sequencing and subsequently nucleotide analysis of all the species were done and also the molecular clock was constructed to locate their position along with their divergence time in correspondence with the other sequences of the allied taxa viz, *Limulus polyphemus*, *Tachypleus gigas* etc. It is found that the *Carcinoscorpius rotundicauda* are more closely related to its allied taxon *Tachypleus gigas* as compared to *Limulus poluphemus* and their divergence period is calculated which is supposed to be the 550 million years ago. Thus, DNA barcoding approach is a useful technique to properly identify species and to construct phylogenetic relationship among the species and subsequent assessment of the species divergence time.

B. Dhar · A. Ghose · S. Kundu · S. Malvika · N. Neelima Devi · S.K. Ghosh (✉)
Department of Biotechnology, Assam University, Silchar, Assam, India
e-mail: drsankarghosh@gmail.com

A. Choudhury
Department of Marine Science, S.D. Marine Biological Research Institute, Sunderban, West Bengal, India

S. Ghorai
Department of Zoology, Egra SSB College, Vidyasagar University, Midnapore, West Bengal, India

S. Trivedi
Department of Biology, University of Tabuk, Tabuk 71491, Saudi Arabia

S. Trivedi et al. (eds.), *DNA Barcoding in Marine Perspectives*,
DOI 10.1007/978-3-319-41840-7_12

Keywords Biodiversity · DNA barcode · Horseshoe crab · Living fossils

1 Introduction

1.1 Horseshoe Crab: The Living Fossils

A living fossil is a living species (or clade) of organism that appears to be similar to a species, otherwise known only from fossils, typically with no close living relatives. Normally the similarity is only by an imagined physical resemblance, between two different species, one extinct, the other extant. Horseshoe crabs are marine chelicerates belonging to the family Limulidae and order Xiphosura or Xiphosurida closely related to arachnids (Masta et al. 2009; Sharma et al. 2015). There are four extant species distributed across the coastal zones of the continental shelves in North America and South East Asia (Kin and Blazejowski 2014). The American horseshoe crab *Limulus polyphemus* Linnaeus has a distribution along the eastern coast of North America and a genetically and geographically isolated population on the northern east coast of the Yucatán Peninsula in Mexico (Dunlop et al. 2003; Faurby et al. 2010; Vasquez et al. 2015). The three remaining species *Tachypleus gigas* Müller, *Tachypleus tridentatus* Leach, and *Carcinoscorpius rotundicauda* Latreille inhabit the shallow waters in the Indo-Pacific. In this area two species, *Tachypleus gigas* and *C. rotundicauda* have a largely overlapping distribution ranging from the Andaman Sea to the South China Sea, but inhabit different habitats (Vasquez et al. 2015). While *T. gigas* lives in sandy and shallow near-coast habitats, *C. rotundicauda* mostly inhabits estuaries and mangroves. The third Asian species *T. tridentatus* lives in shallow coastal zones of Southern China from Vietnam to Japan (Vasquez et al. 2015).

Horseshoe crabs have puzzled evolutionary biologists for centuries and are often used as prime example for organisms that survived long time periods without any significant changes in their anatomy, earning them the name of ‘living fossils’ (Carroll 1997) and ‘phylogenetic relicts’ (Crow and Dove 2000). In the fossil record ancient horseshoe crabs are already known from the Ordovician period (Legg 2014; Van Roy et al. 2010), and modern forms which are indistinguishable from recent species appear during Upper Jurassic (Baek et al. 2014; Briggs et al. 2005). Horseshoe crabs, the closest living relatives of the trilobites (Obst et al. 2012), have persisted for more than 200 million years (Nossa et al. 2014). It seems like these forms maintained a static morphology for at least 150 million years (Schoenemann and Clarkson 2013). This lack of morphological disparity over time is accompanied by low levels of species diversity throughout the entire evolutionary history of horseshoe crabs (Rehm et al. 2012). Even at the peak of their diversification during Carboniferous period the group consisted of not more than a few dozens of documented species (Legg et al. 2012). Following the Permian–Triassic extinction the diversity of the group was greatly reduced and left behind the only recent group of horseshoe crabs, the Limulidae (Fig. 1).

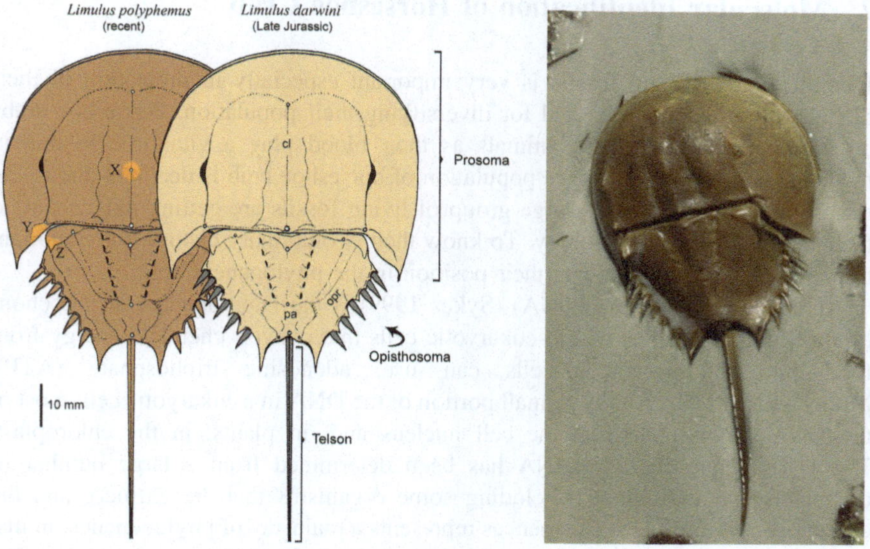

Fig. 1 Anatomy of recent *Limulus polyphemus* (*left*) and oldest known member of the genus *Limulus darwini* (*right*) from Corbulomima horizon of unit III from Late Jurassic sedimentary sequence (central Poland). (X), (Y) and (Z)—details emphasized, are most substantial morphological difference between both these forms. (cl)—cardiac lobe; (opr)—opisthosomal rim; (pa)—posterior area (Kin and Blazejowski 2014)

The anatomical similarity among living horseshoe crabs has greatly impaired the elucidation of the phylogenetic relationships among the recent lineages. Although a wide range of non-morphological data has been studied to date (Silvestro et al. 2015), the only consensus reached so far is that the Atlantic species, *L. polyphemus*, is a sister taxon to the three Indo-Pacific species. The phylogenetic relationship among the three Indo-Pacific species, however, remains unresolved. Cladistic approaches including fossil evidence resulted in low support for a relationship between *Tachypleus gigas* with *T. Tridentatus* (Kin and Blazejowski 2014). In contrast to this an amino acid analysis by Shishikura et al. (1982) (Srimal et al. 1985) suggested *C. rotundicauda* and *T. tridentatus* to be closely related, while *C. rotundicauda* and *T. Gigas* were suggested as sister groups in an analysis of mitochondrial and nuclear genes in the absence of any support measures (Xia 2000). Other approaches such as interspecific hybridization experiments (Shishikura and Sekiguchi 1984), two-dimensional electrophoresis and analyses of two mitochondrial genes (Avise and Bowen 1994) yielded unresolved genealogies for the Asian species. Such an array of conflicting results has led some authors to the conclusion that the three species constitute a phylogenetically unresolvable trichotomy, resulting from a cladogenetic process in which all three Indo-Pacific species formed within a short geological time (Avise and Bowen 1994).

2 Molecular Identification of Horseshoe Crab

Identification of living fossils is very important especially in protection of their biodiversity, their lineages and for diversifying their population. Horseshoe crabs are economically important animals as their blood play a vital role in human medicine, and therefore a large population of horseshoe crab is declined due to the misuse of these animals. A large group of living fossils are getting extinct due to their misapprehend morphology. To know their evolutionary history it is important to identify them and to locate their position in the phylogenetic tree.

Mitochondrial DNA (mtDNA) (Sykes 1999) is the DNA located in mitochondria, cellular organelles within eukaryotic cells that convert chemical energy from food into a form that cells can use, adenosine triphosphate (ATP). Mitochondrial DNA is only a small portion of the DNA in a eukaryotic cell; most of the DNA can be found in the cell nucleus and, in plants, in the chloroplast. The DNA sequence of mtDNA has been determined from a large number of organisms and individuals (including some organisms that are extinct), and the comparison of those DNA sequences represents a mainstay of phylogenetics, in that it allows biologists to elucidate the evolutionary relationships among species. The mitochondrial genome in eukaryotes encodes a total of 37 genes, 22 of which encode transfer RNA (tRNA) molecules, two encode ribosomal RNA (rRNA) molecules and the other 13 encode proteins involved primarily with the process of oxidative respiration. Animal DNA barcodes, 600- to 800-base-pair segments of the mitochondrial gene cytochrome oxidase I, have been proposed as a means to quantify global biodiversity.

Although mtDNA has a long history of use at the species level, recent analyses suggest that the use of a single gene, particularly mitochondrial, is unlikely to yield data that are balanced, universally acceptable, or sufficient in taxonomic scope to recognize many species lineages. Mitochondrial and nuclear genomes have different patterns of evolution and modes of inheritance, which can result in very different assessments of biodiversity. The ramifications of choosing a particular definition of species (species concept) need to be carefully considered because current efforts have designated DNA barcodes as the universal species concept without demonstrating its superiority over pre-existing concepts (Rubinoff 2006).

2.1 Cytochrome c Oxidase *Subunit I*

Cytochrome c oxidase I (*COI*) also known as mitochondrial encoded cytochrome c oxidase I (MT-CO1) is a protein that in humans is encoded by the *mt-CO1* gene. In other eukaryotes, the gene is called *COX1*, or *COI* (Kosakyan et al. 2012) Cytochrome c oxidase I is the main subunit of the cytochrome c oxidase complex. Cytochrome c oxidase subunit I (*CO1* or MT-CO1) is one of three mitochondrial DNA (mtDNA) encoded subunits (MT-CO1, MT-CO2, MT-CO3) of respiratory complex IV. Complex IV is the third and final enzyme of the electron transport chain of mitochondrial oxidative phosphorylation (Garcia-Horsman et al. 1994). The mitochondrial gene cytochrome *c* oxidase I (*COI*) can serve as the core of a global bio-identification system for animals.

There is no compelling a priori reason to focus analysis on a specific gene, but the *cytochrome c oxidase I* gene (COI) does have two important advantages. First, the universal primers for this gene are very robust, enabling recovery of its 5′ end from representatives of most, if not all, animal phyla (Folmer et al. 1994; Zhang and Hewitt 1997). Second, COI appears to possess a greater range of phylogenetic signal than any other mitochondrial gene. In common with other protein coding genes, its third-position nucleotides show a high incidence of base substitutions, leading to a rate of molecular evolution that is about three times greater than that of 12S or 16S rDNA (Dick et al. 2009). In fact, the evolution of this gene is rapid enough to allow the discrimination of not only closely allied species, but also phylogeographic groups within a single species (Cox and Hebert 2001; Wares and Cunningham 2001). Although COI may be matched by other mitochondrial genes in resolving such cases of recent divergence, this gene is more likely to provide deeper phylogenetic insights than alternatives such as cytochrome b (Simmons and Weller 2001) because changes in its amino-acid sequence occur more slowly than those in this, or any other, mitochondrial gene (Lynch and Jarrell 1993). As a result, by examining amino acid substitutions, it may be possible to assign any unidentified organism to a higher taxonomic group (e.g. phylum, order), before examining nucleotide substitutions to determine its species identity.

DNA barcoding has recently emerged as a rapid method for species discovery and biodiversity assessment (Hajibabaei et al. 2006; Borisenko et al. 2008; Stoeckle and Hebert 2008). For animal taxa, the majority of these studies have used a short section of mitochondrial DNA (mtDNA), namely the first ~650 bp of the 5′-end of the cytochrome oxidase I gene (*COI*) (Hebert et al. 2003) (Fig. 2).

DNA barcoding is a new taxonomic tool for identifying biological specimens and managing species diversity (Bhattacharjee et al. 2012; Chakraborty and Ghosh 2014;

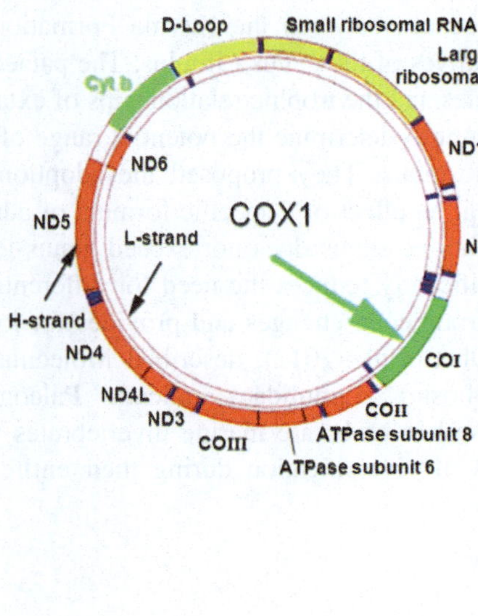

Fig. 2 Mitochondrial genome showing the COX1 gene with the *green arrow*

Trivedi et al. 2014, 2015, 2016a, b, c). Intraspecific variation in this sequence is an order of magnitude less than that observed inter-specifically and this provides the means by which species are differentiated. It is being used as a research tool for refining our understanding of biological diversity, and as a system for assigning biological samples to their species of origin. It is a really generalist identification method and could be pivotal for the identification of certain life stages (e.g. eggs, larvae, nymphs or pupae), which are often impossible to identify otherwise.

There are 3 aspects of barcoding: taxon identification, species delimitation, and phylogenetic placement. DNA barcoding is a technique in which species identification is performed by using DNA sequences from a small fragment of the genome, with the aim of contributing to a wide range of ecological and conservation studies in which traditional taxonomic identification is not practical (Laskar et al. 2013; Mahadani and Ghosh 2013; Dhar and Ghosh 2015). DNA barcoding is well established in animals, but there is not yet any universally accepted barcode for plants. DNA barcoding is based on the premise that a short standardized sequence can distinguish individuals of a species because genetic variation between species exceeds that within species (Hebert et al. 2003). The barcode sequence from each unknown specimen is then compared with a library of reference barcode sequences derived from individuals of known identity. A specimen is identified if its sequence closely matches one in the barcode library. Otherwise, the new record can lead to a novel barcode sequence for a given species (i.e. a new haplotype or geographical variant), or it can suggest the existence of a newly encountered species.

2.2 Current Phylogenetic Status of Horseshoe Crab

Kin and Blazejowski (2014), described a new horseshoe crab species, *Limulus darwini*, from the uppermost Jurassic (ca. 148 Ma) near-shore sediments of the Kcynia Formation, central Poland. The only extant species *Limulus polyphemus* (Linnaeus) inhabits brackish-marine, shallow water environments of the east coast of the United States. It was shown that there were no important morphological differences between the Kcynia Formation specimens and extant juvenile representatives of the genus Limulus. The palaeoecological setting inhabited by the new species and the trophic relationships of extant horseshoe crabs were discussed in an attempt to determine the potential range of food items ingested by these Mesozoic xiphosurans. They proposed the adoption of a new term stabilomorphism, this being: an effect of a specific formula of adaptive strategy among organisms whose taxonomic status does not exceed genus-level. A high effectiveness of adaptation significantly reduces the need for differentiated phenotypic variants in response to environmental changes and provides for long-term evolutionary success.

Obst et al. (2012), described Molecular phylogeny of extant horseshoe crabs (Xiphosura, Limulidae) indicates Paleogene diversification of Asian species. Horseshoe crabs are marine invertebrates well known for their exceptionally low rates of diversification during their entire evolutionary history. Despite the low species diversity in the group, the phylogenetic relationships among the extant species, especially among the three Asian species are still unresolved. They applied a new set of molecular genetic data in combination with a wide geographic sampling of the intra-specific diversity to reinvestigate the evolutionary history among the four living limulid xiphosurans. Their analysis of the intraspecific diversity revealed low levels of connectivity among *Carcinoscorpius rotundicauda* lineages, which can be explained by the estuarine-bound ecology of this species. Moreover, a clear genetic break across the Thai–Malay Peninsula suggests the presence of cryptic species in *C. rotundicauda*. The limulid phylogeny found strong support for a monophyletic genus *Tachypleus* and a diversification of the three Asian species during the Paleogene period, with speciation events well separated in time by several million years. The tree topology suggested that the three Asian species originated in central South East Asia from a marine stem group that inhabited the shallow coastal waters between the Andaman Sea, Vietnam, and Borneo. In this region *C. rotundicauda* probably separated from the *Tachypleus* stem group by invading estuarine habitats, while *Tachypleustridentatus* most likely migrated northeast along the Southern coast of China and towards Japan.

Staton et al. (1997) described Mitochondrial Gene Arrangement of the Horseshoe Crab *Limulus polyphemus*. Numerous complete mitochondrial DNA sequences had been determined for species within two arthropod groups, insects and crustaceans, but there were none for a third, the chelicerates. Most mitochondrial gene arrangements reported for crustacean and insect species are identical or nearly identical to that of Drosophila yakuba. Sequences across 36 of the gene boundaries in the mitochondrial DNA (mtDNA) of a representative chelicerate, *Limulus polyphemus* L., also revealed an arrangement like that of *Drosophila yakuba*. Only the position of the tRNA Leu(UUR) gene differs; in Limulus it is between the genes for tRNA Leu(CUN) and NDl. This positioning was also found in onychophorans, molluscs, and annelids, but not in insects and crustaceans, and indicates that tRNALeU(CUN)- tRNALeU(UUR)-NDw1 as the ancestral gene arrangement for these groups, as suggested earlier. There were no differences in the relative arrangements of protein-coding and ribosomal RNA genes between *Limulus* and *Drosophila*, and none had been observed within arthropods. The high degree of similarity of mitochondrial gene arrangements within arthropods was striking, since some taxa last shared a common ancestor before the Cambrian, and contrasts with the extensive mtDNA rearrangements occasionally observed within some other metazoan phyla (e.g. molluscs and nematodes).

Saitou and Nei (1987) proposed a new method called the neighbor-joining method is proposed for reconstructing phylogenetic trees from evolutionary distance data. The principle of this method is to find pairs of operational taxonomic units (OTUs [=neighbors]) that minimize the total branch length at each stage of clustering of OTUs starting with a star-like tree. The branch lengths as well as the topology of a parsimonious tree can quickly be obtained by using this method. Using computer simulation, we studied the efficiency of this method in obtaining the correct unrooted tree in comparison with that of five other tree-making methods: the unweighted pair group method of analysis, Far-r-is's method, Sattath and

Tversky's method, Li's method, and Tateno et al.'s modified Fan-is method. The new, neighbor-joining method and Sattath and Tversky's method are shown to be generally better than the other methods.

3 Molecular Identification of Indian Horseshoe Crab

Occurrence: The Indian Horseshoe crab is naturally occurring in the Sundarbans Delta of the Bengal region comprising Bangladesh and Eastern India. It is the largest single block of tidal halophytic mangrove forest in the world. The Sundarbans covers approximately 10,000 km^2 (3900 m^2) of which 60 % is in Bangladesh with the remainder in India (Manna et al. 2010). The Indian Sundarbans Delta (ISD) is part of the delta of the Ganga-Brahmaputra-Meghna (GBM) basin in Asia. The Sundarbans, shared between India and Bangladesh is home to one of the largest mangrove forest in the world. It also supports diverse biological resources including fish, birds, mammals, and amphibian species. Crustaceans include Crabs, Squilla, Prawn and Shrimps. Varieties of crabs like the most commonly sighted Fiddler crab, Red crab, Hermit crab, Tree crab, Mud crab are displayed in specimen jars. Above all, one gets the rare opportunity to discover the two varieties of the endangered living fossil Horseshoe Crab, which are found in Sundarban.

4 DNA Barcoding Approach for Species Identification, a Case Study

The comprehensive species identification was made through standard protocol of DNA barcoding. The different samples of Horseshoe crab were collected in different periods and vouchered in the museum. The tissue from the collected specimen were collected aseptically and further processed for DNA extraction following Standard protocol. It was then subjected to amplification of CO1 fragment "barcode segment" followed by DNA sequencing. In this study we amplified Barcode segment using published primers of Folmer et al. (1994).

The PCR reaction was set with an initial denaturation temperature of 94 °C host start for 3 min and subsequently, 94 °C for 1 min denaturation, 54 °C for 45 s for annealing and, 72 °C for 45 s for extension primer annealing for 30 cycles followed by 72 °C for 10 min for final extension using gradient thermal cycler. All the fragments were then sequenced bi-directionally using Automated DNA sequencing technology.

4.1 Outgroup

Lingula is a genus of brachiopods within the class Lingulata. *Lingula* is known to have existed possibly since the Cretaceous or at least the Tertiary. Like its relatives, it has two unadorned phosphatic valves and a long fleshy stalk. *Lingula* lives in burrows in barren sandy coastal seafloor and feeds by filtering detritus from the water. It can be detected by a short row of three openings through which it takes in water (sides) and expels it again (middle). *Lingula* has long been considered as an example of a living fossil; in fact, the perceived longevity of this genus led Darwin to coin this concept. This status is based on the shape of the shell only, and it has been shown that this shape corresponds to a burrowing lifestyle, occurring in different brachiopod lineages, with different and evolving internal structures (Komiya et al. 1980) (Fig. 3).

Lingula is considered a "living-fossil" based on its supposed lengthy morphological conservatism owing to its absence of evolution, and its remarkable survival for more than 550 million years. This conclusion is based on the typical apparently unchanged "linguliform" shape of the shell (Freeman 1999). Over 12,000 fossil species are recognized, (Manna et al. 2010) grouped into over 5000 genera. While the largest modern brachiopods are 100 mm (3.9 in) long (Freeman 1999) a few fossils measure up to 200 mm (7.9 in) wide (Freeman 1999). The earliest confirmed brachiopods have been found in the early Cambrian, inarticulate forms appearing first, followed soon after by articulate forms (Folmer et al. 1994). Three unmineralized species have also been found in the Cambrian, and apparently represent two distinct groups that evolved from mineralized ancestors (Freeman 1999).

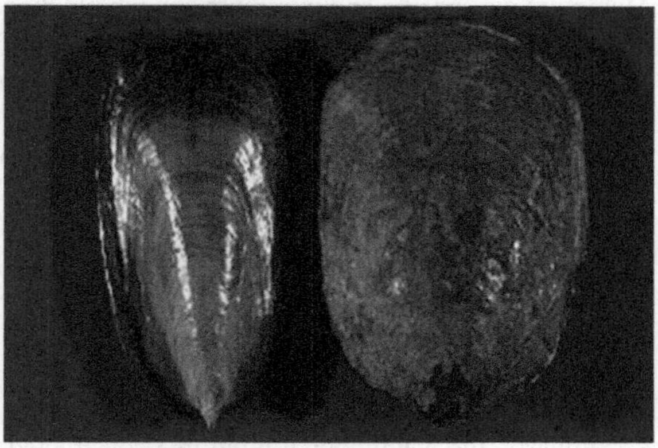

Fig. 3 *Lingula sp.* with its fossil evidence (http://www.geojeff.org/course-materials/historical-geology-lab/lab-6-fossils/brachiopoda/)

4.2 Species Identification Through Similarity Match Approach

In this study we sequenced both the Horseshoe crabs found in Indian sub-continent as well as some other available crustacean and molluscs as OTU or out groups. All the sequence data of the barcode region was subjected to similarity match with the global database. The sequence similarity search for species identification was done in two public databases, viz., BOLD and GenBank. The highest percent pair wise identity for each sequence blasted at NCBI was compared with the percent similarity scores of the same sequence within the BOLD-IDS (BOLD Identification System). The query species that matched either with the same or different species in the databases has been termed as 'specific' or 'non-specific' respectively. The similarity range of 97–100, 92–96 and ≤91 % between the query and the database sequence have been expressed as significant, moderate and insignificant respectively. Table 1 shows the similarity match of these sequences to those of the sequences submitted in GenBank and BOLD databases.

4.3 Identification of Species: Kimura 2-Parameter (K2P) Distance and Neighbour-Joining (NJ) Tree

The calculations of congeneric (between species) and conspecific (within species) genetic distance and Neighbour Joining (NJ) tree construction were done using the computer program MEGA V6.0 (Tamura et al. 2013), selecting the Kimura 2-parameter (K2P) (Kimura 1980) model for phylogenetic reconstruction. Maximum conspecific and minimum congeneric divergences have been determined considering the genetic distance data. A threshold of species boundary or barcode gap was considered by taking the difference of minimum congeneric K2P distance with the maximum conspecific distances (Fig. 4). This threshold was able to distinguish the entire studied specimen. The maximum K2P distance of individuals within species was calculated to be 0.1550100948. The minimum K2P distance of individuals between species was estimated to be 0.200.

The maximum conspecific divergence (0.1550100948, blue line) and the minimum congeneric divergence (0.200, red line) represent the threshold level of conspecific and congeneric divergence respectively. Data that is marked by 'blue cubes' represent the conspecific divergence whereas the 'red plus' represents the congeneric divergence.

Similarly, the Neighbour-Joining tree constructed showed cohesive cluster with the individual of same species and remained distinct from other species. Thus, it was congruent with the genetic distance data and was successful in delineating individuals into respective species (Fig. 5).

Table 1 Identification of the studied species based on similarity match with pre-existing species sequences of BOLD Identification System and Genbank

Sl. No.	Sample ID	Supplied specimen	BOLD-IDS		GenBank (BLASTN similarity in %)	Identified as species
			Species level barcode records	Public record barcode database		
1	Cu1bR, Cu1R	*Limula sp*	*Carcinoscorpius rotundicauda* (98.59 %)	*Carcinoscorpius rotundicauda* (98.59 %)	*Carcinoscorpius rotundicauda* (98)	*Carcinoscorpius rotundicauda (Horseshoe crab)*
			(GBA10392-13)	(GBA10392-13)		
2	PalaR, GcaR	*Panulirus sp*	*Panulirus polyphagus* (99.19 %)	*Panulirus polyphagus* (99.19 %)	*Panulirus polyphagus* (99)	*Panulirus polyphagus (spiny lobster)*
			(GBCMD7090-13)	(GBCMD7090-13)		
3	SGC 03 R	Ucasp Type II	*Uca annulipes* (100 %)	*Uca annulipes* (100 %)	*Uca annulipes* (100)	*Uca annulipes* (Fidder crab)
			(GBCMD5643-13)	(GBCMD5643-13)		
4	CraR, Cru2R	*Crassostrea sp*	Nearest to *Crassostrea belcheri* (97.36 %)	Nearest to *Crassostrea belcheri* (97.36 %)	*Crassostrea cuttackensis* (99)	Tentatively identified as *Crassostrea belcheri*
			(GBMLB4919-13)	(GBMLB4919-13)	*Crassostrea belcheri* (99)	
5	SGC 01	Red crab	No match	No match	*Ocypode ceratophthalmus* (89)	Unidentified
6	SGC 04	Scylla	No match	No match	No match	Unidentified
7	SGL 02	*Lingula sp*	No match	No match	*Lingula adamsi* (80)	Unidentified
8	SGL 01	*Lingula sp*	No match	No match	*Lingula adamsi* (80)	Unidentified

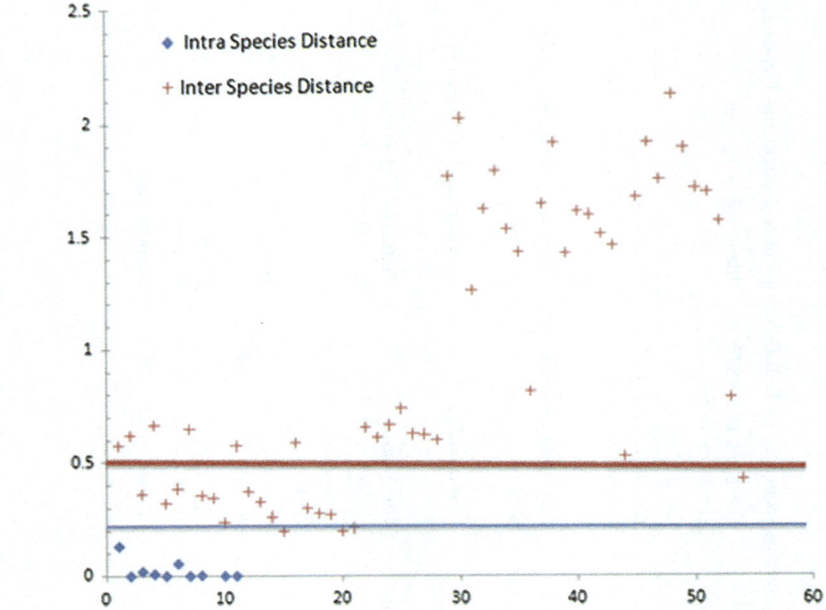

Fig. 4 Distribution of conspecific and congeneric K2P mean divergence of 21 species collected from Sundarbans

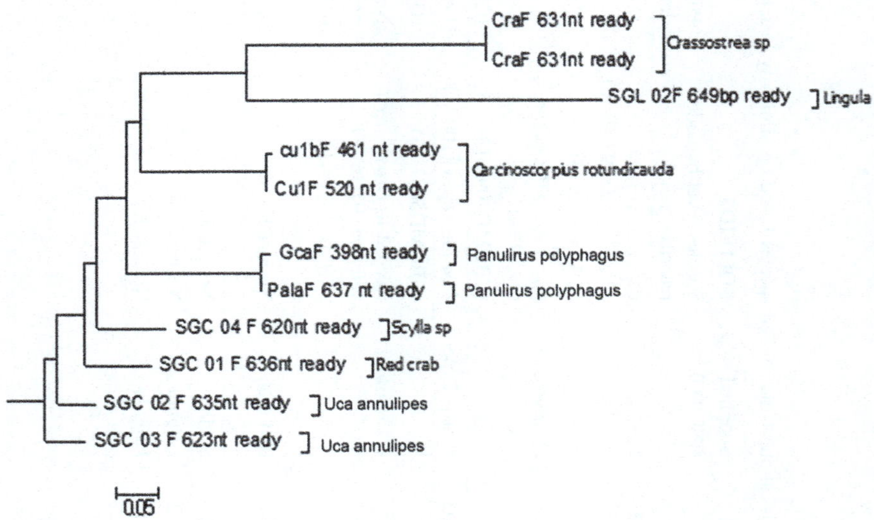

Fig. 5 Neighbour-joining tree construction of the studied species of Horseshoe crab as well as other arthropods and molluscs of Sundarbans area included in the study. All the conspecific sequences clustered cohesively whereas different species clustered distinctly

5 Positioning of the Living Fossils (Horseshoe Crab: *Carcinoscorpius rotundicauda*) in Correspondence with the Global Data

5.1 Maximum Likelihood Tree Construction

To estimate the evolutionary distance and to locate the positions of the living fossil Horseshoe crab viz. *Carcinoscorpius rotundicauda* firstly model test was carried out and subsequently Maximum-likelihood (ML) tree was constructed by taking conspecific sequences from the database (where ever possible) as barcode replicates and other species covering same family or phylum for maximum accuracy and robustness of the ML tree. The best fit model based on model test was selected as summarized in Table 2. Models with the lowest BIC scores (Bayesian Information Criterion) are considered to describe the substitution pattern the best. For each model, AICc value (Akaike Information Criterion, corrected), Maximum Likelihood value (lnL), and the number of parameters (including branch lengths) are also presented (Masta et al. 2009). Non-uniformity of evolutionary rates among sites may be modelled by using a discrete Gamma distribution (+G) with 5 rate categories and by assuming that a certain fraction of sites are evolutionarily invariable (+I). Whenever applicable, estimates of gamma shape parameter and/or the estimated fraction of invariant sites are shown. Assumed or estimated values of transition/transversion bias (R) are shown for each model, as well. They are followed by nucleotide frequencies (f) and rates of base substitutions (r) for each nucleotide pair. Relative values of instantaneous r should be considered when evaluating them. For simplicity, sum of r values is made equal to 1 for each model. For estimating ML values, a tree topology was automatically computed. All positions containing gaps and missing data were eliminated. There were a total of 392 positions in the final dataset.

To construct the Maximum Likelihood tree 2 sequences of *Carcinoscorpius rotundicauda*, 2 species of *Lingula sp,* 6 different crab sequence of the current study with many other closely related groups of the phylum arthropod were taken from the GenBank (Fig. 6). Phylogenetic analyses were performed where Gaps were treated as missing data. The ML analysis consisted of 100 independent runs on the non-partitioned alignment using under the GTR+G+I model (re-estimated all free model parameters) with estimated rearrangement settings.

ML tree analyses consistently showed high support for a sister group relationship between *Tachypleus gigas* and *Tachypleus tridentatus*. This ML tree strongly supports that there is a sistergroup relationship between *Tachypleus* and *C. rotundicauda*. The ML tree interprets that the *Carcinoscorpius rotundicauda* shows lesser similarity with *Limulus polyphemus* as compared to *Tachypleus gigas.* Whereas the second living fossil *Lingula sp.* has shown no similarity with *Carcinoscorpius rotundicauda* and clustered separately as no related sequences of this species was found.

Table 2 Model test result for constructing Maximum Likelihood tree

Model	#Param	BIC	AICc	InL	Invariant	Gamma	R
GTR+G+I	103	19077.6441	18271.01536	−9031.935215	0.252165909	0.83210252	1.530281827
GTR+G	102	19097.99054	18299.18219	−9047.029667	n/a	0.387623613	1.572494722
HKY+G+I	99	19134.85513	18359.50922	−9080.225653	0.250869525	0.788529347	1.481676284
T92+G+I	97	19155.65462	18395.95143	−9100.467863	0.245198156	0.737468139	1.511026715
HKY+G	98	19156.82359	18389.29894	−9096.131117	n/a	0.388014194	1.532548399
TH93+G+I	100	19169.44558	18386.27864	−9092.599646	0.254101771	0.867016546	1.102727803
T92+G	96	19177.86235	18425.98084	−9116.492956	n/a	0.377038536	1.594182648
TN93+G	99	19194.9105	18419.5646	−9110.25334	n/a	0.414350642	1.101096762
K2+G+I	96	19702.92968	18951.04817	−9379.026625	0.260261162	1.064119378	1.240507783
K2+G	95	19731.40326	18987.34365	−9398.184647	n/a	0.454427671	1.240507783
JC+G+I	95	19861.76462	19117.70501	−9463.365324	0.26271954	1.199876195	0.5
JC+G	94	19885.4332	19149.1957	−9480.120844	n/a	0.460752702	0.5
GTR+I	102	19947.35137	19148.54302	−9471.71008	0.295918367	n/a	1.035826162
HKY+I	98	20130.01099	19362.48633	−9582.724814	0.295918367	n/a	0.986083564
TN93+I	99	20139.67142	19364.32551	−9582.633797	0.295918367	n/a	0.986977128
T92+1	96	20150.25255	19398.37104	−9602.688056	0.295918367	n/a	0.983707018
K2+l	95	20403.58786	19659.52825	−9734.276946	0.295918367	n/a	1.073590049
JC+I	94	20520.7376	19784.5001	−9797.773044	0.295918367	n/a	0.5
GTR	101	21351.46478	20560.47702	−10178.68801	n/a	n/a	0.936089074
HKY	97	21557.98027	20798.27708	−10301.63069	n/a	n/a	0.903063841
TN93	98	21567.05554	20799.53089	−10301.24709	n/a	n/a	0.902857643
T92	95	21585.41811	20841.35849	−10325.19207	n/a	n/a	0.90348279
K2	94	21780.25138	21044.01388	−10427.52994	n/a	n/a	1.073590049
JC	93	21878.34839	21149.93322	−10481.49967	n/a	n/a	0.5

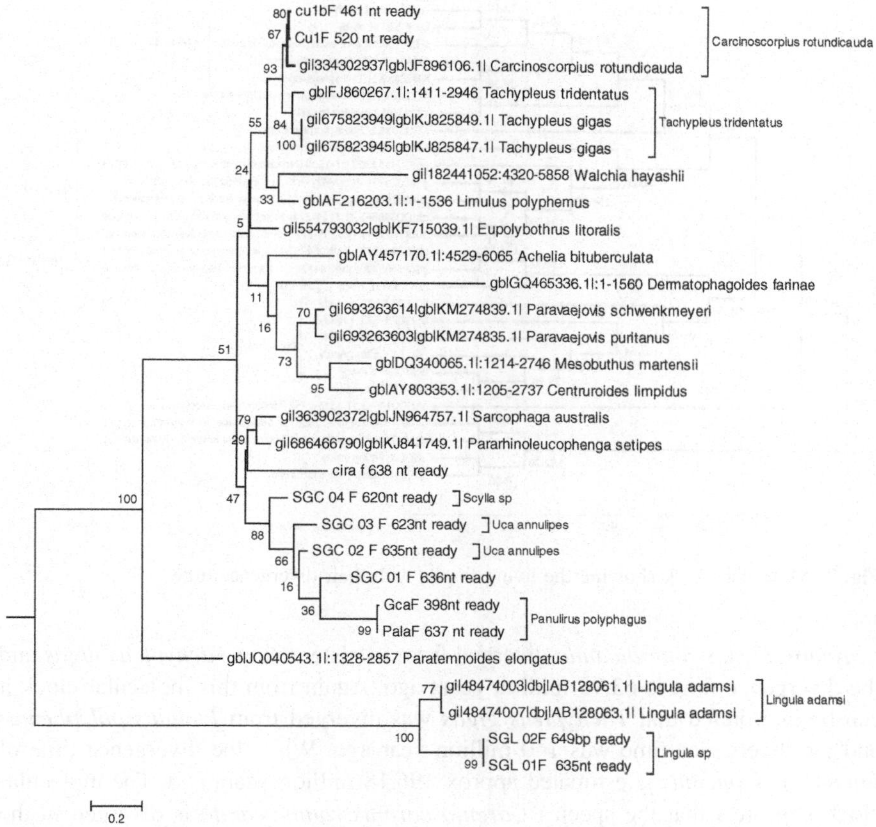

Fig. 6 Maximum-Likelihood tree from the combined analysis based on all molecular data (ML) of the two living fossils (the two *red marked branches*) viz. *Carcinoscorpius rotundicauda* along with their close relatives. *Lingula sp.* taken here as Outgroups

6 Molecular Clock

To estimate the divergence time of the living fossils *Carcinoscorpius rotundicauda* and *Lingula sp.,* molecular clock was generated. The tree was reconstructed from a ML approach (Tamura et al. 2013) using the best fit model of "General Time reversible with gamma distribution (GTR- G+I)". It is then tested for confidence interval for divergence time. A calibration constrains were put at the MRCA of *Limulus polyphemus* and *Tachypleus gigas* based on the fossil data. Molecular clock of the living fossils is given in Fig. 7.

The divergence time of *Tachypleus gigas*, Asian horseshoe crab was dated approx. 32.08 million year ago with the molecular clock, while the molecular clock dated the divergence time of the *Carcinoscorpius rotundicauda* approx. 12.18 million year ago. From the molecular clock it can be interpreted that

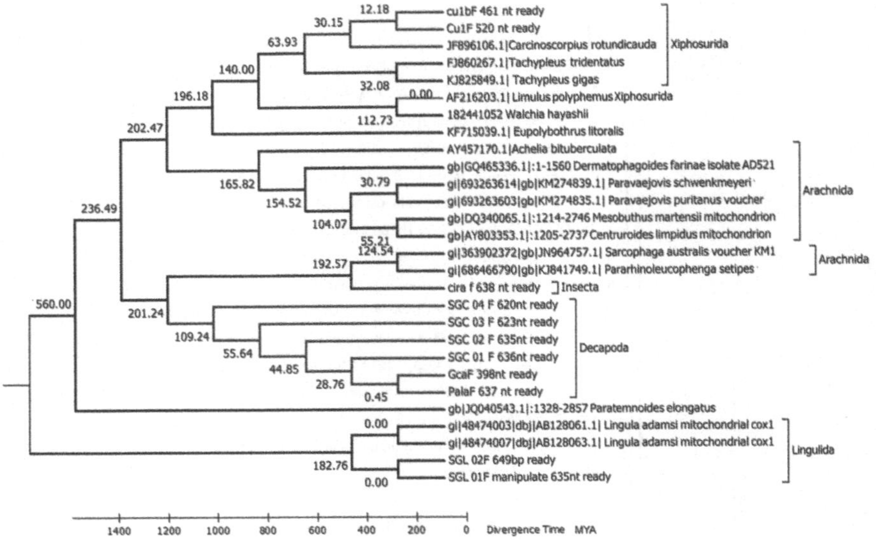

Fig. 7 Molecular clock showing the living fossils with their divergence time

Carcinoscorpius rotundicauda diverged from its sister taxon *Tachypleus gigas* and the divergence time is 63.93 million years ago. Again from this molecular clock it can be interpreted that *Tachypleus gigas* was diverged from *Limulus polyphemus* and the divergence time was 140 million year ago. While the divergence time of *Limulus polyphemus* is estimated approx 196.18 million years ago. The molecular clock interprets that the species *Carcinoscorpius rotundicauda* is diverged in the earth approx. 560 million years ago. The molecular clock interprets that *Lingula sp* diverged approx. 182.76 million years ago.

As earlier studies already indicated (Purvis et al. 2005), the morphostasis of horseshoe crabs cannot be explained by lack of variation on the molecular genetic level. The genetic diversity within *C. rotundicauda* was larger than in *L. Polyphemus* and *T. gigas* which may be explained by the different ecology of this species. As *C. Rotundicauda* often inhabits estuarine habitats and mangroves, and less often enters open oceanic water, the gene flow between populations in this species may be much more restricted. A comparative analysis of the population structure in *C. rotundicauda* and *T. gigas* may reveal contrasting patterns of connectivity and possibly even uncover the presence of reproductively isolated lineages within *C. rotundicauda*. The phylogeographic relationships among the *L. polyphemus* clades were weakly supported and remain unresolved in the current study while three *C. rotundicauda* clades showed a well-supported subdivision of the geographic locations of the Sundarbans and this finding was similar to the findings of *Carcinoscorpius rotundicauda*on each side of the Thai-Malay Peninsula (Obst et al. 2012). Obst et al. (2012) suggests that the diversification of Asian horseshoe crabs most likely occurred during the Paleogene era (app. 65–23 Mya), with speciation events well separated by several million years which is similar to this current study where it is suggested that the diversification of Asian horseshoe crabs occurred during the Paleogene era (app. 63.93–30 Mya).

A possible explanation for the current distribution of recent horseshoe crabs may be that *L. polyphemus* originated in the western Tethys sea and moved westwards with the North American continent during the Atlantic opening. The three Asian species probably originated in the Eastern parts of Tethys sea and survived the eradication of shallow water habitats in central Eurasia following the collision of the continent with the African and Indian plate during the Cenozoic period (Lee 1999). *C. rotundicauda* separated from the *Tachypleus* stem group by invading eustarine habitats, while *T. tridentatus* probably originated by northeast migration along the Southern coast of China and towards Japan. Even though many horseshoe crab fossils suggest that they lived in brackish and freshwater habitats (Briggs et al. 2012).

References

Avise JC, Bowen BW (1994) Investigating sea turtle migration using DNA markers. Curr Opin Genet Dev 4:882–886

Baek SY, Choi EH, Jang KH, Ryu SH, Park SM et al (2014) Complete mitochondrial genomes of Carcinoscorpius rotundicauda and Tachypleus tridentatus (Xiphosura, Arthropoda) and implications for chelicerate phylogenetic studies. Int J Biol Sci 10:479–489

Bhattacharjee MJ, Laskar BA, Dhar B, Ghosh SK (2012) Identification and re-evaluation of freshwater catfishes through DNA barcoding. PLoS ONE 7:e49950

Borisenko AV, Lim BK, Ivanova NV, Hanner RH, Hebert PD (2008) DNA barcoding in surveys of small mammal communities: a field study in Suriname. Mol Ecol Resour 8:471–479

Briggs DE, Moore RA, Shultz JW, Schweigert G (2005) Mineralization of soft-part anatomy and invading microbes in the horseshoe crab Mesolimulus from the Upper Jurassic Lagerstatte of Nusplingen, Germany. Proc Biol Sci 272:627–632

Briggs DE, Siveter DJ, Sutton MD, Garwood RJ, Legg D (2012) Silurian horseshoe crab illuminates the evolution of arthropod limbs. Proc Natl Acad Sci USA 109:15702–15705

Carroll RL (1997) Patterns and processes of vertebrate evolution. Cambridge University Press, New York. xvi, 448 pp

Chakraborty M, Ghosh SK (2014) An assessment of the DNA barcodes of Indian freshwater fishes. Gene 537:20–28

Cox AJ, Hebert PD (2001) Colonization, extinction, and phylogeographic patterning in a freshwater crustacean. Mol Ecol 10:371–386

Crow JF, Dove WF (2000) Perspectives on genetics: anecdotal, historical, and critical commentaries, 1987–1998. University of Wisconsin Press, Madison. xiii, 723 pp

Dhar B, Ghosh SK (2015) Genetic assessment of ornamental fish species from North East India. Gene 555:382–392

Dick MH, Lidgard S, Gordon DP, Mawatari SF (2009) The origin of ascophoran bryozoans was historically contingent but likely. Proc Biol Sci 276:3141–3148

Dunlop JA, Anderson LI, Kerp H, Hass H (2003) Palaeontology: preserved organs of Devonian harvestmen. Nature 425:916

Faurby S, King TL, Obst M, Hallerman EM, Pertoldi C et al (2010) Population dynamics of American horseshoe crabs—historic climatic events and recent anthropogenic pressures. Mol Ecol 19:3088–3100

Folmer O, Black M, Hoeh W, Lutz R, Vrijenhoek R (1994) DNA primers for amplification of mitochondrial cytochrome c oxidase subunit I from diverse metazoan invertebrates. Mol Mar Biol Biotechnol 3:294–299

Freeman G (1999) Regional specification during embryogenesis in the inarticulate brachiopod Discinisca. Dev Biol 209:321–339

Garcia-Horsman JA, Barquera B, Rumbley J, Ma J, Gennis RB (1994) The superfamily of heme-copper respiratory oxidases. J Bacteriol 176:5587–5600

Hajibabaei M, Janzen DH, Burns JM, Hallwachs W, Hebert PD (2006) DNA barcodes distinguish species of tropical Lepidoptera. Proc Natl Acad Sci USA 103:968–971

Hebert PD, Cywinska A, Ball SL, deWaard JR (2003) Biological identifications through DNA barcodes. Proc Biol Sci 270:313–321

Kimura M (1980) A simple method for estimating evolutionary rates of base substitutions through comparative studies of nucleotide sequences. J Mol Evol 16:111–120

Kin A, Blazejowski B (2014) The horseshoe crab of the genus Limulus: living fossil or stabilomorph? PLoS ONE 9:e108036

Komiya H, Shimizu N, Kawakami M, Takemura S (1980) Nucleotide sequence of 5S ribosomal RNA from Lingula anatina. A study on the molecular evolution of 5S ribosomal RNA from a living fossil. J Biochem 88:1449–1456

Kosakyan A, Heger TJ, Leander BS, Todorov M, Mitchell EA et al (2012) COI barcoding of Nebelid testate amoebae (Amoebozoa: Arcellinida): extensive cryptic diversity and redefinition of the Hyalospheniidae Schultze. Protist 163:415–434

Laskar BA, Bhattacharjee MJ, Dhar B, Mahadani P, Kundu S et al (2013) The species dilemma of northeast Indian mahseer (Actinopterygii: Cyprinidae): DNA barcoding in clarifying the riddle. PLoS ONE 8:e53704

Lee MS (1999) Molecular clock calibrations and metazoan divergence dates. J Mol Evol 49:385–391

Legg DA (2014) Sanctacaris uncata: the oldest chelicerate (Arthropoda). Naturwissenschaften 101:1065–1073

Legg DA, Sutton MD, Edgecombe GD, Caron JB (2012) Cambrian bivalved arthropod reveals origin of arthrodization. Proc Biol Sci 279:4699–4704

Lynch M, Jarrell PE (1993) A method for calibrating molecular clocks and its application to animal mitochondrial DNA. Genetics 135:1197–1208

Mahadani P, Ghosh SK (2013) DNA Barcoding: a tool for species identification from herbal juices. DNA Barcodes 1:35–38

Manna S, Chaudhuri K, Bhattacharyya S, Bhattacharyya M (2010) Dynamics of Sundarban estuarine ecosystem: eutrophication induced threat to mangroves. Saline Syst 6:8

Masta SE, Longhorn SJ, Boore JL (2009) Arachnid relationships based on mitochondrial genomes: asymmetric nucleotide and amino acid bias affects phylogenetic analyses. Mol Phylogenet Evol 50:117–128

Nossa CW, Havlak P, Yue JX, Lv J, Vincent KY et al (2014) Joint assembly and genetic mapping of the Atlantic horseshoe crab genome reveals ancient whole genome duplication. Gigascience 3:9

Obst M, Faurby S, Bussarawit S, Funch P (2012) Molecular phylogeny of extant horseshoe crabs (Xiphosura, Limulidae) indicates Paleogene diversification of Asian species. Mol Phylogenet Evol 62:21–26

Purvis A, Gittleman JL, Brooks T (2005) Phylogeny and conservation. Cambridge University Press, Cambridge. xiii, 431 pp

Rehm P, Pick C, Borner J, Markl J, Burmester T (2012) The diversity and evolution of chelicerate hemocyanins. BMC Evol Biol 12:19

Rubinoff D (2006) Utility of mitochondrial DNA barcodes in species conservation. Conserv Biol 20:1026–1033

Saitou N, Nei M (1987) The neighbor-joining method: a new method for reconstructing phylogenetic trees. Mol Biol Evol 4:406–425

Schoenemann B, Clarkson EN (2013) Discovery of some 400 million year-old sensory structures in the compound eyes of trilobites. Sci Rep 3:1429
Sharma PP, Tarazona OA, Lopez DH, Schwager EE, Cohn MJ et al (2015) A conserved genetic mechanism specifies deutocerebral appendage identity in insects and arachnids. Proc Biol Sci 282:20150698
Shishikura F, Nakamura S, Takahashi K, Sekiguchi K (1982) Horseshoe crab phylogeny based on amino acid sequences of the fibrino-peptide-like peptide C. J Exp Zool Part A 223(1):89–91
Shishikura F, Sekiguchi K (1984) Studies on perivitelline fluid of horseshoe crab embryo. II. Purification of agglutinin-binding substance from the perivitelline fluid of Tachypleus gigas embryo. J Biochem 96:629–636
Silvestro D, Antonelli A, Salamin N, Quental TB (2015) The role of clade competition in the diversification of North American canids. Proc Natl Acad Sci USA 112:8684–8689
Simmons RB, Weller SJ (2001) Utility and evolution of cytochrome b in insects. Mol Phylogenet Evol 20:196–210
Srimal S, Miyata T, Kawabata S, Iwanaga S (1985) The complete amino acid sequence of coagulogen isolated from Southeast Asian horseshoe crab, Carcinoscorpius rotundicauda. J Biochem 98:305–318
Staton JL, Daehler LL, Brown WM (1997) Mitochondrial gene arrangement of the horseshoe crab Limulus polyphemus L.: conservation of major features among arthropod classes. Mol Biol Evol 14:867–874
Stoeckle MY, Hebert PD (2008) Barcode of life. Sci Am 299:82–86, 88
Sykes B (1999) The molecular genetics of European ancestry. Philos Trans R Soc Lond B Biol Sci 354:131–138. Discussion 138–139
Tamura K, Stecher G, Peterson D, Filipski A, Kumar S (2013) MEGA6: molecular evolutionary genetics analysis version 6.0. Mol Biol Evol 30:2725–2729
Trivedi S, Affan R, Allesa AHA, Ansari AA, Dhar B, Mahadani P, Ghosh SK (2014) DNA BARcoding of Red Sea fishes from Saudi Arabia—the first approach. DNA Barcodes 2:17–20
Trivedi S, Aloufi AA, Ansari AA, Ghosh SK (2015) Molecular phylogeny of oysters belonging to the genus Crassostrea through DNA barcoding. J Entomol Zool Stud 3(1):21–26
Trivedi S, Aloufi AA, Ansari AA, Ghosh SK (2016a) Role of DNA barcoding in marine biodiversity assessment and conservation: an update. Saudi J Biolog Sci 23(2):161–171
Trivedi S, Aloufi AA, Rehman H, Saggu S, Ghosh SK (2016b) DNA barcoding: tool for assessing species identification in Reptilia. J Entomol Zool Stud 4(1):132–137
Trivedi S, Ansari AA, Rehman H, Saggu S, Abbas ZK, Ghosh SK (2016c) DNA barcoding as a molecular tool for the assessment of plant biodiversity. In: Ansari AA, Gill SS (eds) Plant biodiversity: monitoring assessment and conservation. CABI Publications, UK (in press)
Van Roy P, Orr PJ, Botting JP, Muir LA, Vinther J et al (2010) Ordovician faunas of Burgess Shale type. Nature 465:215–218
Vasquez MC, Murillo A, Brockmann HJ, Julian D (2015) Multiple stressor interactions influence embryo development rate in the American horseshoe crab, Limulus polyphemus. J Exp Biol 218:2355–2364
Wares JP, Cunningham CW (2001) Phylogeography and historical ecology of the North Atlantic intertidal. Evolution 55:2455–2469
Xia X (2000) Phylogenetic relationship among horseshoe crab species: effect of substitution models on phylogenetic analyses. Syst Biol 49:87–100
Zhang DX, Hewitt GM (1997) Assessment of the universality and utility of a set of conserved mitochondrial COI primers in insects. Insect Mol Biol 6:143–150

Part III
DNA Barcoding of Marine Fishes

Mitochondrial DNA Diversity of Wild and Hatchery Reared Strains of Indian *Lates calcarifer* (Bloch)

Prasanna Kumar, B. Akbar John and V. Kanagasabapathy

Abstract *Lates calcarifer*, locally known as seabass in Asia and barramundi in Australia, is a large, euryhaline member of the family Centropomidae that is widely distributed in the Indo-West Pacific region. Its hardy nature, high tolerance to wide range physiological condition and high commercial value has made it a good candidate species for aquaculture practices. In this study we compared the mtDNA diversity of hatchery reared and wild *Lates calcarifer* using universal DNA barcode gene (Cytochrome Oxidase C subunit 1 gene) to assess the genetic health of *L. calcarifer* hatchery practices in India. Sampling stations were randomly chosen to cover both East and West coasts of India. The phylogram constructed with COI sequences (n = 88) of *L. calcarifer* revealed that geographic distributions of clades are not restricted to any particular sampling stations. Gene flow appeared to have transported haplotypes between the clades from their likely origins across the sampled range. Both Nucleotide (π) and haplotyte (h) diversity of wild *L. calcarifer* was higher in East coast samples compared to West coast samples. The comparative genetic diversity analysis assessed through COI sequences between hatchery reared and wild catches of *L. calcarifer* showed that the nucleotide diversity of hatchery strains was 2.7 times lesser than that of wild strains, demanding immediate action plans to restore genetic diversity in *L. calcarifer* hatchery practices in India.

Keywords *Lates calcarifer* · COI gene · DNA barcoding · Haplotype diversity · Genetic diversity

---

P. Kumar
Centre of Advanced Study in Marine Biology, Annamalai University, Parangipettai, India

B. Akbar John (✉)
INOCEM Research Station (IRS), Kulliyyah of Science,
International Islamic University of Malaysia, Kuantan, Pahang, Malaysia
e-mail: akbarjohn50@gmail.com

V. Kanagasabapathy
Central Institute of Brackish Water Aquaculture (ICAR), Chennai, India

V. Kanagasabapathy
Unit of Live Feed Culture, Department of Zoology, University of Madras, Chennai, India

S. Trivedi et al. (eds.), *DNA Barcoding in Marine Perspectives*,
DOI 10.1007/978-3-319-41840-7_13

1 Introduction

Lates calcarifer (Bloch), more commonly known as barramundi or Asian sea bass (Pusey et al. 2004), is a large, protandrous hermaphroditic, centropomid, fish distributed widely throughout the Indo-West Pacific region, including northern regions of Australia. It is a commercially exploitable species whose reproductive physiology is reasonably well understood, with the life cycle fully closed and hatchery production of juveniles routinely achieved (Macbeth et al. 2002). The seabass (*Lates calcarifer*) is a potential candidate for farming in India, because of its fast growth rate, tolerance to wide environmental conditions and its demand in domestic and export markets. The FAO report suggested that annual seabass production has been relatively static since 1998, at ~20,000–27,000 tonnes. A notable increase in its production has observed in year 2008 and 2010, at ~34,000–49,000 tonnes respectively (FAO 2011).

Extensive research on genetic diversity of this species was deeply investigated in Australian (Chenoweth et al. 1998; Ward et al. 2008), Singapore, Taiwan, Malaysian and Indonesian waters (Wang et al. 2007; Yue et al. 2002). The genetic diversity of wild populations and cultured stocks of Asian seabass has been studied using mitochondrial DNA sequences (Doupe and Recher 1999; John et al. 2010), allozymes (Keenan 1994) and microsatellites (Yue et al. 2002; Zhu et al. 2006). Ward et al. (2008) has used COI gene fragments in assessing the genetic population of *L. calcarifer*, but the study faced serious problem of sampling size (n = 10). Almost all the previous studies on assessing genetic diversity of *L. calcarifer* did not yield stable refined conclusion, as different markers were used for the estimation. It should also be noted that hatchery fish stocking for stock enhancement has been operated at a massive and global scale. However, the use of hatchery fish as a means of stock enhancement is highly controversial, and little is known about its effects on wild stock and consequences for stock enhancement (Araki and Schmid 2010).

Though the species was recovered heavily from wild catches and hatchery cultures in India, the studies on assessing of genetic diversity of *L. calcarifer* from Indian waters is still scanty. Due to an emerging DNA barcoding technology using universal barcode gene as a reference source has attracted us to explore the genetic structure and mtDNA diversity of wild and hatchery reared strains of Indian *Lates calcarifer*. The objectives of the present study was to (i) estimate genetic diversity of Indian *L. calcarifer* using single standard COI marker gene (DNA barcode) and (ii) compare the genetic diversity of hatchery reared and wild catches of the species.

2 Materials and Methods

2.1 Sampling Design

The spatial distribution of mitochondrial DNA lineages of *Lates calcarifer* in Indian waters was characterized by sampling stations in East and West coast of the country. Five sampling stations 2 from Arabian Sea, 1 in Indian Ocean and 2 from Bay of Bengal was selected for the study. The sapling stations were PatwadiFish Landing Centre (PFLC) (19° 03′N, 72° 48′E) in Maharashtra; Mangalore Fishery Harbor (MFHK) (12° 53′N, 74° 49′E) in Karnataka; Kanyakumari Fishery Harbor (KFHT) (8° 05′N, 77° 33′E) in Tamil Nadu, Kakinada Fishing Harbor (KFHA) (16° 59′N, 82° 17′E) in Andra Pradesh and Roychowk Fishing Harbor (RFHW) (22° 11′ N, 88° 03′E) in West Bengal (Fig. 1). A total of 20 individual samples were collected from each station constituting 100 specimens for DNA sequencing. Besides wild specimens, *Lates calcarifer* in different life stages were procured from 2 hatcheries (Rajiv Gandhi Centre for Aquaculture (RGCA) (12° 56′N, 80° 15E) in Tamil Nadu (in East coast) and Takave farms (MH) (18° 32′N, 73° 53′E) in Mumbai (West coast of India). Hatchery collection includes 25 specimens for the DNA sequencing.

2.2 DNA Barcoding

We screened 650 bp PCR product of mitochondrial Cytochrome C Oxidae subunit I gene (COI) of 125 individuals of *L. calcarifer* using agarose gel electrophoresis and automated DNA sequencing methodology. The sequences obtained below the recommended length (i.e., <650 bps) (Hebert et al. 2003) were not considered for further analysis. Khan et al. (2010) and Kumar et al. (2011) methods were adopted for DNA extraction, amplification, screening and sequencing. DNA from eggs was isolated by modifying the protocols of Kumar et al. (2011). The modification was in terms of adjustments with the volume of lysis buffer (reduced to half of its original amount) and high salt solution. Sequencing was performed using MegaBace sequencer and all sequences were deposited in NCBI nucleotide database and could be accessed through GenBank accession numbers JF919740-JF919828.

2.3 Statistical Analysis

Neighbor-Joining phylogenetic trees were constructed based on corrected average pair-wise difference distance matrices using the program Molecular Evolutionary Genetics Analysis (MEGA) (Tamura et al. 2007). Multiple alignment of the sequences was performed using Clustal X (Larkin et al. 2007). Haplotype diversity

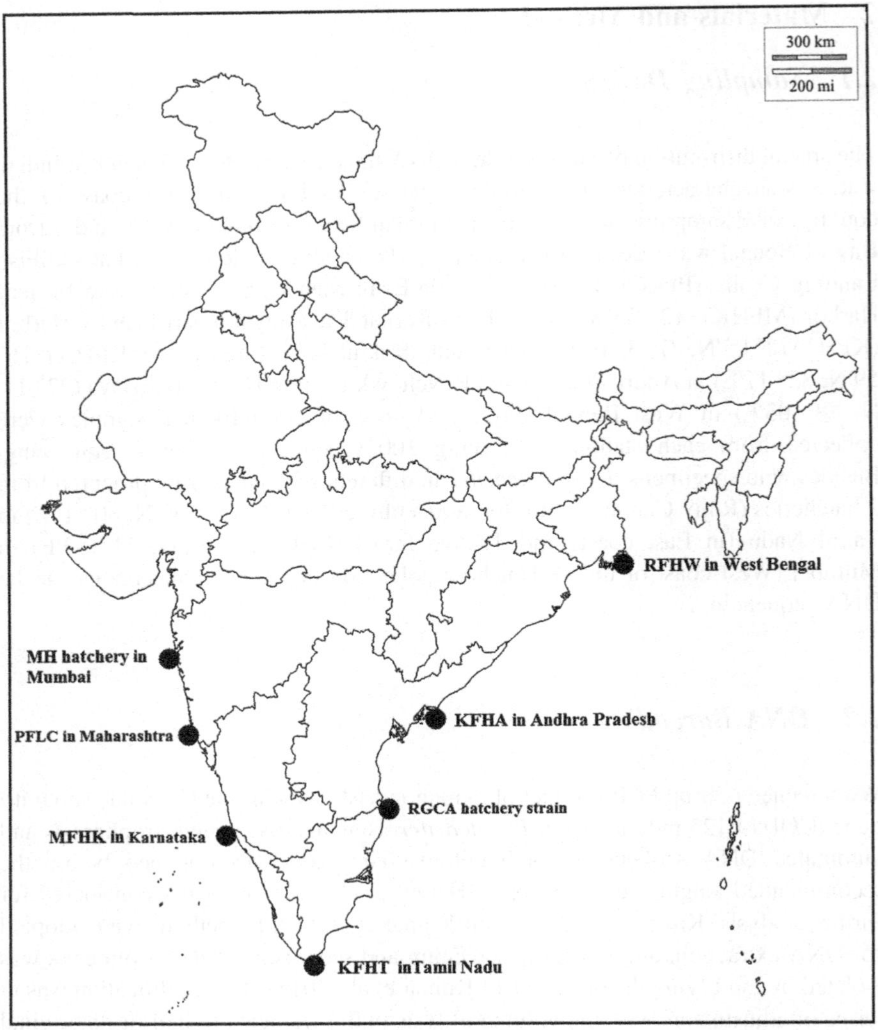

Fig. 1 Location of the sampling sites covering both East and West coasts of India

(Nei 1987) was calculated using the DnaSP package (Librado and Rozas 2009), nucleotide diversity statistics and standard error were calculated using of (Nei and Jin 1989) algorithm, patristic distances generated by the neighbor-joining method (Saitou and Nei 1987). Codon usage of translated conceptual amino acids of COI region was calculated using BioEidt (Hall 1999). Transition/transversion ratios were estimated by direct counts.

3 Result and Discussion

3.1 Maximum Likelihood Tree

From 125 PCR products sequenced, 88 sequences were found to contain nucleotides of more than 650 bps length and were taken for further analysis. From the 88 individuals screened, 40 putative haplotyes were resolved. The maximum likelihood phylogeny reconstruction yielded strong support for 4 distinct clades (named as Clade A, Clade B, Clade C and Clade D) differed by approximately 3 % sequence divergence. Clade with bootstrap less than 45 was fused (Fig. 2). Geographical distributions of Clade A, B, C and D were not restricted to any particular sides of sampled stations. In clade A majority of the sequences (25 %) were contributed by strains collected from Roychowk fishing harbor West Bengal (RFHW) and (20 %) second majority were from strains of Kanyakumari fishing harbor Tamil Nadu (KFHT). In clade B, majority of the sequences (41 %) was from Kakinada fish harbor Andra Pradesh (KFHA) and 28 % were from Patwadi fish landing centre Maharashtra (PFLC). Least of 9 % of sequences were contributed by hatchery strains of Mumbai (MH). Majority of sequences (53 %) of clade C was contributed by Mangalore fishing harbor Karnataka (MFHK) and 35 % was from Kakinada fishing harbor Andra Pradesh (KFHA). Least of 12 % was contributed by hatchery strains of Rajiv Gandhi Centre for Aquaculture (RGCA) in clad

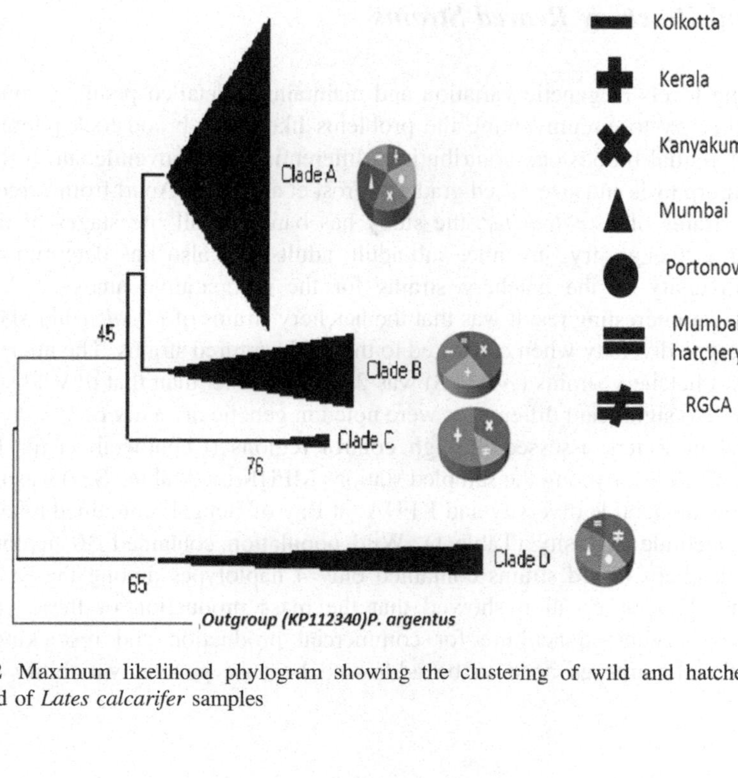

Fig. 2 Maximum likelihood phylogram showing the clustering of wild and hatchery reared srtaind of *Lates calcarifer* samples

C. In Clade D, the majority of the sequences (39 %) were from RGCA hatchey strains followed by the strains of Roychowk fishing Harbor, West Bengal (RFHW) (30 %). Least of 17 and 13 % were contributed by strains from Mumbai hatchery stains (MH) and and Kanyakumari fishery harbor, Tamil Nadu (KFHT) respectively.

Marine fishes generally show low levels of genetic differentiation among geographic regions due to higher dispersal potential during planktonic egg, larval, or adult history stages coupled with an absence of physical barriers to movement between ocean basins or adjacent continental margins (Grant and Bowen 1998; Hewitt 2000). The patterns of Indian-Pacific phylogenetic breaks observed in marine species like the coconut crab, *Birguslatro*, which exhibits only 2 % total mtDNA break between Pacific and Indian Ocean populations (Lavery et al. 1996) were not observed in this study due to high dispersal ability of seabass along the sampling stations. Gene flow appears to have transported haplotypes between the clades from their likely origins across the sampled range. Direct investigations of *L. calcarifer* movements using mark-recapture techniques suggested a mean single generation dispersal distance of 15.3 km in Coral Sea populations (Russell and Garrett 1988). Although this figure may have a large variance and is not exactly estimating gene flow, it can be used as a crude measure to determine if the secondary integration can be explained.

3.2 Nucleotide and Haplotype Diversity of Wild and Hatchery Reared Strains

Monitoring levels of genetic variation and maintaining detailed pedigrees on progeny is the key to circumventing the problems like small broodstock population sizes, differential broodstock contribution, differential larval/juvenile survival during metamorphosis and size-based grading (Frost et al. 2006). Apart from barcoding the wild strains of *L. calcarifer*, the study has barcoded all life stages of the *L. calcarifer* viz, eggs, fry, juvenile, sub-adult, adults and also has determined the genetic diversity of the hatchery strains for the comparative analysis. As we expected, yet interesting result was that the hatchery strains of *L. calcarifer* showed low haplotype diversity when compared to the wild captured strains. The nucleotide diversity of hatchery strains ($\sim$0.008) was 2.7 times lesser than that of wild strains ($\sim$0.046). No significant differences were noted in genetic diversity of *L. calcarifer* of Australian waters assessed through control regions (Chenoweth et al. 1998; Doupe et al. 1999). Among the sampled stations MFHK (at Arabian Sea) was found to have low nucleotide diversity and KFHA (at Bay of Bengal) contained relatively higher nucleotide diversity (Table 1). Wild population contained 36 haplotypes whereas hatchery reared strains contained only 4 haplotypes among the sampled specimens. This observation showed that the mass production of these valued species for serving aquaculture for commercial production and restocking for conservational purposes need to be addressed because genetic variability is an

Table 1. mtDNA Genetic diversity including nucleotide and haplotype diversity and the putative haplotypes of *Lates calcarifer* samples sampled from wild and hatchery reared strains

Sampled stations	Genetic diversity		No. of Haplotypes
	Nucleotide diversity (π)	Haplotype diversity (h)	
PFLC (Maharashtra)	0.049 ± 0.005	0.872 ± 0.048	7
MFHK (Karnataka)	0.041 ± 0.008	0.859 ± 0.023	5
KFHT (Tamil Nadu)	0.051 ± 0.001	0.921 ± 0.041	9
KFHA (Andhra Pradesh)	0.058 ± 0.009	0.949 ± 0.076	3
RFHW (West Bengal)	0.053 ± 0.009	0.936 ± 0.062	13
MH (Takave farms in Mumbai)	0.008 ± 0.001	0.018 ± 0.021	2
RGCA (Hatchery in Tamil Nadu)	0.009 ± 0.004	0.019 ± 0.081	2
Average	0.074 ± 0.012	0.653 ± 0.041	40

important attribute in domesticating better productive traits (Tave 1993). In addition, fixed haplotype differences observed among wild stock populations (e.g., KFHT, KFHA and RFHW in East coast and PFLC and MFHK in West coast) are potentially useful for communal rearing experiments, for monitoring the genetic effects of selection during selective breeding programs, brood stock management, and for developing markers to assist selection (Liu et al. 1999; Cross et al. 2000). Genetic variation is pivotal for populations to adapt to a changing environment or demographic events. The efficacy of an aquaculture operation or are stocking program is influenced by the genetic variation of the brood stock and associated propagation practices (Allendorf and Ryman 1987; Ferguson et al. 1991). More often than not, the variability in hatchery brood stock is found to be lower than in wild populations (as it is observed in our study) due to founder effects and inbreeding accumulated over generations, as observed in closed breeding systems where progeny become future brood stock. Genetic changes in captivity may lead to an alteration in genetic composition of the wild counter parts vi are stocking programs (Hindar et al. 1991; Crozier 1993; Carr et al. 1997).

3.3 *Intra-specific Sequence Variation*

The full recommended barcode length of 88 specimens was subjected to clustal analysis for determining variable and conserved regions. It was found that the entire alignment contained 7.5 % variable sites and 92.4 % conserved regions. Among the variable region, 5.5 % were parsimonious informative and 2 % were singleton sites. The analysis for relative codon usage revealed that the codon "CCG" which codes for Proline has been used frequently next to "UUC" (Phenylalanine). Among 650 bps analyzed, 50 segregating sites were detected. Overall the sequence variation within the barcode length was low.

4 Conclusion

The phylogram constructed with COI sequences of *L. calcarifer* revealed that geographic distributions of clades are not restricted to any particular sampling stations. Gene flow appears to have transported haplotypes between the clades from their likely origins across the sampled range. The comparative analysis of genetic diversity assessed through COI sequences between hatchery reared and wild catches of *L. calcarifer* states that the nucleotide diversity of hatchery strains was 2.7 times lesser than that of wild strains, demanding immediate action plans to restore genetic variability of *L. calcarifer* in hatchery practices in India. COI gene has more parsimonious site for genetic diversity assessment than CytB gene (data not included). We strongly recommend that COI gene could be used for further genetic analysis of *L. calcarifer* for sustainable hatchery practices.

Acknowledgments This work was partly supported by UGC fellowship. Our special thanks are due to the Dean of Centre of Advanced Studies in Marine Biology (CASMB), Annamalai University for his constant encouragement and University officials for providing the necessary facilities. Our special thanks to RGCA, Tamil Nadu and Takave farms, Maharastra for their helpfulness in sampling. We acknowledge the technical support extended by Priority Life Sciences, Coimbatore, India and Macrogen Inc., North Korea.

References

Allendorf FW, Ryman N (1987) Genetic management of hatchery stocks. In: Ryman N, Utter F (eds) Population genetics and fisheries management. University of Washington Press, Seattle, pp 141–159

Araki H, Schmid C (2010) Is hatchery stocking a help or harm? Evidence, limitations and future directions in ecological and genetic surveys. Aquaculture 308:S2–S11

Carr JW, Anderson JM, Whoriskey FG, Dilworth T (1997) The occurrence and spawning of cultured Atlantic salmon (Salmosalar) in a Canadian river. ICESJ Mar Sci 56:064–1073

Chenoweth SF, Hughes JM, Keenan CP, Lavery S (1998) When Oceans meet: a teleost shows secondary intergradation at an India-Pacific interface. Proc R Soc Lond B 265:415–420

Cross T, Dillance E, Galvin P (2000) Which molecular markers should be chosen for different specific applications in fisheries and aquaculture? National University of Ireland. http://www.ucc.ie/ucc/research/adc/molmark/index.html

Crozier W (1993) Evidence of genetic interaction between escaped farmed salmon and wild Atlantic salmon (*Salmosalar* L.) in a Northern Irish river. Aquaculture 113:19–29

Doupe R, Recher H (1999) Gene pool management of hatchery Barramundi *Lates calcarifer* for production and stock augmentation programmes. Pac Conserv Biol 5:73–75

Doupe R, Horwitz P, Lymbery A (1999) Mitochondrial genealogy of western Australian barramundi: application of inbreeding coefficient coalescent analysis for separating temporal population process. Jr Fish Boil 54:1197–1209

FAO (2011) Asia-Pacific fishery commission—Report of the thirty-first session. RAP Publication 2010/14

Ferguson MM, Ihssen PE, Hynes JD (1991) Are culture stocks of brown trout (*Salmotruta*) and rainbow trout (*Oncorhynchusmykiss*) genetically similar to their source populations? Can J Fish Aquat Sci 46:149–158

Frost LA, Evans BS, Jerry DR (2006) Loss of genetic diversity due to hatchery culture practices in Barramundi (*Lates calcarifer*). Aquaculture 261:1056–1064

Grant WS, Bowen BW (1998) Shallow population histories in deep evolutionary lineages of marine fishes: insights from sardines and anchovies and lessons for conservation. J Hered 89:415–426

Hall TA (1999) BioEdit: a user-friendly biological sequence alignment editor and analysis program for windows 95/98/NT. Nucl Acids Symp Ser 41:95–98

Hebert PDN, Cywinska A, Ball SL, Ward JR (2003) Biological identification through DNA barcode. Proc R Soc Lond B 270:313–321

Hewitt GM (2000) The genetic legacy of the Quaternary ice ages. Nature 405:907–913

Hindar K, Ryman N, Utter F (1991) Genetic effects of cultured fish on natural fish populations. Can J Fish Aquat Sci 48:945–957

John BA, Kumar CP, Lyla PS, Khan AS, Jalal KCA (2010) DNA barcoding *Lates calcarifer* (Bloch 1970). Res J Biol Sci 5(6):414–419

Keenan CP (1994) Recent evolution of population-structure in Australian barramundi, *Lates calcarifer* (Bloch)—an example of isolation by distance in one-dimension. Aust J Mar Fresh Res 45:1123–1148

Khan S, Lyla PS, John BA, Kumar CP, Murugan S, Jalal KCA (2010) DNA barcoding of *Stolephorus indicus, Stolephorus commersonnii* and *Terapon jarbua* of Parangipettai coastal waters. Biotechnology 9(3):373–377

Kumar CP, John BA, Khan AS, Lyla PS, Murugan S, Rozihan M, Jalal KCA (2011) Efficiency of universal barcode gene (cox1) on morphologically cryptic Mugilidae fishes delineation. Tr Appl Sci Res 6(9):1028–1036

Larkin MA, Blackshields G, Brown NP, Chenna R, McGettigan PA, McWilliam H, Valentin F, Wallace IM, Wilm A, Lopez R, Thompson JD, Gibson TJ, Higgins DG (2007) Clustal W and Clustal X version 2.0. Bioinformatics 23:2947–2948

Lavery S, Moritz C, Fielder DR (1996) Indo-Pacific population structure and evolutionary history of the coconut crab Birguslatro. Molec Ecol 5:557–570

Librado P, Rozas J (2009) DnaSP v5: a software for comprehensive analysis of DNA polymorphism data. Bioinformatics 25:1451–1452. doi:10.1093/bioinformatics/btp187

Liu ZJ, Li P, Argue BJ, Dunham RA (1999) Random amplified polymorphic DNA markers: usefulness for gene mapping and analysis of genetic variation of catfish. Aquaculture 174: 59–68

Macbeth M, O'Brien L, Palmer P, Lewer R, Garrett R, Wingfield M, Knibb W (2002) Selective breeding in Barramundi: technical report for the Australian Barramundi Farmers Association, Department of Primary Industries

Nei M, Jin L (1989) Variances of the average numbers of nucleotide substitutions within and between populations. Molec Biol Evol 6:290–300

Nei M (1987) Evolutonary molecular genetics. University of Columbia Press, New York

Pusey B, Kennardm M, Arthington A (2004) Freshwater fishes of North-Eastern Australia. CSIRO Publishing, Collingwood

Russell DJ, Garrett RN (1988) Movements of Juvenile Barramundi, *Lates calcarifer* (Bloch), in North-Eastern Queensland. Aust J Mar Freshwater Res 39:117–123

Saitou N, Nei M (1987) The neighbour-joining method: a new method for reconstructing phylogenetic trees. Molec Biol Evol 4:406–425

Tamura K, Dudley J, Nei M, Kumar S (2007) MEGA4: Molecular Evolutionary Genetics Analysis (MEGA) software version 4.0. Mole Biol Evol 24:1596–1599

Tave D (1993) Genetics for fish hatchery managers. Van Nostr and Reinhold, New York, 415p

Wang CM, Zhu ZY, Lo LC, Feng F, Lin G, Yang WT, Li J, Yue GH (2007) A microsatellite linkage map of Barramundi, Lates calcarifer. Genetics 175:907–915

Ward RD, Holmes BH, Yearsley GK (2008) DNA barcoding reveals likely second species of Asian sea bass (Barramundi) (*Lates calcarifer*). J Fish Biol 72:458–463

Yue GH, Li Y, Chao TM, Chao R, Orban L (2002) Novel microsatellites from Asian seabass (*Lates calcarifer*) and their application to broodstock analysis. Mar Biotechnol 4:503–511

Zhu ZY, Lin G, LoL C, Xu YX, Renee C, Yue GH (2006) Genetic analyses of Asian seabass stocks using novel polymorphic microsatellites. Aquaculture 256:167–173

Barcoding Antarctic Fishes: Species Discrimination and Contribution to Elucidate Ontogenetic Changes in Nototheniidae

E. Mabragaña, S.M. Delpiani, J.J. Rosso, M. González-Castro, M. Deli Antoni, R. Hanner and J.M. Díaz de Astarloa

Abstract Fish species richness in the Southern Ocean accounts for approximately 2 % of the world's ocean species, with more than 370 species registered and several awaiting for formal description. Here we explore on the use of DNA barcoding to discriminate fishes from Antarctic Peninsula by compiling our results and placing them into a comparative framework with other previous studies to provide a comprehensive review of available barcodes for Antarctic fishes. A total of 275 specimens, belonging to 36 different putative species were barcoded. Nearly all species exhibit unique barcodes or clusters of closely related haplotypes, and only four species lacked genetic resolution using Barcode Index Numbers (BINs). Thus, ~90 % of the species barcoded in this study could be identified at species level with accuracy using BINs. However the use of nucleotic diagnostic character allowed us to discriminate the remaining species. Compiling our results with previous studies, about 80 species inhabiting the Antarctic Peninsula were already barcoded, representing approximately 60 % of the species occurring in the area. Finally, we highlighted ontogenetic morphological traits observed in some Nothothenidae, which may lead to misidentification of juveniles. DNA Barcoding was a cornerstone element for obtaining a reliable identification of these specimens. These results are crucial for management and conservation purposes since an accurate species-level resolution of juveniles is necessary to determine nursery areas and to clarify species distributions.

Keywords DNA barcodes · Fishes · Antarctic Peninsula · Barcode Index Number · Species discrimination · Ontogenetic variation

E. Mabragaña (✉) · S.M. Delpiani · J.J. Rosso · M. González-Castro · M. Deli Antoni · J.M. Díaz de Astarloa
Laboratorio de Biotaxonomía Morfológica y Molecular de Peces,
Instituto de Investigaciones Marinas y Costeras,
Consejo Nacional de Investigaciones Científicas y Técnicas (CONICET),
Universidad Nacional de Mar del Plata, Mar del Plata, Argentina
e-mail: emabraga@mdp.edu.ar

R. Hanner
Biodiversity Institute of Ontario, Department of Integrative Biology,
University of Guelph, Guelph, Canada

S. Trivedi et al. (eds.), *DNA Barcoding in Marine Perspectives*,
DOI 10.1007/978-3-319-41840-7_14

1 Introduction

Fish species richness in the Southern Ocean accounts for approximately 2 % of the world's ocean species, with more than 370 species registered and several awaiting for formal description (Duhamel et al. 2014). The Notothenioidei (with representatives from six families), liparids and zoarcids are the dominant components of the Antarctic fish fauna (Eastman 2005), being snail fishes (Liparidae) the most speciose Antarctic fish family.

Eastman (2005) suggests that after a century of ichthyofaunal research in Antarctic waters, the fauna is fairly well known. However, not all groups are completely understood as revealed by the number of species recently described for notothenioids and other highly diversified fish families such as, Liparidae and Zoarcidae (Duhamel et al. 2010, 2014; Stein 2012). Moreover, although several taxonomic studies have been carried out on Antarctic fishes, the number of species is probably underestimated as some taxa and regions have not been deeply explored. Indeed, the asymptotic level in species richness has not yet been reached (Duhamel et al. 2014). Several families (Rajidae, Muraenolepididae, Harpagiferidae, among others) still require thorough taxonomic revision due to the lack of detailed species diagnoses or because of misidentifications in the scientific record (Duhamel et al. 2014). On the other hand, most taxonomic studies do not usually cover different ontogenetic stages, lacking information about the morphological identification of juveniles.

The Antarctic fish fauna has a remarkably high level of endemism (c.a 90 % of the species recorded are found only in Antarctic waters, Eastman 2005; Smith et al. 2012). This feature, not restricted to fishes, increases the concern about the potential influence of global warming, habitat loss, UV exposure and ocean acidification in the southern ocean ecosystem (Clarke et al. 2005; Thatje 2005; Aronson et al. 2009; Turner et al. 2009; Cook et al. 2005, 2010; Constable et al. 2014), specially because some changes are already visible in the Antarctic Peninsula (Steig et al. 2009; Naish et al. 2009). Therefore, a more comprehensive knowledge of their biodiversity is required.

Over the last few decades several molecular studies have been conducted on Antarctic fishes in order to support morphological research. Some of these works includes the use of DNA Barcoding as a standarized molecular taxonomic approach (Lautrédou et al. 2010; Smith et al. 2008, 2011, 2012; Rock et al. 2008; Rey et al. 2011; Dettai et al. 2011; Duhamel et al. 2010). The Fish Barcode of Life initiative (FISH-BOL; Ward et al. 2009) seeks to establish a mitochondrial 5′ cytochrome *c* oxidase subunit I (COI) reference sequence library for the molecular identification of fishes worldwide, following a common protocol that includes links to voucher specimens (Steinke and Hanner 2010). The use of DNA barcoding in fishes can facilitate subsequent species identification by non-specialists, help highlight specimens that represent a range expansion of known species, flag previously unrecognized (e.g. cryptic) species, and enable identifications where traditional methods are not applicable (e.g. fillets, eggs and larvae). As of July 2010, nearly 7800 fish species had been barcoded, including at least one species for ~90 % of all families (Becker et al. 2011). Five years later, this number has risen to nearly 11,000 species (www.fishbol.org).

Many studies have demonstrated the usefulness of DNA barcoding to discriminate marine and freshwater fishes around the world (Pegg et al. 2006; Steinke et al. 2009; Ward and Holmes 2007; Ward et al. 2008; Huber et al. 2008; Valdez-Moreno et al. 2009; Mabragaña et al. 2011; Rosso et al. 2012; Knebelsberger et al. 2014, among others). Rock et al. (2008) were the first to provide barcodes from Antarctic fishes, analyzing DNA barcoding of 34 putative species representing seven different families, collected in the Scotia Sea. Duhamel et al. (2010) conducted a survey in the eastern sector of the Southern Ocean and provide the first molecular data (COI) for 13 species of liparids, allowing the identification of most species. Lautrédou et al. (2010) analysed the boundaries between 12 species of *Trematomus* collected in different sectors of the southern ocean. Rey et al. (2011) analysed the difference between two species of *Gymnodraco* off Terre Adélie based on morphology and DNA barcoding. Dettai et al. (2011), provided barcodes from 57 species in the Eastern part of the Antarctic continental shelf. Finally, Smith et al. (2012) provided an overview of barcode records for the Ross Sea fishes and a comparison of genetic divergence within the Ross Sea and between this and other regions of the Southern Ocean. They found that DNA barcoding could discriminate 87.5 % of Antarctic species. Nevertheless, these molecular studies showed incomplete species discrimination within some representatives of Notothenidae, Artedidraconidae and Liparidae.

As part of a global project conducted by Consejo Nacional de Investigaciones Científicas y Técnicas (CONICET) on the systematics and biology of Antarctic organisms, DNA barcodes of fish species around Antarctic Peninsula and adjacent Islands were obtained and their concordance with traditional morphological identification was explored. The overarching objetive of this study was to further extend the use of DNA barcoding to discriminate Antarctic fishes by compiling the results of our survey and placing them into a comparative framework with other previous studies to provide a comprehensive review of available barcodes for Antarctic fishes. We also highlighted ontogenetic morphological traits observed in some species of Notothenidae, which may lead to misidentification of juveniles through the use of taxonomical keys.

2 Materials and Methods

2.1 Study Area

The sampling region encompasses part of the Argentine Antarctic Sector including the South Shetland Islands, the South Orkney Islands and the north of the Antarctic Peninsula (Fig. 1). Around the South Shetland Islands the shelf break lies at depths between 225 and 380 m in the north-east (Elephant Island) and between 250 and 450 m in the rest of the archipelago. The Islands are located along 481 km of shelf in a NE–SW direction. Around the South Orkney Islands the shelf is very narrow to the north and a broad plain to the south, breaking mainly below the 500 m isobath (Acosta et al. 1989; Jones 2000). The Antarctic Peninsula is separated from the

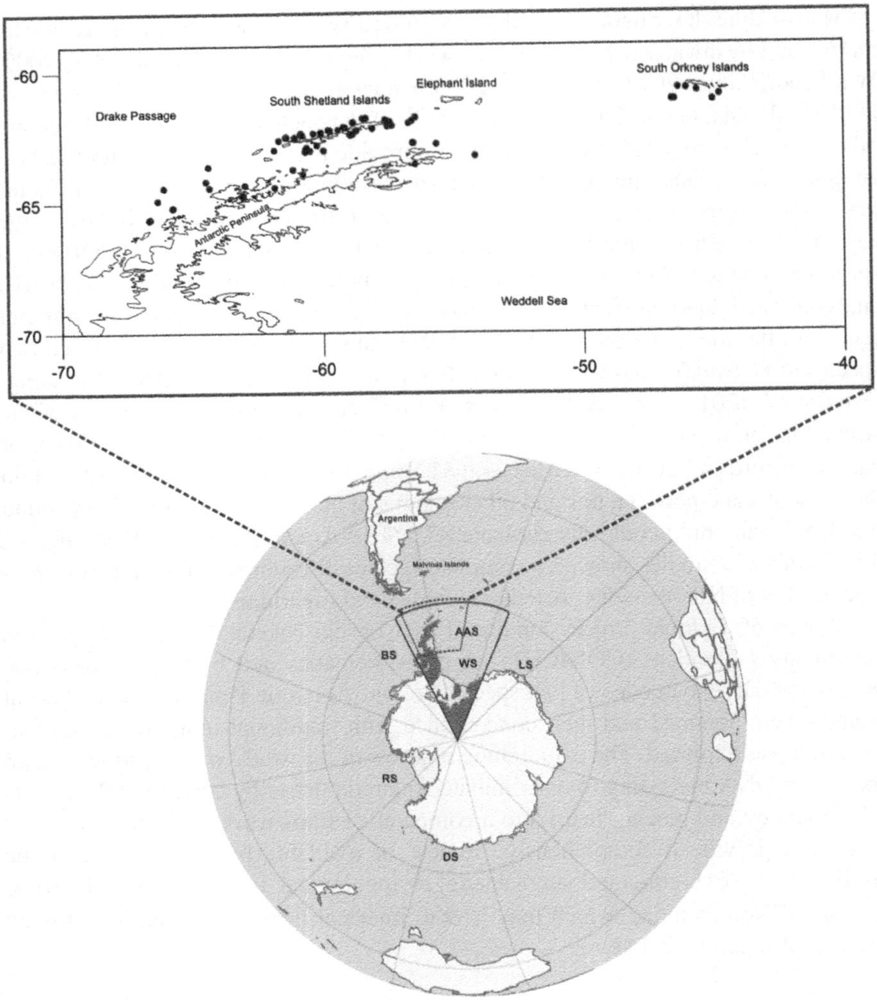

Fig. 1 Collection sites (*black circles*) for specimens examined in this study. *AAS* Argentine Antarctic Sector, *BS* Bellingshausen Sea, *DS* Durmont d'Urville Sea, *LS* Lazarev Sea, *RS* Ross Sea, *WS* Weddel Sea

Shetland Islands by the deep waters of Bransfield Strait (Acosta et al. 1989; Kock et al. 2000). The bottom topography in the west and north of the Antarctic Peninsula might be described as a shelf surrounded by islands, communicating with the open sea by troughs of varying depths (Barrera-Oro 2002). The circulation in the region is controlled by an eastward component of the Antarctic Circumpolar Current which balances the waters coming from the Antarctic Peninsula, the Weddell Sea and waters from the Bellingshausen Sea (Gordon 1988; Barrera-Oro 2002; Turner et al. 2009). In the study area 131 species have been reported (Gon and Heemstra 1990; Andriashev 1998; Matallanas and Pequeño 2000; La Mesa et al. 2002; Kock 2005; Chernova 2006; Matallanas 2009, 2011; Balushkin and Prirodina 2010; Balushkin 2012).

2.2 *Sample Collection*

Overall 5297 specimens belonging to 40 species and 9 families (Artedidraconidae, Bathydraconidae, Channichthyidae, Harpagiferidae, Myctophidae, Nototheniidae, Liparidae, Rajidae and Zoarcidae) were collected on board the Oceanographic Vessel "Puerto Deseado" of CONICET during summers of 2011, 2012 and 2013. This number represents almost the 30 % of the species reported for the surveyed area. A total of 70 stations were performed (Fig. 1). Fish were collected using two bottom trawls, including a shrimp net (50 mm mesh in the wings, and 20 mm in the cod end; vertical height 1 m, horizontal opening 4 m), and a bottom trawl net (135 mm mesh in the wings, and 60 mm in the cod end; vertical height 3.7 m, horizontal opening 10 m). We also used "long liner gears" in shallow areas. Specimens were identified on board using diagnostic keys (and reexamined in laboratory when necessary, after results of molecular analysis) and measured (total and standard length). Vouchers were morphologically identified following the identification reliability level 2 according to the Fish-BOL collaborator's protocol (Steinke and Hanner 2010): "specimen identified by a trained identifier who had prior knowledge of the group in the region or used available literature to identify the specimen". We followed Eschmeyer (2015) for species names and its higher classification.

2.3 *DNA Analysis*

A portion of tissue was taken from representatives of each species and preserved in 96 % ethanol for subsequent molecular analysis. The voucher specimens were labelled, photographed, formalin fixed (with further alcohol long-term preservation) and deposited as vouchers in the fish collection of "Instituto de Investigaciones Marinas y Costeras (IIMyC)- CONICET- Universidad Nacional de Mar del Plata", Argentina.

DNA extraction, polymerase chain reaction (PCR), and sequencing of the 5′ region of the COI gene were performed following standard DNA barcoding protocols (Ivanova et al. 2006) coupled with primers and primer cocktails developed for fishes (Ward et al. 2005; Ivanova et al. 2007). Extraction and amplification were performed in two International Barcode of Life reference Laboratories of CONICET in Argentina, one located at the Museo Argentino de Ciencias Naturales and the other one at the IIMyC. Sequencing was performed in Advanced Analysis Center´s Genomics Facility (College of Biological Sciences, University of Guelph, Ontario Canadá) and in the Canadian Centre for DNA Barcoding (CCDB) at the Biodiversity Institute of Ontario, (University of Guelph, Ontario, Canada).

Amplification of the 5′ region of COI, corresponding to base positions 6474 to 7126 of the *Danio rerio* mitochondrial genome (Broughton et al. 2001), was first attempted using FF2d_t1/FR1d_t1 primer combination and C_FishF1t1/C_FishR1t1 primer cocktails (Ivanova et al. 2006). The primer combinations C_FishF1t1 and C_FishR1t1 both contained two primers (FishF2_t1/VF2_t1 and FishR2_ t1/FR1d_t1, respectively). PCR reactions were performed in 96-well plates. The reaction master mix consisted of 825 μl water, 125 μl 106 buffer, 62.5 μl MgCl2 (25 mM), 6.25 μl dNTP (10 mM), 6.25 μl each primer (0.01 mM) and 6.25 μl Taq DNA polymerase (5 U/μl). This mixture was prepared for each plate, and each well contained 10.5 μl of solution and 2 μl of genomic DNA. The PCR reaction profile was comprised of an initial step of 2 min at 95 °C, and 35 cycles of 30 s at 94 °C, 40 s at 52 °C, and 1 min at 72 °C, with a final extension at 72 °C for 10 min. For specimens that failed to amplify using the primer combinations above, the primer combinations C_VF1LFt1/C_ VR1LRt1 (Ivanova et al. 2007) consisting of VF1_t1/VF1d_t1/LepF1_t1/VFli_t1 and VR1_t1/VR1d_t1/LepR1_t1/VRli_t1 primer sets respectively were tried. All primers were appended with M13 tails to facilitate sequencing.

Amplicons were visualized on a 2 % agarose E-GelH 96-well system (Invitrogen). Sequencing reactions applied M13 forward and reverse primers using the BigDyeH Terminator v.3.1 Cycle Sequencing Kit (Applied Biosystems Inc.), and the reaction profile was comprised of an initial step of 2 min at 96 °C and 35 cycles of 30 s at 96 ° C, 15 s at 55 °C, and 4 min at 60 °C. Products were directly sequenced using an ABI 3730 capillary sequencer, according to manufacturer's instructions.

2.4 Data Analysis

DNA sequences were aligned with SeqScape v.2.1.1 software (Applied Biosystems, Inc.) and further double-checked visually. Barcode sequences were subjected to distance-based, diagnostic character (Maximum Likelihood) and spectral clustering (BIN) analyses.

Sequence divergences were calculated using the Kimura two parameter (K2P) distance model (Kimura 1980), and Neighbor-joining (NJ) trees of K2P distances were created to provide graphic representations of divergence between species, using the software MEGA v5.0 (Tamura et al. 2011). The *p* distance model was also tested. Differences in distance estimates and tree topology between *p* distance and K2P models were minimal. Consequently, K2P model was chosen for comparison purposes, as it is commonly used for describing differences among species in DNA barcoding studies. Nevertheless, K2P has been described as a poorly fitting model at the species level (Collins et al. 2012). Moreover, distance-based models erase all character-based information (DeSalle 2006). As a result, K2P/NJ clusters of taxonomical units with either high intra-specific or low interspecific divergences were more closely inspected by a subsequent character-based analysis. For this purpose, the best nucleotide substitution model was selected to perform a maximum likelihood (ML) analysis using MEGA v5.0 (Tamura et al. 2011).

The Barcode Index Number (BIN) was used to estimate the number of species directly from the barcode records and congruence of these estimates with the distance based and character based approaches were evaluated. BINs is "an online framework that clusters barcode sequences algorithmically, generating a web page for each cluster. Since clusters show high concordance with species, BINs can be used to verify species identifications as well as document diversity when taxonomic information is lacking" (see boldsystems.org and Ratnasingham and Hebert 2013 for further details on BINs). The public library of BINs in BOLD, was also used to scrutinize whether the literature about DNA Barcoding of Antarctic fishes had incorporated different BINs under this unique nominal taxa (i.e. to ascertain taxonomic conflicts among barcode studies conducted by different teams of researchers). For those species sharing the same BIN, we additionally explored their COI sequences for diagnostic characters with the tool available in BOLD.

The nearest-neighbour distance (NND) distribution analysis, that is, the minimum genetic distance between a species and its closest neighbour-species, was also performed. BOLD was also used to explore the genetic divergence between barcode records of given species of the Argentine Antarctic Peninsula with other available barcode sequences for the corresponding species from other sector of Antarctica.

All sequence assemblies, electropherogram (trace) files, primer sequences and specimen provenance data were deposited in the "Argentinean Antarctic Fishes phase I" (Code AAFI) on the Barcode of Life Database (BOLD, Ratnasingham and Hebert 2007). This included digital images of morphological voucher specimens, sex and ontogenetic stage (juvenile or adult), total and standard length as well as GPS coordinates for all specimen collection localities. Sequence data are also available on GenBank (Accession numbers pending).

3 Results and Discussion

A total of 275 specimens, belonging to 36 different putative species from 8 families, including Artedidraconidae, Bathydraconidae, Channichthyidae, Harpagiferidae, Myctophidae, Nototheniidae, Liparidae and Zoarcidae were successfully barcoded (Table 1). Representatives of family Arhyncobatidae (*Bathyraja maccaini* and *B. murrayi*), as well as some species within families Bathydraconidae (*Bathydraco marri*), Liparidae (*Paraliparis trilobodon*) and Zoarcidae (*Pachycara brachycephalum*) did not amplify following the same protocols. No stop codons, insertions or deletions were found in any of the amplified sequences, showing that all of them constitute functional mitochondrial COI sequences. Four species were represented by only one sequence. The analysis of COI sequences, with variable levels of divergence, revealed that interspecific divergence (D) was relatively high except in some species within Artedidraconidae, Liparidae, Channichthidae and Zoarcidae ($D < 3$ %) (Fig. 2 and Table 2). The K2P genetic distances averaged 0.27 % within species, 8.96 % within genera, and 14.1 % within families (Table 3). The full K2P/NJ tree also showed that nearly all species exhibit unique barcodes or clusters of closely related haplotypes. The spectral cluster (BIN) did not agree with current taxonomic classification of our specimens in 100 % of cases. BIN analysis

Table 1 Species barcoded during the surveys 2011–2013 on Antarctic Peninsula

Family	Species	Nº	Range TL (mm)
Artedidraconidae	*Artedidraco skottsbergi* (Lönnberg 1905)	14	45–95
	Pogonophryne scotti (Regan 1914)	5	350
	Pogonophryne permitini (Andriashev 1967)	1	166
	Dolloidraco longedorsalis (Roule 1913)	1	155
Bathydraconidae	*Gymnodraco acuticeps* (Boulenger 1902)	4	285
	Parachaenichthys charcoti (Vaillant 1906)	12	118–496
	Prionodraco evansii (Regan 1914)	6	109–189
Channichthyidae	*Chaenocephalus aceratus* (Lönnberg 1906)	7	232
	Chionodraco rastrospinosus (De Witt and Hureau 1979)	5	255–387
	Chaenodraco wilsoni (Regan 1914)	6	163–320
	Champsocephalus gunnari (Lönnberg 1905)	2	321–445
	Cryodraco antarcticus (Dollo 1900)	2	140–343
	Pagetopsis macropterus (Boulenger 1907)	4	188–237
	Pseudochaenichthys georgianus (Norman 1937)	3	177–369
Harpagiferidae	*Harpagifer antarcticus* (Nybelin 1947)	9	48–117
Myctophidae	*Electrona antárctica* (Günther 1878)	1	81
Nototheniidae	*Gobionotothen gibberifrons* (Lönnberg 1905)	13	70–395
	Notothenia coriiceps (Richardson 1844)	6	360–411
	Notothenia rossii (Richardson 1844)	5	277–359
	Lepidonotothen squamifrons (Günther 1880)	21	223–295
	Lindbergichthys nudifrons (Lönnberg 1905)	36	58–182
	Nototheniops larseni (Lönnberg 1905)	20	53–203
	Pleuragramma antárctica (Boulenger 1902)	11	167–188
	Trematomus newnesi (Boulenger 1902)	4	133–196
	Pseudotrematomus scotti (Boulenger 1907)	13	60–180
	Pseudotrematomus hansoni (Boulenger 1902)	1	222–371
	Pseudotrematomus eulepidotus (Regan 1914)	9	130–237
	Pseudotrematomus bernacchii (Boulenger 1902)	4	112–136
Liparidae	*Paraliparis antarcticus* (Regan 1914)	4	199
	Paraliparis sp.	4	61–104
	Careproctus sp.	7	36–93
Zoarcidae	*Ophtalmolycus* cf. *amberensis* (Tomo, Marschoff and Torno 1977)	7	269
	Lycenchelys nigripalatum (DeWitt and Hureau 1979)	7	135–196
	Lycenchelys tristichodon (DeWitt and Hureau 1980)	1	269
	Lycenchelys sp.	1	146–181

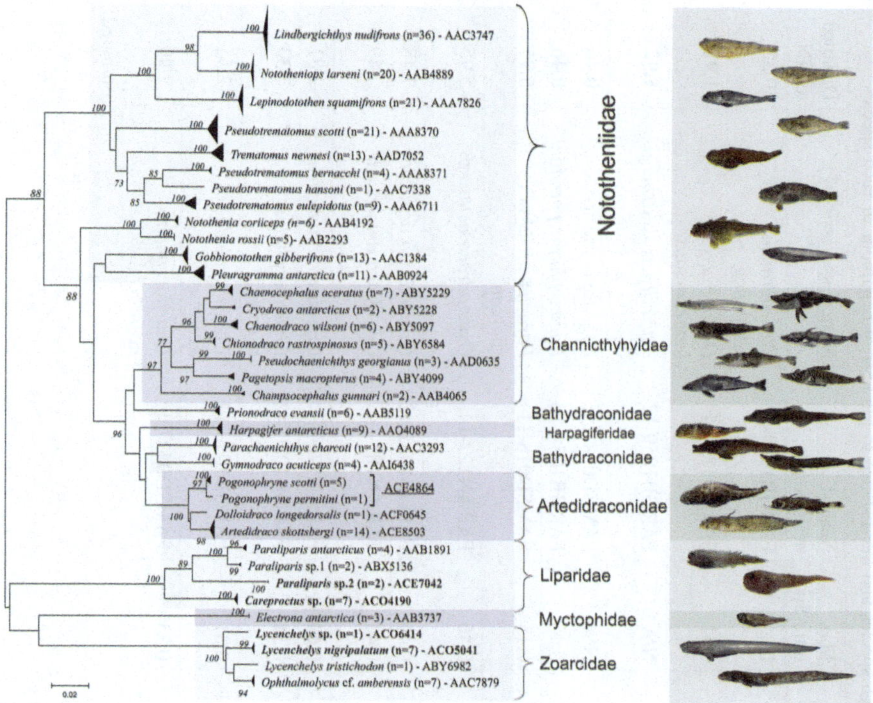

Fig. 2 Neighbour-Joining tree based on K2P distances. *Numbers after taxa* indicate the corresponding BIN. *Solid triangles* represent clusters of multiple specimens, with the vertical dimension proportional to the number of specimens, and the horizontal depth proportional to the genetic variation within that cluster. *Number at nodes* represent bootstrap values, (only values greater than 70 are given). Underlined BIN include more than one species. In *bold* those species that represent new BINs for BOLD. Figures contain representatives of each genera

recognized 35 taxonomic units from the 36 putative species (Fig. 2). Indeed albeit forming distinctive clusters in the K2P/NJ tree, *Pogonophryne scotti* and *P. permitini* showed low (0.6 %) genetic divergence (Fig. 2) and were included in the same BIN. Our work yielded barcodes for 13 species of the Antarctic Peninsula region for the first time. Sequence divergence between these specimens and those (public or published) from other regions were compared (Table 4). Most widely distributed species showed little or no sequence divergences among regions, as was observed by Smith et al. 2012.

3.1 *Nototheniidae*

This family is represented in the Antarctic waters by c.a 38 species, from which 21 were reported for the northern Antarctic Peninsula and adjacent waters (AP).

Table 2 Barcode gap analysis showing the distribution of distances within each species and the distance to the nearest neighbour of each species

Order	Family	Species	Mean intra-Sp	Max intra-Sp	Nearest neighbour	Nearest species	Distance to NN
Myctophiformes	Myctophidae	*Electrona antarctica*	0	0	FARGB1134-12	*Pleuragramma antarctica*	20.35
Perciformes	Artedidraconidae	*Artedidraco skottsbergi*	0.26	0.62	FARAN183-14	*Dolloidraco longedorsalis*	**1.92**
Perciformes	Artedidraconidae	*Dolloidraco longedorsalis*	N/A	N/A	FARGB1059-12	*Artedidraco skottsbergi*	**1.92**
Perciformes	Artedidraconidae	*Pogonophryne permitini*	N/A	N/A	FAMSD064-11	*Pogonophryne scotti*	**0.46**
Perciformes	Artedidraconidae	*Pogonophryne scotti*	0.25	0.46	FARGB1054-12	*Pogonophryne permitini*	**0.46**
Perciformes	Bathydraconidae	*Gymnodraco acuticeps*	0.05	0.15	FAMSD018-11	*Parachaenichthys charcoti*	5.95
Perciformes	Bathydraconidae	*Parachaenichthys charcoti*	0.13	0.46	FARAN127-14	*Gymnodraco acuticeps*	5.95
Perciformes	Bathydraconidae	*Prionodraco evansii*	0.2	0.46	FARGB1031-12	*Gymnodraco acuticeps*	7.48
Perciformes	Channichthyidae	*Chaenocephalus aceratus*	0.19	0.46	FARAN185-14	*Cryodraco antarcticus*	2.49
Perciformes	Channichthyidae	*Chaenodraco wilsoni*	0.21	0.64	FARAN176-14	*Cryodraco antarcticus*	2.81
Perciformes	Channichthyidae	*Champsocephalus gunnari*	0.31	0.31	FARGB1050-12	*Chaenocephalus aceratus*	7.33
Perciformes	Channichthyidae	*Chionodraco rastrospinosus*	0.15	0.31	FARGB1050-12	*Chaenocephalus aceratus*	2.51
Perciformes	Channichthyidae	*Cryodraco antarcticus*	0.33	0.33	FAMSD042-11	*Chaenocephalus aceratus*	2.49

(continued)

Table 2 (continued)

Order	Family	Species	Mean intra-Sp	Max intra-Sp	Nearest neighbour	Nearest species	Distance to NN
Perciformes	Channichthyidae	*Pagetopsis macropterus*	0.31	0.62	FARAN154-14	*Pseudochaenichthys georgianus*	5.11
Perciformes	Channichthyidae	*Pseudochaenichthys georgianus*	0.11	0.16	FARAN054-12	*Pagetopsis macropterus*	5.11
Perciformes	Harpagiferidae	*Harpagifer antarcticus*	0.3	0.62	FARGB1095-12	*Artedidraco skottsbergi*	7.35
Perciformes	Nototheniidae	*Gobionotothen gibberifrons*	0.28	0.77	FARGB1136-12	*Pleuragramma antarctica*	9.38
Perciformes	Nototheniidae	*Lepidonotothen squamifrons*	0.09	0.46	FARGB627-12	*Nototheniops larseni*	8.98
Perciformes	Nototheniidae	*Lindbergichthys nudifrons*	0.22	0.62	FARGB627-12	*Nototheniops larseni*	5.91
Perciformes	Nototheniidae	*Notothenia coriiceps*	0.2	0.46	FARGB980-12	*Notothenia rossii*	3.74
Perciformes	Nototheniidae	*Notothenia rossii*	0	0	FARGB839-12	*Notothenia coriiceps*	3.74
Perciformes	Nototheniidae	*Nototheniops larseni*	0.24	0.77	FARGB1020-12	*Lindbergichthys nudifrons*	5.91
Perciformes	Nototheniidae	*Pleuragramma antarctica*	0.39	0.94	FARGB838-12	*Gobionotothen gibberifrons*	9.38
Perciformes	Nototheniidae	*Pseudotrematomus bernacchii*	0.08	0.15	FARGB1071-12	*Pseudotrematomus hansoni*	4.81
Perciformes	Nototheniidae	*Pseudotrematomus eulepidotus*	0.54	1.24	FARGB1051-12	*Pseudotrematomus bernacchii*	5.63
Perciformes	Nototheniidae	*Pseudotrematomus hansoni*	N/A	N/A	FARGB1051-12	*Pseudotrematomus bernacchii*	4.81
Perciformes	Nototheniidae	*Pseudotrematomus scotti*	0.43	1.09	FARGB1071-12	*Pseudotrematomus hansoni*	9.43

(continued)

Table 2 (continued)

Order	Family	Species	Mean intra-Sp	Max intra-Sp	Nearest neighbour	Nearest species	Distance to NN
Perciformes	Nototheniidae	*Trematomus newnesi*	0.69	1.55	FAMSD036-11	*Pseudotrematomus bernacchii*	7.31
Perciformes	Zoarcidae	*Lycenchelys nigripalatum*	0.13	0.46	FARGB963-12	*Lycenchelys tristichodon*	2.66
Perciformes	Zoarcidae	*Lycenchelys sp.*	N/A	N/A	FARGB963-12	*Lycenchelys tristichodon*	2.7
Perciformes	Zoarcidae	*Lycenchelys tristichodon*	N/A	N/A	FAMSD028-11	*Ophthalmolycus sp.*	2.35
Perciformes	Zoarcidae	*Ophthalmolycus* sp.	0.17	0.46	FARGB963-12	*Lycenchelys tristichodon*	2.35
Scorpaeniformes	Liparidae	*Careproctus georgianus*	0.05	0.2	FARGB971-12	*Paraliparis* sp.	8.27
Scorpaeniformes	Liparidae	*Paraliparis antarcticus*	0.15	0.31	FARAN129-14	*Paraliparis* sp.	**1.55**
Scorpaeniformes	Liparidae	*Paraliparis* sp.	4.51	6.81	FARGB1046-12	*Paraliparis antarcticus*	**1.55**

Where the species is a singleton, N/A is displayed for intra-specific values. Distances are highlighted if the nearest neighbour is less than 2 % divergent, or when the distance to the nearest neighbour is less than the max intra-specific distance

Table 3 Summary of distribution of sequence divergence at each taxonomic level

	n	Taxa	Comparisons	Min Dist (%)	Mean Dist (%)	Max Dist (%)	SE Dist (%)
Within species	270	30	1877	0	0.27	6.81	0
Within Genus	69	5	409	0.46	8.96	11.6	0.01
Within Family	263	6	11545	1.92	14.1	23.7	0

Table 4 Intraspecific K2P divergences within specimens from Antarctic Peninsula and adjacent Islands (AP) and among representative of AP and other regions of the Southern Ocean (OR)

Species	D (%) AP	D (%) OR
Electrona antárctica	0	0–0.78
Pogonophryne permittini	–	No specimens
Dolloidraco longedorsalis	–	0.16–0.48
Careproctus sp.	0–0.2	No specimens
Paraliparis antarcticus	0–0.31	0–0.64
Paraliparis sp1.*	–	0.93–1.1 %
Gymnodraco acuticeps	0–0.15	0–0.33 %
Chaenodraco wilsoni	0–0.64 %	0–0.77 %
Cryodraco antarcticus	–	0.16–0.33 %
Pagetopsis macropterus	0–0.62 %	0.31–0.46 %
Lycenchelis tristichodon	–	0.62–0.8 %
Ophtalmolycus cf. *amberensis*	0–0.46 %	0–0.48 % Ross Sea/1.63–2.18 % AAT
Pleuragramma antárctica	0–0.94 %	0–1.1 %

Data included only species that were not previously barcoded in AP. *AAT* Australian Antarctic Territory

Recently Duhamel et al. (2014) proposed the inclusion of several families (Artedidraconidae, Bathydraconidae, Channichthyidae, Harpagiferidae) as sub-families within Nototheniidae, based on previous molecular analysis. Although this seems to be a reasonably well-supported hypothesis concerning their classification, we followed the accepted classification presented by Eschmeyer (2015) and treated each separately. Several barcoding studies were made on Antarctic representatives of this family (Rock et al. 2008; Lautrédou et al. 2010; Dettai et al. 2011; Smith et al. 2012). These studies included 22 species of genera *Cryothenia*, *Dissostichus*, *Gobionotothen*, *Lepidonotothen*, *Lindbergichthys*, *Notohtenia*, *Nototheniops*, *Pleuragramma*, *Pseudotrematomus* and *Trematomus*, showing that COI provided effective species-level discrimination for nearly all species. The exception appeared within the genus *Trematomus* in which lack of COI divergence was reported for

T. loennbergi and *T. lepidorhinus* (Lautrédou et al. 2010; Dettai et al. 2011; Smith et al. 2012). Public data, available on BOLD, reveal that BINs are congruent with all previous results, placing all species, except *T. loennbergi* and *T. lepidorhinus* into different and exclusive BINs. In the present study specimens belonging to 12 species of Nototheniidae were barcoded (Table 1). All these species were previously barcoded. However, no barcode sequences of *Pleuragramma antarctica* from AP were published yet (Rock et al. 2008; Lautrédou et al. 2010; Dettai et al. 2011; Smith et al. 2012). *Pleuragramma antarctica* showed little or no sequence divergences among AP and other regions, sharing haplotypes (similarity ranged 98.9 to 100 %), as was observed by Smith et al. (2012) for other Notothenids.

3.1.1 The Application of Barcode to Elucidate Ontogenetic Changes in Diagnostic Characters

The diagnostic features used for species identification within the family Nototheniidae rely on a specimen range in size from about 10 cm to over 2 m (DeWitt et al. 1990). In that key, the size range of fish used for the diagnosis of some species is not specified. Ontogenetic changes are discussed only for some structures such as otoliths (DeWitt et al.1990) or in body morphometrics for punctual species (Piacentino and Barrera-Oro 2009). In this respect, we observed that juveniles of several species of Nototheniidae lacked some diagnostic characteristics (e.g. pattern of head squamation) commonly employed in available diagnostic keys as well as species descriptions. Therefore, these juveniles could be erroneously assigned to a different species or even different genus. In this respect, DNA barcodes were useful to clarify the inconsistence and allowed us to properly match each juvenile with the corresponding species.

Notably, we found that juveniles a priori identified as *Gobionotothen* sp., were assigned to *Pseudotrematomus scotti* by BIN analysis after the barcode sequences were obtained. According to the key to genera of Nototheniidae, presented in Dewitt et al. (1990), the presence of scales in the preorbital separates (among others) *Pseudotrematomus* (at the time of publication valid as *Trematomus*) from *Gobionotothen*, being naked in the last genus. The morphological description of *P. scotti* further characterized this species by a fully scaled occipital and interorbital regions, as well as cheeks and opercles. We noted that in juveniles (51–56 mm TL) of *Pseudotrematomus scotti* the head is mostly naked with the sole presence of a few scales in the occipital region (Fig. 3). On the other hand we have note that juveniles (61–74 mm TL) of *Lepidonotothen squamifrons*, lacked scales on dorsal part of eyes while they are present in adults (Fig. 4). Finally, some juveniles identified as *Trematomus* sp. (63–68 mm TL) corresponded to *Nototheniops larseni* (at the time of publication valid as *Lepidonotothen larseni*). The incongruence here was that these specimens presented the pre-orbital region naked whereas this part of the body is scaled in adults (Fig. 5). Diagnostic meristic counts (dorsal-fin, anal-fin and pectoral-fin rays and tubular scales along the upper lateral line) have overlapping ranges between some genera, therefore the presence/absence of scales in

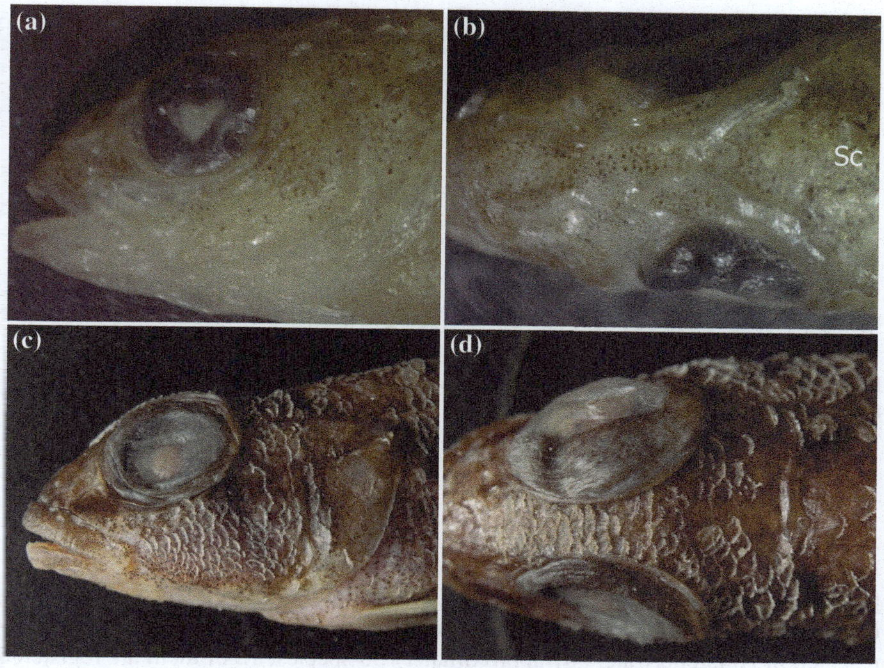

Fig. 3 Ontogenetic changes observed in specimens of *Pseudotrematomus scotti*. Lateral and dorsal view of juveniles **a** and **b** and adults **c** and **d**

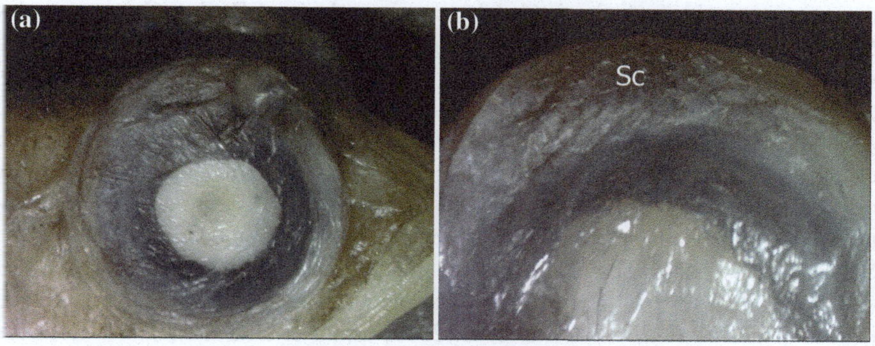

Fig. 4 Ontogenetic changes observed in dorsal squamation of eyes in specimens of *Lepidonotothen squamifrons*. **a** Juveniles and **b** adults. *Sc* Scales

preorbital and occipital regions are the diagnostic features that primarily distinguish them (Dewitt et al. 1990) and as was mentioned this characteristic was variable among juveniles and adults.

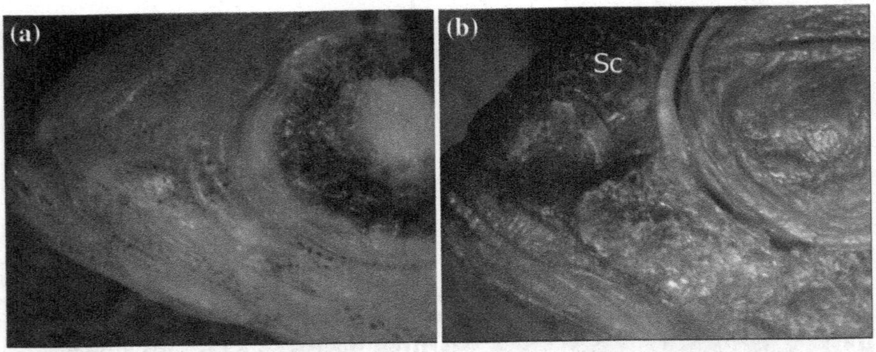

Fig. 5 Ontogenetic changes observed in specimens of *Nototheniops larseni.* (**a**) juveniles **b** Adults. *Sc* Scales

3.2 *Myctophidae*

Myctophids constitutes the dominant fish family of the mesopelagic and bathypelagic Antarctic waters in terms of their species richness, abundance and biomass (Donnelly et al. 1990; Donnelly and Torres 2008; Koubbi et al. 2011). It is represented in Antarctic waters by approximately 36 species, 12 of them are registered in AP (Gon and Heemstra 1990). At least, 9 species of lanternfishes were previously barcoded (Rock et al. 2008; Dettai et al. 2011; Smith et al. 2012). These studies showed that COI provides effective species-level discrimination and also highlighted possible new species (Smith et al. 2012). In turn, BINs mostly supported these findings with each species possessing an exclusive BIN, but *Gymnoscopelus bolini* which harbored different BINs, suggesting the existence of potentially cryptic species (Smith et al. 2012). During our surveys only *Electrona antarctica* was collected and barcoded. *Electrona antarctica* is typically found south of the Antarctic Polar Front (APF). This species has a wide distribution in the southern ocean. Biogeographic data indicate that *Electrona antarctica* has a circumpolar distribution mainly between the Antarctic Slope Front (as delimited by the continental 500 m isobath) and the APF, although small specimens can be taken in the Sub-Antarctic Zone. The southernmost record for the species is at 74.67 °S (Duhamel et al. 2014). There are published barcode records from all Antarctic regions (Rock et al. 2008; Dettai et al. 2011; Smith et al. 2012). Those from the west Antarctic correspond to the Georgias Islands (Rock et al. 2008). Our results showed that specimens from AP presented shallow intra-specific divergences with those from other regions, sharing haplotypes (COI Similarity 99.21–100 %), as was observed by Smith et al. 2012.

3.3 *Artedidraconidae*

Artedidracons, known as plunderfishes, are the less well-known notothenioid fishes (Eakin et al. 2009). This family is composed of approximately 26 species, 10 of which have been recorded in the AP (Eakin 1990). The genus *Pogonophryne* has a circum-Antarctic distribution with some species extending as far north as the South Orkney Islands, and in depths ranging from 100 m to more than 2500 m (Duhamel et al. 2014). The genus comprises nearly 70 % of the diversity within the family. Currently, five species groups are recognize within this genus based on differences in spotting patterns and meristics: the *P. mentella* group, the *P. scotti* group, the *P. barsukovi* group, the *P. marmorata* group and the *P. albipinna* group (Eakin et al. 2009). A molecular phylogenetic analysis of this family was provided by Eakin et al. (2009), showing low genetic divergences among species and limited phylogenetic resolution among the five species groups. Species of the genus *Pogonophyryne* are especially difficult to identify because there is scarce meristic separation (Eakin et al. 2009). The mental barbel in *Pogonophryne*, which has been used to distinguish species appears to be highly variable in shape within the various species (Eakin et al. 2001) and makes species identification difficult (Duhamel et al. 2014).

Twenty species of plunderfishes were previously barcoded showing a lack of resolution in many of them (Rock et al. 2008; Dettai et al. 2011; Smith et al. 2012). Based on relatively few specimens, Rock et al. (2008) found that barcodes failed to discriminate *Artedidraco loennbergi* from *A. skottsbergi*. In contrast, Dettai et al. (2011), using more than 100 specimens from 7 species, found that species formed unique molecular clusters excepting some *Pogonophryne* species. However, almost all interspecific differences were smaller than 2 % and several less than 0.8 %. Finally, Smith et al. (2012), analysed barcode data from 4 species of Artedidraconidae and found low divergences among species and highlighted a lack of region-specific haplotypes. Moreover, other molecular markers showed low genetic divergences suggesting recent radiation (Lecointre et al. 2011). In concordance with all these results, when exploring public barcode data only 5 different BINs were obtained within this family: *Artedidraco loennbergi*, *Artedidraco skottsbergi*, *Artedidraco mirus*, *Artedidraco orianae*, and two BINs with several species each, one containing *Dolloidraco longedorsalis*, *Artedidraco shackletoni* and *Histiodraco velifer* and another one with all the species of *Pogonophryne*.

Four species were barcoded in the present study including *Artedidraco skottsbergi, Pogonophryne scotti, P. permittini* and *Dolloidraco longedorsalis.* All these species but *P. permittini* were previously barcoded but there are no published barcode records for *D. longedorsalis* from the AP. Different BINs were detected for *Artedidraco skottsbergi,* and *Dolloidraco longedorsalis* whereas *Pogonophryne scotti* and *P. permittini,* are included in the same BIN (Fig. 2 and Table 5). Nevertheless, discrimination between these two species was supported by both the

Table 5 Summary of species collected and main information containing each BIN (Barcode Index Number)

Species	BIN	Average/max distance p-dist (%)	Distance to the nearest p-dist (%)	Nearest member
Electrona antárctica	AAB3737	0.08/0.81	2.39	*Symbolophorus veranyi*
Artedidraco skottsbergi	ACE8503	0.27/0.8	1.28	*Artedidraco loennbergi*
Dolloidraco longedorsalis	ACF0645	0.84/2.25	1.28	*Artedidraco orianae*
Pogonophryne scotti	ACE4864	0.54/1.29	1.44	*Artedidraco loennbergi*
Pogonophryne permitini	ACE4864	0.54/1.29	1.44	*Artedidraco loennbergi*
Gymnodraco acuticeps	AAI6438	0.06/0.47	4.46	*Cygnodraco mawsoni*
Parachaenichthys charcoti	AAC3293	0.17/1.08	2.25	*Parachaenichthys georgianus*
Prionodraco evansii	AAB5119	0.19/0.77	4.65	*Racovitzia glacialis*
Chaenocephalus aceratus	ABY5229	0.2/0.49	2.09	*Cryodraco antarcticus*
Chaenodraco wilsoni	ABY5097	0.32/0.96	2.09	*Chionodraco myersi*
Champsocephalus gunnari	AAB4065	0.55/1.62	6.74	*Chionodraco myersi*
Chionodraco rastrospinosus	ABY6584	0.11/0.32	1.77	*Chionodraco myersi*
Cryodraco antarcticus	ABY5228	0.14/0.49	1.46	*Chionobathyscus dewitti*
Pagetopsis macropterus	ABY4099	0.26/0.62	2.09	*Pagetopsis maculatus*
Pseudochaenichthys georgianus	AAD0635	0.06/0.17	4.38	*Neopagetopsis ionah*
Harpagifer antarcticus	AAO4089	0.47/1.44	6.74	*Artedidraco skottsbergi*
Gobionotothen gibberifrons	AAC1384	0.22/1.12	4.34	*Gobionotothen acuta*
Lepidonotothen squamifrons	AAA7826	0.27/1.61	8.07	*Patagonotothen tessellata*
Lindbergichthys nudifrons	AAC3747	0.37/1.63	1.52	*Lindbergichthys mizops*
Notothenia coriiceps	AAB4192	0.22/0.85	3.57	*Notothenia rossii*
Notothenia rossii	AAB2293	0.03/0.32	3.57	*Notothenia coriiceps*

(continued)

Table 5 (continued)

Species	BIN	Average/max distance p-dist (%)	Distance to the nearest p-dist (%)	Nearest member
Nototheniops larseni	AAB4889	0.26/0.96	6.16	*Lindbergichthys mizops*
Pleuragramma antarctica	AAB0924	0.26/1.15	8.63	*Cryothenia peninsulae*
Pseudotrematomus bernacchii	AAA8371	0.18/0.53	4.29	*Cryothenia amphitreta*
Pseudotrematomus eulepidotus	AAA6711	0.8/1.98	4.31	*Trematomus bernacchii*
Pseudotrematomus hansoni	AAC7338	0.52/1.32	4.54	*Trematomus bernacchii*
Pseudotrematomus scotti	AAA8370	0.31/1.12	8.18	*Trematomus* sp.
Trematomus newnesi	AAD7052	0.58/1.87	7.06	*Pagothenia borchgrevinki*
Lycenchelys nigripalatum	ACO5041	0.21/0.55	1.14	*Lycenchelys aratrirostris*
Lycenchelys sp.	ACO6414	–	2.27	*Lycenchelys aratrirostris*
Lycenchelys tristichodon	ABY6982	0.24/0.92	1.94	*Lycenchelys aratrirostris*
Ophthalmolycus cf. *amberensis*	AAC7879	0.14/0.48	1.61	*Ophthalmolycus amberensis*
Careproctus sp.	ACO4190	0.05/0.2	2.53	*Careproctus longipectoralis*
Paraliparis antarcticus	AAB1891	0.27/0.66	1.08	*Paraliparis aff. longipectoralis*
Paraliparis sp.1	ABX5136	0.58/1.14	2.25	*Paraliparis* sp.
Paraliparis sp.2	ACE7042	–	1.13	*Paraliparis aff. longipectoralis*

NJ (Fig. 2) and ML (Fig. 4) analyses and also by NDC. Interestingly, a single BIN includes all species of *Pogonophryne* already barcoded (12 species). Compared with public data on BOLD our specimen identified as *P. permittini* clustered together with specimens of *Pogonophryne barsukovi* (from Smith et al. 2012) and *Pogonophryne* sp1 (from Dettai et al. 2011). The NDC did not discriminate these species but, as a group, two NDC, #97 (A) and #603 (A), allowed to differenciate them from the remaining *Pogonophryne*.

There are no previous barcode records for *Dolloidraco longedorsalis* from the AP. COI similarity with conspecific from other regions (Dettai et al. 2011; Smith et al. 2012) ranged 99.52–99.84 %. As noted above, exploring public BOLD data we found that, along with *Dolloidraco longedorsalis*, two more species are included in the same BIN: *Histiodraco velifer* and *Artedidraco shackletoni*. Interspecific

Table 6 Nucleotide position for each diagnostic character in some of the species analysed in this study

Species/nucleotic position	#39	#321	#336	#351	#468	#492	#495	#531	#621
Dolloidraco longedorsalis	T	C	C	T	T	C	T	T	C
Artedidraco shackletoni	T	C	C	C	C	G	C	T	C
Histiodraco velifer	C	T	T	T	T	C	C	G	T
Species/nucleotic position	#252	#555	#579						
Harpagifer antarcticus	T	T	A						
H. bispinnis/H. paliolatus	C	C	G						

divergence among them is low (<2 %), but specimens from each species yielded a cohesive cluster. In addition, the three species could be discriminating through NDC (Table 6).

3.4 Bathydraconidae

Antarctic dragonfishes are slender-bodied species endemic to the Southern Ocean and live mostly on the shelf and upper slope. They are represented by 15 species in Antarctic waters, 7 of them were recorded in AP (Gon 1990). All species of dragonfishes were previously barcoded. Rock et al. (2008) stated that certain species (they barcoded 3 species of *Bathydraco*) were not resolved by COI, representing one single MOTU. In the same way Smith et al. (2012) reported lack of species resolution within *Bathydraco* species with shared haplotypes among two pair of species. Exploring public BOLD data, 11 different BINs were obtained within this family. Each species, except those of the genus *Bathydraco* formed an exclusive BIN.

Three dragonfishes were barcoded in this study, *Gymnodraco acuticeps*, *Prionodraco evansii* and *Parachaenichthys charcoti.* Each species formed unique BINs. No barcode records of *G. acuticeps* from the AP were published yet. Similarity percentage between ours specimens and those from other regions ranged from 99.67 to 100 % showing shallow intra-specific divergences among them.

3.5 Channichthyidae

Crocodile icefishes are unique among vertebrates in lacking haemoglobin. The family is represented by 18 species in Antarctic waters, 15 of them were registered for AP (Iwami and Kock 1990; Kock 2005; La Mesa et al. 2002). DNA barcoding was applied to 15 species of Channichthyidae (Rock et al. 2008; Dettai et al. 2011;

Smith et al. 2012) and with the exception of *Cryodraco antarcticus*, no problems of resolution with COI were detected (Rock et al. 2008; Dettai et al. 2011; Smith et al. 2012). Smith et al. 2012 found that *C. antarcticus* and *C. atkinsoni* shared haplotypes. In concordance with these studies 14 different public BINs are available on BOLD, all species except *C. antarcticus* and *C. atkinsoni* formed different BINs.

Seven species of crocodile icefishes were succesfully barcoded in this study. All these species were previously barcoded. Each species formed unique BINs (Fig. 2 and Table 5). No published barcode records are available for *Chaenodraco wilsoni, Cryodraco antarcticus* and *Pagetopsis macropterus* in west Antarctic waters. The three species are distributed around the Antarctic continental shelf but also occur in small numbers as far north as the southern Scotia Arc (Kock 1992). Our results show shallow intra-specific divergences among AP and other regions (Table 3).

3.6 *Harpagiferidae*

Spiny plunderfishes are small benthic/epibenthic species, confined to coastal waters from 0 m (under rocks in pools at low tide) to 200 m depth. They are ecologically and morphologically similar species (Eastman 2005). Currently, 10 species of the Genus *Harpagifer* have been described of which 8 are restricted to the sub-Antarctic islands and southern Scotia Arc islands. *Harpagifer antarcticus* is present along the western side of the Antarctic Peninsula. Littoral and sublittoral zones seem inhabited by different species (Neyelov and Prirodina 2006). *H. permitini* seems restricted to South Georgia; *H. crozetensis* and *H. spinosus* to the Crozet Islands; *H. kerguelensis* and *H. nybelini* to Kerguelen Islands and Heard Islands; *H. marionensis* to the Prince Edward Islands; *H. macquarensis* and *H. andriashevi* to Macquarie Island and *H. georgianus* in South Georgia, Prince Edward and Macquarie islands (Duhamel et al. 2014).

No published barcode records are available for any species. In the present study we barcoded specimens of *Harpagifer antarcticus*, which formed a unique BIN (Table 3). The same BIN also contains *H. bispinnis, H. palliolatus, H. georgianus* and *H. spinosus. H. bispinis* and *H palliolatus* are restricted to south Patagonian waters and yielded a unique cluster whereas specimens of *H antarcticus* were not all clustered together. Specimens of *H. antarcticus* showed 0–0.62 % intraspecific divergence and 98.6–99.53 % similitude with those of *H. bispinnis/H. palliolatus*. However, NDC allowed discrimination between *H. bispinnis/H. palliolatus* from *H. antarcticus* (Table 6). The lack of resolution between valid species of Harpagifer using COI is not surprising. Hüne et al. (2014) found low level of genetic divergence between *Harpagifer antarcticus* (from Antarctica) and *H. bispinnis* (from Patagonia) using the mitochondrial control region D-loop (a rapid mitochondrial marker) suggesting a recent (Quaternary) colonization of Patagonia from the Antarctic Peninsula.

3.7 *Zoarcidae*

Eelpouts are an important component of benthic ecosystems in Antarctic waters with approximately 32 reported species, 22 of them have been recorded in AP (Matallanas 2009, 2010, 2011; Matallanas et al. 2012). At least 9 species of eelpouts were previously barcoded. COI showed a high species-level resolution and highlighted possible new species (Rock et al. 2008; Dettai et al. 2011; Smith et al. 2012). Four different species were collected in this study: *Ophtalmolycus* cf. *amberensis*, and three species of *Lycenchelys*, *L. nigripalatum*, *L. tristichodon* and an unidentified species *Lycenchelys* sp. Each species formed unique BINs (Fig. 2 and Table 5). Discrimination among these species was also supported by both NJ (Fig. 2) and ML (Fig. 5) trees. *Ophthalmolycus amberensis* is widely distributed along the Antarctic continent to the Antarctic Peninsula (Duhamel et al. 2014).

The seven specimens of *Ophtalmolycus* cf. *amberensis* were included in the same BIN. There is no previous barcode data of this species from the AP. The BIN also contains specimens of *O. amberensis* from the Ross Sea. Similarity between specimens from both regions ranged 99.84–100 %. The nearest neighbor of this BIN contained *O. amberensis* from the Durmont d'Urville Sea. The analysis of public barcode sequences yielded two different clusters, one containing *O. amberensis* from the Ross Sea and those collected in AP, and the other containing specimens from Australian Antarctic Territory (AAT). Smith et al. (2012), already highlighted the high divergence between specimens from Ross Sea and AAT and suggested that this high level of divergence would be indicative of species-level divergence. The authors stressed the need for analyze additional specimens from the type locality (Ambers Islands, Antarctic Peninsula) to resolve the taxonomic status of specimens from the Ross Sea and AAT. As mentioned above, our specimens collected in the Antarctic Peninsula, matched specimens of *O. amberensis* from the Ross Sea and are included in the same BIN. If the type locality is Antarctic Peninsula, it is probable that specimens from this BIN represent the real *O. amberensis* and those from the AAT be a new species. Further morphological analysis, comparing vouchers from the different regions are needed to corroborate this hypothesis (Fig. 6).

The seven specimens of *Lycenchelys nigripalatum* were clustered together and formed a unique BIN (Fig. 2 and Table 5) which is new for BOLD (ACO5041). There is no previous barcode data of this species from any region of Antarctica. The nearest neighbor is *Lycenchelys aratrirostris* with a percentage similarity of 98.54–98.72 %. Regarding *Lycenchelys tristichodon* the specimens formed a unique BIN. This species was already barcoded (Dettai et al. 2011) but there is no public barcode data of specimens from AP. The similarity percentage with specimens from other regions range 99.22–99.38 %. There were no sharing haplotypes between regions and 3 different NDC were also found: Site#172 (G vs A) Site#181 (C vs T); Site#541 (T vs C). The nearest neighbor (97.98 % similarity) is *L. aratirostris*. Finally *Lycenchelys* sp. constitute a singleton forming a unique and new BIN for BOLD. The distance to the nearest neighbor (*Lycenchelys aratrirostris)* is 2.27 %

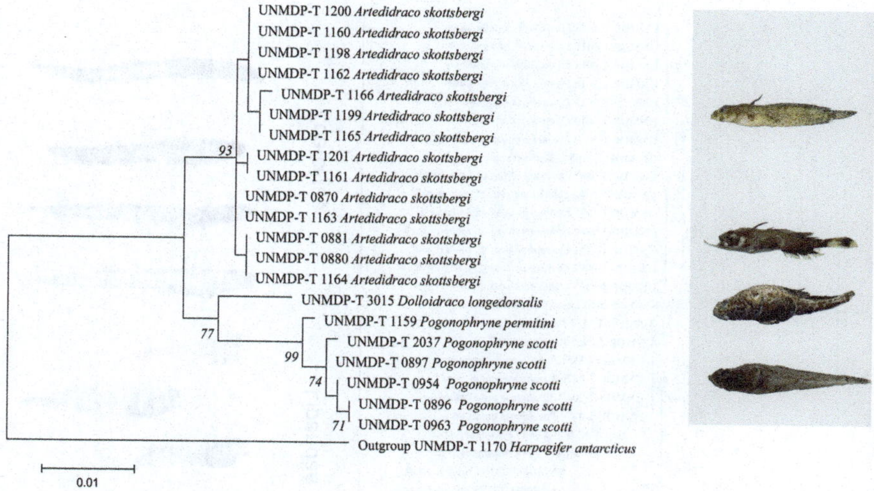

Fig. 6 COI relationships among specimens of Artedidraconidae collected in the Antarctic Peninsula. ML tree rooted with *Harpagifer antarcticus*; number at nodes are bootstrap percentage (>70 %) after 500 replicates based on ML. Scale bar is a K2+G distance reference. Code numbers represent BOLD sample IDs. Photos on the right correspond to each species

(p-dist). The record of zoarcid species and their geographical distribution in the Southern Ocean need to be completed. Species identification is still a problem in some species. For example, very few species have been recorded from East Antarctica (Duhamel et al. 2014).

3.8 *Liparidae*

Snailfishes are the most speciose family of Antarctic fishes with approximately 66 species described, 17 of which were recorded for the AP and adjacent waters (Heemstra and Duhamel 1990; Andriashev 1998; Matallanas and Pequeño 2000; Chernova 2006; Balushkin and Prirodina 2010; Balushkin 2012; Stein 2012). Identification requires a high level of expertise and a thorough study of osteological characters. Given the fact that several species were barcoded (see Duhamel et al. 2010) including species within genera *Careproctus* and *Paraliparis*, and the identification of specimens were made by specialists, assignation of specific names for our specimens were based on results from BINs. At least 13 species of Liparidae were previously barcoded (Rock et al. 2008; Duhamel et al. 2010; Dettai et al. 2011; Smith et al. 2012), showing different degrees of resolution. Duhamel et al. (2010) found that all individuals from a single species are grouped together in the molecular trees. The distance among species is mostly over 2 %, except for a few pairs of *Paraliparis* species (*P. charcoti*-*P. leobergi, P. rosaceus*-*P. neelovi*). Smith

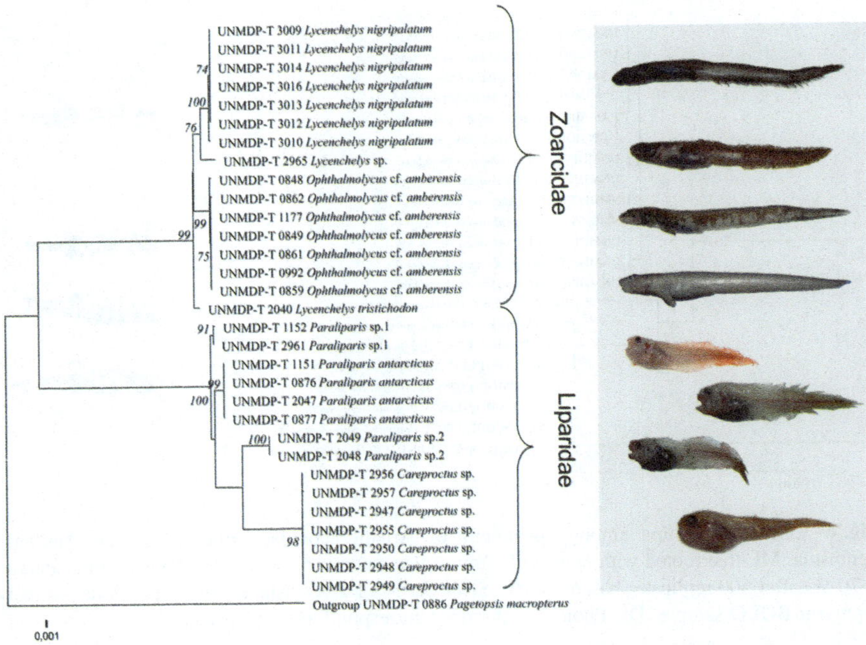

Fig. 7 COI relationships among specimens of Zoarcidae and Liparidae collected in the Antarctic Peninsula. ML tree rooted with *Pagetopsis macropterus*; number at nodes are bootstrap percentage (>70 %) after 500 replicates based on ML. Scale bar is a K2+G distance reference. Code numbers represent BOLD sample IDs. Photos on the right correspond to each species

et al. (2012) noticed lack of resolution among Ross Sea liparids with shallow of zero divergence among recently described species (Stein 2012) of the genus *Paraliparis* (Fig. 7).

In the present study, four different BINs were recorded within this family, corresponding to a species initially identified as *Careproctus georgianus, Paraliparis antarcticus* and another two species that could not be identified to species level and were named *Paraliparis* sp. One of them, corresponding to the BIN ACE7042, showed low (D < 2 %) interspecific divergence (Fig. 2) related to *Paraliparis antarcticus,* but both the NJ (Fig. 2) and ML (Fig. 5) trees, and BIN approaches supported it is a different species. The other cluster of *Paraliparis* sp. (BIN ABX5136) showed a high distance (>8 %).

Seven specimens of *Careproctus* sp. were barcoded. All of them were clustered together and constituted the same BIN (Fig. 2 and Table 5). No previous barcode data were available on this species representing a new BIN for BOLD. All specimens came from AP. The distance (p-dist) to Nearest Neighbor, *Careproctus longipectoralis* (BIN AAI6622), is 2.53 %. The species were initially identified as *C. georgianus*, but the similarity with public records of *C. georgianus* in BOLD, one from Antarctic waters (Rock et al. 2008) and two from the northern Pacific (Steinke el al. 2009) is 93.91–93.93 %, indicating represent another species. For the

West Antarctic Ocean at least 12 species of *Careproctus* have been described (Andriashev and Prirodina 1990; Duhamel 1992; Andriashev and Stein 1998): *C. acifer*, *C. eltianae*, *C. federovi*, *C. georgianus*, *C. improvisus*, *C. lacmi*, *C. leptorhinus*, *C. parviparratus*, *C. polarstein*, *C. rimiventris*, *C. scophopterus* and *C. steini*. There are only barcode data of *C. georgianus*. There are barcodes from species of other Antarctic regions (*C. crozotensis*, *C. continentalis*, *C. discoveryae*, *C. longipectoralis*). Our specimens are most similar to *C. longipectoralis*, although correspond to a different BIN and has a genetic divergence (K2P) >3 %.

Regarding *Paraliparis antarcticus* there is no previous barcode record of this species from the West Antarctic. In our samples, all specimens formed a unique BIN in which conspecifics from other regions were also present (Fig. 2 and Table 5). The similarity percentage with specimens from other regions ranged 99.36–100 %, showing shallow intra-specific divergences among AP and other regions, sharing haplotypes. Two specimens of *Paraliparis* sp1. were barcoded in this study. These specimens formed a unique BIN (Fig. 2 and Table 5) which also contained several *P. mawsoni* from other Antarctic region (Dettai et al. 2011). In the NJ tree these two species clustered separately with a percentage similarity ranging 98.9–99.07 %. Two specimens of *Paraliparis* sp2. were also barcoded in this study. These specimens formed a unique BIN (Fig. 2 and Table 5) which also contained another *Paraliparis* sp. from the same area (Rock et al. 2008). The nearest neighbor was *P. aff longipectoralis* that formed another BIN.

3.8.1 Extending the Distribution of Some Antarctic Fish Species

Most species collected during our survey were previously cited for the study area. However some of them constitute new records or an expansion in the distribution range of the species. Within liparids, *P. antarcticus* was previously cited for the east Antarctic shelf, and southermost Weddell Sea, thus these records extended its distribution to the northern part of the Antarctic Peninsula. Besides, if *Paraliparis* sp1 correspond to *P. mawsoni* (according to BIN), the presence of this species is new for the area because it had been reported only for the south eastern Lazarev Sea (Heemstra and Duhamel 1990), east Antarctic (Terre Adélie and George V Land) (Duhamel et al. 2010), and Weddell Sea (Matallanas, 1999). Although *Paraliparis antarcticus* and *P. mawsoni* have been mentioned to occur in a wider circum-antarctic distribution, knowledge of the distribution of Southern Ocean liparids is still limited due to poor coverage of the deep-sea (Duhamel et al. 2014). *Lycenchelis nigripalatum* and *L. tristichodon* were previously registered off Adelaide Island and Durmont d'Urville Sea (Dettai et al. 2011), thus our records for these species in the study area spread its distribution to the northern part of Antarctic Peninsula.

4 Conclusions

Fishes from Antarctic Peninsula and adjacent waters represent 35 % of fish richness in Antarctic waters. In this study, four new BINs were generated and incorporated in BOLD. They corresponded to *Lycenchelys nigripalatum*, *Lycenchelys* sp., *Careproctus* sp., and a *Paraliparis* sp. Even though barcode records from all the other species barcoded in this study are available at BOLD, this work contributed to the Barcode reference library with samples from an area not fully barcoded (see Rock et al. 2008; Lautrédou et al. 2010; Duhamel et al. 2010; Dettai et al. 2011; Smith et al. 2012).

In this study, 35 different BINs from 36 putative species were obtained, with only *Pogonophryne scotti* and *P. permittini* sharing the same BIN. However, when exploring public barcode data, we found that two more BINs corresponding to *Dolloidraco longedorsalis* and *Harpagifer antarcticus* also included other valid species. Thus, using only a conservative approach (BINs), 88.6 % of the species barcoded in this study could be identified at species level with accuracy. A similar level of species discrimination using BINs was reported for the fish fauna of the Northeast Atlantic (Knebelsberger et al. 2014). However the use of nucleotic diagnostic character allowed us to discriminate the remaining species.

Compiling our results with previous studies, about 80 species inhabiting the Antarctic Peninsula were already barcoded, representing approximately 60 % of the species occurring in the area. Over 75 % of the barcoded species could be identified at species level with accuracy using BINs. Our results suggest that the use of nucleotic diagnostic character may help to improve the level of species discrimination.

In the present work, juveniles of some species of Nototheniiidae were not able to be identified to the species level using external morphology since they did not share with the corresponding adults the same diagnostic features. These results evidenced an important component of ontogenetic variation in the character state of diagnostic features commonly used to construct taxonomic keys for this family. DNA Barcoding was, therefore, a cornerstone element for obtaining a reliable identification of these specimens. These results are very relevant for management and conservation purposes since an accurate species-level resolution of juveniles is necessary to determine nursery areas and to clarify species distributions.

References

Acosta J, Canals M, Herranz P, Sanz J (1989) Informe de resultados "Antártida 8611", investigación geológica-geofísica y sedimentológica en el arco de Escocia y península Antártica. In Resultados de la campaña "Antártida 8611". Publ Espec Inst Esp Oceanogr 2:9–82

Andriashev AP (1998) A review of recent studies of Southern Ocean Liparidae (Teleostei: Scorpaeniformes). Cybium 22(3):255–266

Andriashev AP, Prirodina VP (1990) Review of the Antarctic species of the genus *Careproctus* (Liparididae) with notes on the carcinophilic species of the genus. Vopr Ikhtiol 30(5):709–719

Andriashev AP, Stein DL (1998) Review of the Snailfish genus *Careproctus* (Liparidae, Scorpaeniformes) in Antarctic and adjacent waters. Contrib Sci (Los Ang) 470:1–63

Aronson RB, Moody RM, Ivany LC, Blake DB, Werner JE, Glass A (2009) Climate change and trophic response of the Antarctic bottom fauna. PLoS ONE 4(2):e4385

Balushkin AV (2012) *Volodichthys* gen. nov. new species of the primitive snailfish (Liparidae: Scorpaeniformes) of the southern hemishpere. Description of new species V. *Solovjevae* sp. nov. (Cooperation Sea, the Antarctic). J Ichthyol 52(1):1–10

Balushkin AV, Prirodina VP (2010) A new species of Muraenolepididae (Gadiformes) *Muraenolepis evseenkoi* sp. nova from continental seas of Antarctica. J Ichthyol 50(7):495–502

Barrera-Oro E (2002) The role of fish in the Antarctic marine food web: differences between inshore and offshore waters in the southern Scotia Arc and west Antarctic Peninsula. Antarct Sci 14:293–309

Becker S, Hanner R, Steinke D (2011) Five years of FISH-BOL: brief status report. Mitochondr DNA 22:3–9

Broughton RE, Milam JE, Roe BA (2001) The complete sequence of the zebrafish (*Danio rerio*) mitochondrial genome and evolutionary patterns in vertebrate mitochondrial DNA. Genome Res 11:1958–1967

Chernova NV (2006) New and rare snailfishes (Liparidae, Scorpaeniformes) with the description of four new species from the Southern Hemisphere and Tropical East Pacific. J Ichthyol 46 (1):1–14

Clarke A, Barnes DKA, Hodgson DA (2005) How isolated is Antarctica? Trends Ecol Evol 20:1–3

Collins RA, Boykin LM, Cruickshank RH, Armstrong KF (2012) Barcoding's next top model: an evaluation of nucleotide substitution models for specimen identification. Methods Ecol Evol 3:457–465

Constable A, Costa D, Murphy E, Hofmann E, Schofield O, Press A et al (2014) Assessing status and change in Southern Ocean ecosystems. In: De Broyer C, Koubbi P, Griffiths HJ, Raymond B, Udekem d'Acoz C et al (eds) Biogeographic Atlas of the Southern Ocean. Scientific Committee on Antarctic Research, Cambridge, pp 404–407

Cook AJ, Fox AJ, Vaughan DG, Ferrigno JG (2005) Retreating glacier fronts on the Antarctic Peninsula over the past half century. Science 308:541–544

Cook AJ, Vaughan DG (2010) Overview of areal changes of the ice shelves on the Antarctic Peninsula over the past 50 years. Cryosphere 4:77–98

DeSalle R (2006) Species discovery versus species identification in DNA barcoding efforts: response to Rubinoff. Conserv Biol 20:1545–1547

Dettai A, Lautredou AC, Bonillo C, Goimbault E, Busson F, Causse R et al (2011) The actinopterygian diversity of the CEAMARC cruises: barcoding and molecular taxonomy as a multi level tool for new findings. Deep Sea Res II 58:250–263

DeWitt HH, Heemstra PC, Gon O (1990) Nototheniidae. In: Gon O, Heemstra PC (eds) Fishes of the Southern Ocean. JLB Smith Institute of Ichthyology, Grahamstown, pp 279–331

Donnelly J, Torres JJ (2008) Pelagic fishes in the Marguerite Bay region of the West Antarctic Peninsula shelf. Deep-Sea Res II 55(3–4):523–539

Donnelly J, Torres JJ, Hopkins TL, Lancraft TM (1990) Proximate composition of Antarctic mesopelagic fishes. Mar Biol 106(1):13–23

Duhamel G (1992) Description d'espèces Nouvelles de *Careproctus* et *Paraliparis* et Donnés Nouvelles sur ces Genres et le Genre *Edentoliparis* du l'océan Austral (Cyclopteridae, Liparinae). Cybium 16(3):183–207

Duhamel G, Hautecoeur M, Dettai A, Causse R, Pruvost P, Busson F et al (2010) Liparids from the Eastern sector of Southern Ocean and first information from molecular studies. Cybium 34:319–343

Duhamel G, Hulley P-A, Causse R , Koubbi P, Vacchi, M, Pruvost P et al (2014) Biogeographic patterns of fish. In: De Broyer C, Koubbi P, Griffiths HJ, Raymond B, Udekem d'Acoz C et al (eds) Biogeographic Atlas of the Southern Ocean, Chap 7. Scientific Committee on Antarctic Research, Cambridge, pp 328–362

Eakin R (1990) Artedidraconidae. In: Gon O, Heemstra PC (eds) Fishes of the Southern Ocean. JLB Smith Institute of Ichthyology, Grahamstown, pp 332–356

Eakin RR, Eastman JT, Jones CD (2001) Mental barbell variation in *Pogonophryne scotti* Regan (Pisces: Perciformes: Artedidraconidae). Antarct Sci 13:363–370

Eakin RR, Eastman JT, Near TJ (2009) A new species and a molecular phylogenetic analysis of the Antarctic fish genus *Pogonophryne* (Notothenioidei: Artedidraconidae). Copeia 4:705–713

Eastman JT (2005) The nature of the diversity of Antarctic fishes. Polar Biol 28:93–107

Eschmeyer WN (ed) (2015) Catalog of fishes. California Academy of Sciences. http://research.calacademy.org/research/ichthyology/catalog/fishcatmain.asp.2015. Accessed 03 Jan 2015

Gon O (1990) Bathydraconidae. In: Gon O, Heemstra PC (eds) Fishes of the Southern Ocean. JLB Smith Institute of Ichthyology, Grahamstown, pp 364–380

Gon O, Heemstra PC (1990) Fishes of the Southern Ocean, 1st edn. JLB Smith Institute of Ichthyology, Grahamstown

Gordon AL (1988) Spatial and temporal variability within the Southern Ocean. In: Saharge D (ed) Antarctic Ocean and resources variability. Springer, Berlin, pp 41–56

Heemstra PC, Duhamel G (1990) Congiopodidae. In: Gon O, Heemstra PC (eds) Fishes of the Southern Ocean. JLB Smith Institute of Ichthyology, Grahamstown, pp 229–230

Hubert N, Hanner R, Holm E, Mandrak NE, Taylor E, Burridge M et al (2008) Identifying Canadian freshwater fishes through DNA barcodes. PLoS ONE 3:1–8

Hüne M, González-Wevar C, Poulin E, Mansilla A, Fernández DA, Barrera-Oro E (2014) Low level of genetic divergence between Harpagifer fish species (Perciformes: Notothenioidei) suggests a Quaternary colonization of Patagonia from the Antarctic Peninsula Polar Biol. DOI 10.1007/s00300-014-1623-6

Ivanova NV, de Waard JR, Hebert PDN (2006) An inexpensive, automation friendly protocol for recovering high-quality DNA. Mol Ecol Notes 6:998–1002

Ivanova NV, Zemlak TS, Hanner RH, Hebert PDN (2007) Universal primer cocktails for fish DNA barcoding. Mol Ecol Notes 7:544–548

Iwami T, Kock KH (1990) Channichthyidae. In: Gon O, Heemstra PC (eds) Fishes of the Southern Ocean. JLB Smith Institute of Ichthyology, Grahamstown, pp 381–389

Jones CD, Anderson ME, Balushkin AV, Duhamel G, Eakin RR, Eastman JT, Kuhn KL, Lecointre G, Near TJ, North AW, Stein DL, Vacchi M, Detrich HW (2008) Diversity, relative abundance, new locality records and population structure of Antarctic demersal fishes from the northern Scotia Arc islands and Bouvetøya. Polar Biol 31:1481–1497

Kimura M (1980) A simple method for estimating evolutionary rate of base substitutions through comparative studies of nucleotide sequences. J Mol Evol 16:111–120

Knebelsberger T, Landi M, Neumann H, Kloppmann M, Sell AF, Campbell PD et al (2014) A reliable DNA barcode reference library for the identification of the North European shelf fish fauna. Mol Ecol Res 14:1060–1071

Kock KH (1992) Antarctic fish and fisheries. Cambridge University Press, Cambridge

Kock KH (2005) Antarctic icefishes (Channichthyidae): a unique family of fishes. Rev Part I Polar Biol 28(11):862–895

Kock KH, Jones CD, Wilhelms S (2000) Biological characteristics of Antarctic fish stocks in the southern Scotia Arc region. CCAMLR Sci 7:1–41

Koubbi P, Moteki M, Duhamel G, Goarant A, Hulley PA, O'Driscoll R, Ishimaru T, Pruvost P, Tavernier E, Hosie G (2011) Ecoregionalization of myctophid fish in the Indian sector of the Southern Ocean: results from generalized dissimilarity models. Deep Sea Res II 58:170–180

La Mesa M, Vacchi M, Iwami T, Eastman JT (2002) Taxonomic studies of the Antarctic icefish genus *Cryodraco* Dollo, 1900 (Notothenioidei: Channichthyidae). Polar Biol 25:384–390

Lautrédou A-C, Bonillo C, Denis G, Cruaud C, Ozouf-Costaz C, Lecointre G, Dettai A (2010) Molecular taxonomy and identification within the Antarctic genus *Trematomus* (Notothenioidei, Teleostei): how valuable is barcoding with COI? Polar Sci 4:333–352

Lecointre G, Gallut C, Bonillo C, Couloux A, Ozouf-Costaz C, Dettai A (2011) The Antarctic Fish Genus Artedidraco is paraphyletic (Teleostei, Notothenioidei, Artedidraconidae). Polar Biol 34:1135–1145

Mabragaña E, de Astarloa JMD, Hanner R, Zhang J, González Castro M (2011) DNA barcoding identifies Argentine fishes from marine and brackish waters. PLoS ONE 6:e28655

Matallanas J (1999) New and rare snailfish genus *Paraliparis* from the Weddell Sea with the description of two new species. J Fish Biol 54:1017–1028

Matallanas J (2009) Description of a new genus and species of zoarcid fish, *Bellingshausenia olasoi*, from the Antarctic Ocean. Polar Biol 32:873–878

Matallanas J (2010) Description of two new genera, *Santelmoa* and *Bentartia* and two new species of Zoarcidae (Teleostei, Perciformes) from the Southern Ocean. Polar Biol 33(5):659–672

Matallanas J (2011) Description of two new species of *Ophthalmolycus* (Teleostei: Zoarcidae) from the Southern Ocean and key to species of the genus. J Mar Biol Ass UK 91:561–567

Matallanas J, Pequeño G (2000) A new snailfish species, *Paraliparis orcadensis* sp. nov (Pisces: Scorpaeniformes) from the Scotia Sea (Southern Ocean). Polar Biol 23:298–300

Matallanas J, Corbella C, Møller PR (2012) Description of two new species of *Santelmoa* (Teleostei, Zoarcidae) from the Southern Ocean. Polar Biol 35(9):1395–1405

Naish T, Powell R, Levy R, Wilson G, Scherer R, Talarico F et al (2009) Obliquity-paced Pliocene West Antarctic ice sheet oscillations. Nature 458:322–328

Neyelov AV, Prirodina VP (2006) Description of *Harpagifer permitini* sp. nova (Harpagiferidae) from the sublittoral zone of South Georgia and redescription of the littoral *H. georgianus* Nybelin. J Ichthyol 46(1):1–12

Pegg GG, Sinclair B, Briskey L, Aspden WJ (2006) MtDNA barcode identification of fish larvae in the southern Great Barrier Reef. Australia Sci Mar 70:7–12

Piacentino GLM, Barrera-Oro E (2009) Phenotypic plasticity in the Antarctic fish *Trematomus newnesi* (Nototheniidae) from the South Shetland Islands. Polar Biol 32:1407–1413

Ratnasingham S, Hebert PDN (2007) BOLD: the Barcode of Life Data System. Mole Ecol Notes 7:355–364. http://www.barcodinglife.org

Ratnasingham S, Hebert PDN (2013) DNA-based registry for all animal species: the Barcode Index Number (BIN) system. PLoS ONE 8(8):e66213

Rey O, Bonillo C, Gallut C, Cruaud C, Dettaï A, Ozouf-Costaz C, Lecointre G (2011) Naked dragonfishes *Gymnodraco acuticeps* and *G. victori* (Bathydraconidae, Notothenioidei) off Terre Adélie are a single species. Cybium 35(2):111–119

Rock J, Costa FO, Walker DI, North AW, Hutchinson WF, Carvalho GR (2008) DNA barcodes of fish of the Scotia Sea, Antarctica indicate priority groups for taxonomic and systematics focus. Antarct Sci 20:253–262

Rosso JJ, Mabragaña E, Gonzalez Castro M, de Astarloa JMD (2012) DNA barcoding Neotropical fishes: recent advances from the Pampa Plain. Argentina Mol Ecol Resour 12:999–1011

Smith PJ, Steinke D, McVeagh SM, Stewart AL, Struthers CD, Roberts CD (2008) Molecular analysis of Southern Ocean skates (*Bathyraja*) reveals a new species of Antarctic skate. J Fish Biol 73:1170–1182

Smith PJ, Steinke D, McMillan PJ, Stewart AL, McVeagh SM, Diaz de Astarloa JM, Welsford D, Ward RD (2011) DNA barcoding highlights a cryptic species of grenadier (genus *Macrourus*) in the Southern Ocean. J Fish Biol 78:355–365

Smith PJ, Steinke D, Dettai A, McMillan P, Welsford D, Stewart A, Ward RD (2012) DNA barcodes and species identifications in Ross Sea and Southern Ocean Fishes. Pol Biol 35 (9):1297–1310

Steig EJ, Schneider DP, Rutherford SD, Mann ME, Comiso JC, Shindell DT (2009) Warming of the Antarctic ice-sheet surface since the 1957 international geophysical year. Nature 457:459–462

Stein DL (2012) Snailfishes (family Liparidae) of the Ross Sea, Antarctica, and closely adjacent waters. Zootaxa 3285:1–120

Steinke D, Hanner R (2010) The FISH-BOL collaborators' protocol. Mitochon DNA 21:1–5

Steinke D, Zemlak TS, Boutillier JA, Hebert PDN (2009) DNA barcoding of Pacific Canada's fishes. Mar Biol 156:2641–2647

Tamura K, Peterson D, Peterson N, Stecher G, Nei M, Kumar S (2011) MEGA5: molecular evolutionary genetics analysis using maximum likelihood, evolutionary distance, and maximum parsimony methods. Mol Biol Evol 28:2731–2739

Thatje S (2005) The future fate of the Antarctic marine biota? Trends Ecol Evol 20:418–419

Turner J, Bindschadler R, Convey P, Di Prisco G, Fahrbach E, Gutt L, Hodgson D, Mayewski P, Summerhayes C (2009) Antarctic climate change and the environment. Scientific Committee on Antarctic Research, Cambridge

Valdez-Moreno M, Ivanova NV, Elías-Gutiérrez M, Contreras-Balderas S, Hebert PDN (2009) Probing diversity in freshwater fishes from Mexico and Guatemala with DNA barcodes. J Fish Biol 74:377–402

Ward RD, Zemlak TS, Innes BH, Last PR, Hebert PDN (2005) DNA barcoding Australia's fish species. Philos T R Soc B 360:1847–1857

Ward RD, Holmes BH (2007) An analysis of nucleotide and amino acid variability in the barcode region of cytochrome c oxidase I (cox1) in fishes. Mol Ecol Notes 7:899–907

Ward RD, Holmes BH, White WT, Last PR (2008) DNA barcoding Australasian chondrichthyans: results and potential uses in conservation. Mar Freshw Res 59:57–71

Ward RD, Hanner R, Hebert PDN (2009) The campaign to DNA barcode all fishes, FISH-BOL. J Fish Biol 74:329–356

Barcoding of Indian Marine Fishes: For Identification and Conservation

V.S. Basheer, Labrechai Mog Chowdhury, C. Mohitha and K.K. Bineesh

Abstract India has a rich natural heritage and nurtures a unique bio-diversity, placing it among the 12 most biodiverse countries. Globally the number of valid fish species recorded so far is more than 31000, with the addition, at an average, of 100–150/year. Among these 2,508 are indigenous to Indian subcontinent (877 freshwater, 113 brackish water and 1,518 marine species). DNA barcoding is a molecular method for species identification and classification of biological organisms based on the analysis of short, standardized gene sequences. In most animals, the fragment of mitochondrial gene cytochrome c oxidase subunit I (COI) has been used as the target sequence. This novel system is designed to provide rapid, accurate, and automatable species identifications by using short, standardized gene regions as internal species tags. Of this rich natural biodiversity, comprising over 1518 native marine species, at present barcodes of about 500 marine fish species are available, which is approximately 33 % of total Indian marine fish diversity. Whereas major portion of registered marine fishes remain untouched. Hence more emphasis should be given to DNA barcoding, with mandate of barcoding all the species to establish global comprehensive reference libraries. The traditional taxonomists will play a vital role in completing such a global database; hence there is a pressing need to make a integration of DNA barcoding with traditional taxonomy. In a nutshell, it can be said that DNA barcoding can be taken up as pragmatic approach for resolving unambiguous identification of the fish fauna which can play a crucial role in biodiversity assessment and conservation of marine ecosystem of country.

Keywords Biodiversity · Conservation · DNA barcode · Fish · Marine fish species

V.S. Basheer (✉) · L.M. Chowdhury · C. Mohitha · K.K. Bineesh
Peninsular and Marine Fish Genetic Resources Centre, NBFGR, CMFRI Campus, Kochi 682018, Kerala, India
e-mail: vsbasheer@gmail.com

S. Trivedi et al. (eds.), *DNA Barcoding in Marine Perspectives*,
DOI 10.1007/978-3-319-41840-7_15

1 Introduction

Since the Convention of Biological Diversity summit at Rio-de-Janeiro in 1992, all over the world, there has been lot of interest in identification of biological resources. Identification of species in the classic way, using morphological characters takes longer time and manpower, hence scientists developed an alternative method in conjunction with the genetic make up of the species. DNA barcoding is a molecular method for species identification and classification of biological organisms based on the analysis of short, standardized gene sequences (Hebert et al. 2003). In most animals, the fragment of mitochondrial gene cytochrome c oxidase subunit I (COI) has been used as the target sequence. This novel system is designed to provide rapid, accurate, and automatable species identifications by using short, standardized gene regions as internal species tags. DNA barcoding is now gaining more attention in the field of assessment of cryptic species, taxon diagnosis, authentication and safety assessment of seafood, wildlife forensics, conservation genetics and detection of invasive alien species from whole fish, fillets, fins, fragments, juveniles, larvae, eggs, any properly preserved tissue or environment sample (Trivedi et al. 2016). The concerted global research FISH-BOL (Fish Barcode of Life) on DNA barcoding was launched in 2005, with the goal of collection and assembling specific DNA barcode sequences and associated voucher provenance data in accurate reference sequence library to aid the molecular identification of all fish species.

India has a rich natural heritage and nurtures a unique bio-diversity, placing it among the 12 biodiversity rich countries. Globally the number of valid fish species recorded so far is more than 31000, with the addition, at an average, of 100–150/year (Eschmeyer et al. 2010). Among these 2,508 are indigenous to Indian subcontinent (877 freshwater, 113 brackish water and 1,518 marine species) (NBFGR, 2013). Out of which about 500 marine species has been barcoded (Tables 1, 2 and Fig. 1), covering carangids, scombrids, serranids, scianids, polynemids, nemipterids, pomacanthids, gobiids, clupeids, mugilids, sharks, rays, skates and other taxonomically important groups, under national programme on DNA barcoding which was launched in 2005 with initiative of NBFGR and other research programmes. In view of the growing importance of fish DNA barcoding, Nagpure et al. (2012) developed Fish Barcode Information System (FBIS) database on Indian fishes. FBIS represent barcode sequences for Indian fishes, reported in India as well as from other countries. Presently, it comprises barcode sequence of 1083 marine, 349 Freshwater and 49 brackish water with 23429 specimen sequences (http://mail.nbfgr.res.in/fbis/) belonging to 213 families and 48 orders.

Table 1 List of Indian marine fish species barcoded (Bony fishes)

Sl. No.	Order	Family	Species	Acessions
1	Perciformes	Carangidae	*Carangoides armatus*	FJ459577
2			*Carangoides chrysophrys*	FJ237546
3			*Carangoides ferdau*	EU514505
4			*Carangoides malabaricus*	FJ347935
5			*Carangoides praeustus*	KC508506
6			*Carangoides talamparoides*	KC508507
7			*Caranx caranges*	EU514501
8			*Alectis indica*	FJ347894
9			*Alectis ciliaris*	EU514500
10			*Alepes djedaba*	EF609498
11			*Alepes kleinii*	FJ237545
12			*Atropus atropos*	EF609506
13			*Atule mate*	EU514511
14			*Parastromateus niger*	EF609570
15			*Megalaspis cordyla*	EF609548
16			*Caranx hippos*	FJ347906
17			*Caranx ignobilis*	EU014221
18			*Caranx sexfasciatus*	EU514509
19			*Gnathanodon speciosus*	EU148562
20			*Elagatis bipinnulata*	EU014215
21			*Decapterus russelli*	EF609508
22			*Decapterus macrosoma*	EU514515
23			*Decapterus macarellus*	EU514517
24			*Scomberoides lysan*	FJ347900
25			*Scomberoides commersonnianus*	EU514496
26			*Selar boops*	FJ347890
27			*Selar crumenophthalmus*	FJ347941
28			*Selaroides leptolepis*	EU514521
29			*Seriolina nigrofasciata*	EU014235
30			*Trachinotus blochii*	EU148559
31			*Uraspis helvola*	EU514510
32		Scombridae	*Auxis rochei*	FJ226520
33			*Auxis thazard*	FJ226525
34			*Euthynnus affinis*	EU148529
35			*Katsuwonus pelamis*	EU014258
36			*Rastrelliger brachysoma*	KJ590064
37			*Rastrelliger faughni*	KJ590073
38			*Rastrelliger kanagurta*	FJ237548

(continued)

Table 1 (continued)

Sl. No.	Order	Family	Species	Acessions
39			*Sarda orientalis*	EF609591
40			*Scomberomorus commerson*	KM677209
41			*Scomberomorus guttatus*	EU541328
42			*Thunnus albacares*	EF609629
43			*Thunnus tonggol*	FJ226524
44		Serranidae	*Aethaloperca rogaa*	KM226213
45			*Cephalopholis argus*	FJ237556
46			*Cephalopholis aurantia*	KM226217
47			*Cephalopholis boenak*	FJ237553
48			*Cephalopholis formosa*	KF268156
49			*Cephalopholis urodeta*	FJ459565
50			*Cephalopholis sonnerati*	FJ237541
51			*Cephalopholis miniata*	FJ237607
52			*Cephalopholis nigripinnis*	KM226226
53			*Chelidoperca investigatoris*	KP009558
54			*Chelidoperca occipitalis*	JX185306
55			*Chelidoperca maculicauda*	JX185308
56			*Liopropoma randalli*	KF814980
57			*Epinephelus areolatus*	JX674967
58			*Epinephelus bleekeri*	JX674971
59			*Epinephelus chlorostigma*	EU392203
60			*Epinephelus coeruleopunctatus*	KF268167
61			*Epinephelus coioides*	KJ755858
62			*Epinephelus diacanthus*	EF609517
63			*Epinephelus macrospilos*	JX675007
64			*Epinephelus longispinis*	HQ658119
65			*Epinephelus latifasciatus*	EU014219
66			*Epinephelus fuscoguttatus*	JX674997
67			*Epinephelus flavocaeruleus*	KM226266
68			*Epinephelus faveatus*	JX674974
69			*Epinephelus fasciatus*	EU392208
70			*Epinephelus malabaricus*	FJ237599
71			*Epinephelus merra*	FJ237598
72			*Epinephelus melanostigma*	KM226281

(continued)

Table 1 (continued)

Sl. No.	Order	Family	Species	Acessions
73			*Epinephelus morrhua*	EU392189
74			*Epinephelus miliaris*	KM226282
75			*Epinephelus polylepis*	KM226288
76			*Epinephelus polyphekadion*	KM226293
77			*Epinephelus radiatus*	KM226297
78			*Epinephelus quoyanus*	KM226294
79			*Epinephelus spilotoceps*	KM226298
80			*Epinephelus tauvina*	EU148566
81			*Epinephelus undulosus*	JX675024
82			*Hyporthodus octofasciatus*	KM226304
83			*Odontanthias perumali*	KR105805
84			*Plectropomus areolatus*	KJ607966
85			*Plectropomus maculatus*	JX123681
86			*Plectropomus leopardus*	KM226309
87			*Plectropomus laevis*	KM226311
88			*Sacura boulengeri*	KR105842
89			*Variola louti*	FJ459559
90			*Variola albimarginata*	KM226312
91		Scianidae	*Dendrophysa russelii*	EU148580
92			*Johnius belangerii*	FJ347918
93			*Johnius borneensis*	FJ347922
94			*Johnius carutta*	FJ265843
95			*Johnius dussumieri*	FJ347915
96			*Johnius elongatus*	EF534123
97			*Nibea maculata*	EU014249
98			*Nibea soldado*	HQ219159
99			*Otolithes cuvieri*	FJ347924
100			*Otolithes ruber*	FJ237586
101			*Otolithoides biauritus*	EF534127
102			*Panna microdon*	JX983436
103			*Macrospinosa cuja*	JX260908
104			*Protonibea diacanthus*	EF528229
105		Sphyraendiae	*Sphyraena acutipinnis*	JX260977
106			*Sphyraena barracuda*	FJ265849
107			*Sphyraena jello*	EF609620
108			*Sphyraena obtusata*	FJ265848

(continued)

Table 1 (continued)

Sl. No.	Order	Family	Species	Acessions
109		Polynemidae	*Filimanus similis*	KJ468470
110			*Filimanus heptadactyla*	EF609523
111			*Leptomelanosoma indicum*	EF609539
112			*Polydactylus microstomus*	KJ468474
113			*Polydactylus plebeius*	KC576978
114			*Polydactylus sexfilis*	KJ468467
115			*Polydactylus sextarius*	EU392177
116			*Eleutheronema tetradactylum*	EF609512
117			*Polynemus paradiseus*	HQ219165
118		Leiognathidae	*Equulites leuciscus*	FJ265836
119			*Equulites lineolatus*	FJ237600
120			*Eubleekeria splendens*	FJ384712
121			*Gazza minuta*	EU148512
122			*Leiognathus equulus*	FJ347946
123			*Karalla daura*	EU392205
124			*Photopectoralis bindus*	EF609534
125			*Secutor ruconius*	EF609612
126			*Secutor insidiator*	FJ265837
127		Mullidae	*Parupeneus barberinus*	EU148577
128			*Parupeneus forsskali*	FJ347965
129			*Parupeneus macronemus*	KJ632829
130			*Parupeneus pleurostigma*	FJ237573
131			*Parupeneus trifasciatus*	FJ459576
132			*Upeneus sulphureus*	EF609637
133			*Upeneus vittatus*	FJ347944
134			*Mulloidichthys auriflamma*	EU014232
135		Nemipteridae	*Nemipterus bipunctatus*	HQ423413
136			*Nemipterus japonicus*	EF609554
137			*Nemipterus mesoprion*	EF609559
138			*Nemipterus nematophorus*	JN992286
139			*Nemipterus zysron*	JN992287
140			*Nemipterus hexodon*	EF609414
141			*Nemipterus furcosus*	EF609413
142			*Nemipterus virgatus*	EJ237835
143			*Nemipterus peronii*	EF609415
144			*Parascolopsis boesemani*	KR105824
145			*Parascolopsis eriomma*	KR105820
146			*Parascolopsis aspinosa*	KR105815

(continued)

Table 1 (continued)

Sl. No.	Order	Family	Species	Acessions
147		Apogonidae	*Pristiapogon kallopterus*	EU392192
148			*Apogon quadrifasciatus*	EU148585
149			*Apogon norfolcensis*	FJ237579
150		Chaetodontidae	*Chaetodon collare*	FJ237559
151			*Chaetodon decussatus*	FJ237562
152			*Chaetodon trifascialis*	FJ237610
153			*Heniochus singularius*	JX669540
154			*Heniochus acuminatus*	EU014239
155			*Roa jayakari*	KF268183
156		Gerreidae	*Gerres erythrourus*	KC774649
157			*Gerres filamentosus*	KC774650
158			*Gerres oyena*	JX260873
159			*Pentaprion longimanus*	EU392182
160		Lethrinidae	*Lethrinus conchyliatus*	EU148536
161			*Lethrinus miniatus*	EU148533
162		Lutjanidae	*Lutjanus bohar*	EU541340
163			*Lutjanus fulviflamma*	EU541339
164			*Lutjanus johnii*	EU148538
165			*Lutjanus kasmira*	HQ658118
166			*Lutjanus lutjanus*	EU148541
167			*Lutjanus malabaricus*	EU014231
168			*Lutjanus russellii*	EU148540
169			*Macolor niger*	KJ425304
170			*Pinjalo pinjalo*	EU541341
171			*Pristipomoides multidens*	FJ237568
172		Pomacentridae	*Amphiprion ephippium*	JX987299
173			*Amphiprion frenatus*	JX901062
174			*Amphiprion clarkii*	JX573169
175			*Amphiprion ocellaris*	JX454573
176			*Amphiprion polymnus*	JX975292
177			*Amphiprion sebae*	KJ397926
178			*Amphiprion percula*	JX573170
179			*Amphiprion nigripes*	JX573171
180			*Amphiprion melanopus*	JX548321
181			*Amphiprion sandaracinos*	JX548320
182			*Amphiprion perideraion*	JX548324
183			*Amphiprion akallopisos*	JX975291
184			*Premnas biaculeatus*	JX548322
185			*Abudefduf septemfasciatus*	KJ129002

(continued)

Table 1 (continued)

Sl. No.	Order	Family	Species	Acessions
186			*Abudefduf vaigiensis*	FJ237570
187			*Chrysiptera unimaculata*	KF268162
188			*Dascyllus aruanus*	HQ589913
189		Pomacanthidae	*Apolemichthys xanthurus*	KC626014
190			*Pomacanthus annularis*	KF268138
191		Terapontidae	*Terapon puta*	KC774676
192			*Terapon theraps*	FJ347958
193			*Terapon jarbua*	FJ347885
194		Trichiuridae	*Lepturacanthus savala*	EF609542
195			*Trichiurus lepturus*	FJ347953
196			*Trichiurus russelli*	FJ265829
197			*Trichiurus auriga*	KR105923
198			*Aphanopus intermedius*	KP244485
199		Rachycentridae	*Rachycentron canadum*	EF609584
200		Scatophagidae	*Scatophagus argus*	EF609604
201		Priacanthidae	*Priacanthus hamrur*	EF609576
202			*Priacanthus prolixus*	KF815020
203			*Priacanthus tayenus*	FJ265857
204			*Pristigenys refulgens*	KF815040
205			*Priacanthus sagittarius*	KF815027
206			*Priacanthus blochii*	KF815022
207			*Cookeolus japonicus*	KF815042
208		Lactariidae	*Lactarius lactarius*	FJ347949
209		Ephippidae	*Ephippus orbis*	EU014240
210			*Platax teira*	KJ129011
211		Sparidae	*Acanthopagrus berda*	EU014244
212			*Acanthopagrus latus*	JX983210
213			*Argyrops spinifer*	EU541345
214		Ariommatidae	*Ariomma indicum*	KP244487
215		Blennidae	*Petroscirtes variabilis*	EU148523
216		Pempheridae	*Pempheris mangula*	KJ020193
217		Centrolophidae	*Psenopsis cyanea*	EU392194
218		Menidae	*Mene maculata*	FJ347939
219		Acanthuridae	*Acanthurus mata*	FJ459542
220			*Acanthurus triostegus*	KF434770
221		Gobiidae	*Acentrogobius chlorostigmatoides*	JX193727
222			*Acentrogobius audax*	JX193752
223			*Acrygobius baliurus*	JX193733
224			*Bathygobius fuscus*	JX193747

(continued)

Table 1 (continued)

Sl. No.	Order	Family	Species	Acessions
225			*Oxyurichthys ophthalmonema*	JX193748
226			*Oligolepis acutipennis*	JX193738
227			*Odontamblyopus rubicundus*	JX193743
228			*Obliquogobius cometes*	KP244597
229			*Exyrias puntang*	KC788177
230			*Periophthalmus novemradiatus*	KM229327
231			*Psammogobius biocellatus*	JX193732
232			*Trypauchen vagina*	JX193742
233			*Boleophthalmus boddarti*	KM229329
234			*Glossogobius giuris*	FJ459498
235		Labridae	*Halichoeres zeylonicus*	FJ158563
236			*Halichoeres timorensis*	KF422721
237			*Thalassoma lunare*	FJ237565
238			*Thalassoma jansenii*	FJ459567
239			*Thalassoma quinquevittatum*	FJ459571
240		Eleotridae	*Butis butis*	JX193740
241			*Eleotris fusca*	JX193751
242		Latidae	*Lates calcarifer*	FJ384689
243			*Psammoperca waigiensis*	FJ237578
244		Caesionidae	*Caesio caerulaurea*	FJ237594
245		Coryphaenidae	*Coryphaena hippurus*	FJ237540
246		Haemulidae	*Diagramma picta*	FJ237604
247			*Pomadasys kaakan*	JX260937
248		Drepaneidae	*Drepane longimana*	FJ459579
249			*Drepane punctata*	EU541347
250		Stromateidae	*Pampus argenteus*	FJ226532
251			*Pampus chinensis*	FJ226529
252		Siganidae	*Siganus canaliculatus*	KJ679902
253			*Siganus javus*	KJ679903
254		Sillaginidae	*Sillago sihama*	EF609615
255			*Sillago vincenti*	KC774673
256		Cepolidae	*Sphenanthias whiteheadi*	JN704806
257			*Acanthocepola indica*	KP244472
258		Bathyclupeidae	*Bathyclupea hoskynii*	KP244492
259		Emmelichthyidae	*Erythrocles acarina*	KP244547
260		Malacanthidae	*Hoplolatilus fronticinctus*	KC110755
261		Istiophoridae	*Istiophorus platypterus*	EF609527

(continued)

Table 1 (continued)

Sl. No.	Order	Family	Species	Acessions
262		Pinguipedidae	*Parapercis maculata*	KF876338
263		Gempylidae	*Ruvettus pretiosus*	HM990654
264			*Neoepinnula orientalis*	KP244591
265			*Promethichthys prometheus*	KP244604
266			*Lepidocybium flavobrunneum*	KP244579
267		Xiphiidae	*Xiphias gladius*	KJ739601
268		Percophidae	*Bembrops caudimacula*	KP244495
269		Symphysanodontidae	*Symphysanodon xanthopterygion*	KR105909
270		Ambassidae	*Ambassis ambassis*	KJ614388
271		Bramidae	*Brama dussumieri*	KJ020208
272	Clupeiformes	Clupeidae	*Hilsa kelee*	FJ158559
273			*Tenualosa toli*	EF609624
274			*Tenualosa ilisha*	JX260883
275			*Sardinella albella*	FJ237551
276			*Sardinella gibbosa*	FJ237613
277			*Sardinella longiceps*	EF609594
278			*Nematalosa nasus*	FJ384687
279			*Anodontostoma chacunda*	FJ347933
280		Engraulidae	*Encrasicholina heteroloba*	EU392186
281			*Stolephorus andhraensis*	EU541322
282			*Stolephorus commersonnii*	EU541323
283			*Stolephorus indicus*	FJ347957
284			*Thryssa dussumieri*	JX983289
285			*Thryssa hamiltonii*	EU148567
286			*Thryssa malabarica*	FJ347883
287			*Thryssa setirostris*	EU541324
288		Chirocentridae	*Chirocentrus dorab*	FJ347877
289		Pristigasteridae	*Pellona ditchela*	FJ347928
290		Dussumieriidae	*Dussumieria elopsoides*	FJ347962
291			*Dussumieria acuta*	EU014223
292		Nomeidae	*Psenes cyanophrys*	KJ020212
293			*Psenes arafurensis*	KJ020215
294			*Cubiceps whiteleggii*	KP244519

(continued)

Table 1 (continued)

Sl. No.	Order	Family	Species	Acessions
295	Mugiliformes	Mugilidae	*Ellochelon vaigiensis*	JQ045780
296			*Liza klunzingeri*	JX983355
297			*Liza subviridis*	JQ045782
298			*Valamugil speigleri*	JQ045778
299			*Liza macrolepis*	FJ347967
300			*Liza tade*	JQ045776
301			*Liza parsia*	JQ045779
302			*Liza planiceps*	JQ045784
303			*Mugil cephalus*	FJ347895
304			*Moolgarda cunnesius*	FJ384690
305			*Moolgarda seheli*	JQ045781
306	Siluriformes	Ariidae	*Arius arius*	KF010175
307			*Arius gagora*	JX260835
308			*Arius jella*	FJ265865
309			*Arius maculatus*	FJ869856
310			*Arius subrostratus*	EU148555
311			*Arius maculatus*	FJ403390
312			*Plicofollis platystomus*	KF824838
313			*Plicofollis tenuispinis*	KF824836
314			*Netuma thalassina*	EU014254
315			*Osteogeneiosus militaris*	EF609563
316		Plotosidae	*Plotosus limbatus*	KF824845
317			*Plotosus lineatus*	EU148554
318			*Plotosus canius*	KF824847
319	Pleuronectiformes	Cynoglsidae	*Cynoglossus dubius*	FJ347907
320			*Cynoglossus cynoglossus*	JX983282
321			*Cynoglossus bilineatus*	FJ384697
322			*Cynoglossus puncticeps*	EU541318
323			*Cynoglossus macrostomus*	FJ347954
324			*Cynoglossus lingua*	EU541316
325			*Cynoglossus carpenteri*	KP244525
326		Paralichthyidae	*Pseudorhombus arsius*	JX260939
327			*Pseudorhombus elevatus*	EU541314
328			*Pseudorhombus malayanus*	EU541312
329		Psettodidae	*Psettodes erumei*	EF609580
330		Bothidae	*Laeops macrophthalmus*	KP244572
331			*Chascanopsetta lugubris*	KP244514
332		Soleidae	*Zebrias synapturoides*	FJ347914

(continued)

Table 1 (continued)

Sl. No.	Order	Family	Species	Acessions
333	Beloniformes	Hemiramphidae	*Hemiramphus far*	EU148546
334			*Hyporhamphus dussumieri*	JX983320
335			*Hyporhamphus xanthopterus*	EU148545
336			*Rhynchorhamphus georgii*	JX983484
337			*Rhynchorhamphus malabaricus*	KJ641743
338		Belonidae	*Strongylura leiura*	FJ237566
339			*Strongylura strongylura*	EU014257
340	Aulopiformes	Synodontidae	*Harpadon nehereus*	EU148582
341			*Saurida pseudotumbil*	KF876337
342			*Saurida nebulosa*	KF876020
343			*Saurida tumbil*	EF609603
344			*Saurida undosquamis*	FJ347930
345			*Synodus variegatus*	FJ265846
346			*Saurida* cf. *micropectoralis*	KR105884
347			*Saurida longimanus*	KR105853
348			*Trachinocephalus myops*	EF609630
349		Chlorophthalmidae	*Chlorophthalmus acutifrons*	JX944228
350			*Chlorophthalmus corniger*	JX944224
351	Tetraodontiformes	Tetraodontidae	*Arothron hispidus*	KC409391
352			*Arothron immaculatus*	KC409372
353			*Arothron leopardus*	KJ093731
354			*Arothron stellatus*	KC409388
355			*Chelonodon patoca*	KC409363
356			*Lagocephalus guentheri*	KC409371
357			*Lagocephalus inermis*	JX995942
358			*Lagocephalus spadiceus*	FJ384711
359		Ostraciidae	*Lactoria cornuta*	FJ237606
360			*Tetrosomus gibbosus*	KF268149
361		Balistidae	*Odonus niger*	FJ459554
362			*Rhinecanthus rectangulus*	FJ459548
363		Monacanthidae	*Anacanthus barbatus*	FJ541074

(continued)

Table 1 (continued)

Sl. No.	Order	Family	Species	Acessions
364	Anguilliformes	Muraenidae	*Gymnothorax undulatus*	KF297588
365			*Gymnothorax punctatus*	KF297589
366			*Gymnothorax pictus*	KF297590
367			*Strophidon sathete*	FJ384683
368		Muraenesocidae	*Muraenesox cinereus*	KF297592
369		Nemichthyidae	*Nemichthys acanthonotus*	KP244586
370		Scaridae	*Scarus quoyi*	KF930376
371	Syngnathiformes	Syngnathidae	*Hippocampus trimaculatus*	EU930320
372			*Hippocampus kuda*	FJ541049
373			*Hippocampus kelloggi*	GQ502149
374		Centriscidae	*Centriscus scutatus*	FJ265863
375		Fistulariidae	*Fistularia petimba*	FJ541073
376	Scorpaeniformes	Scorpaenidae	*Pterois russelii*	KF268175
377			*Ebosia falcata*	KP244540
378			*Pontinus nigerimum*	KR105829
379		Dactylopteridae	*Dactyloptena orientalis*	EU148590
380		Peristediidae	*Satyrichthys adeni*	KR105846
381		Setarchidae	*Setarches guentheri*	KR105907
382		Berycidae	*Beryx mollis*	KP244504
383	Beryciformes	Holocentridae	*Myripristis murdjan*	FJ459546
384			*Sargocentron rubrum*	KF442242
385			*Ostichthys kaianus*	KR105810
386		Trachichthyidae	*Gephyroberyx darwinii*	KP244553
387	Gonorynchiformes	Chanidae	*Chanos chanos*	JX260845
388	Lophiiformes	Diceratiidae	*Bufoceratias thele*	KP244512
389		Lophiidae	*Lophiodes lugubris*	KP244581
390			*Lophius indicus*	KP244583
391	Perciformes	Pentacerotidae	*Histiopterus typus*	KP244559
392	Notacanthiformes	Notacanthidae	*Notacanthus indicus*	KR105800
393			*Notacanthus* sp. A	KR105803
394	Ophidiiformes	Ophidiidae	*Neobythites steatiticus*	KP244588
395	Osmeriformes	Platytroctidae	*Normichthys yahganorum*	KR105797
396		Alepocephalidae	*Alepocephalus bicolor*	KP244479
397	Zeiformes	Parazenidae	*Cyttopsis rosea*	KP244533
398		Zeidae	*Zenopsis conchifer*	KR105931

Table 2 List of Indian marine fish species barcoded (Cartilaginous fishes)

Sl No.	Order	Family	Species	Acc. No.
1	Chimaeriformes		*Neoharriotta pinnata*	HM239670
2			*Hydrolagus africanus*	KF89952
3	Hexanchiformes	Hexanchidae	*Heptranchias perlo*	HM239668
4			*Hexanchus griseus*	KF899463
5	Echinorhiniformes	Echinorhinidae	*Echinorhinus brucus*	HM467790
6	Orectolobiformes	Hemiscylliidae	*Chiloscyllium griseum*	KF899626
7		Stegostomatidae	*Stegostoma fasciatum*	KF899644
8		Ginglymostomatidae	*Nebrius ferrugineus*	KM973183
9		Rhincodontidae	*Rhincodon typus*	KF899634
10	Lamniformes	Alopiidae	*Alopias pelagicus*	HM239672
11			*Alopias superciliosus*	KF899554
12			*Alopias vulpinus*	KF899558
13			*Isurus oxyrinchus*	KF899536
14		Odontaspididae	*Odontaspis noronhai*	KF899559
15		Lamnidae	*Isurus oxyrinchus*	KF899541
16			*Isurus paucus*	KF899542
17		Pseudocarchariidae	*Pseudocarcharias kamoharai*	KF899532
18	Carcharhiniformes	Carcharhinidae	*Carcharhinus macloti*	KF913242
19			*Carcharhinus longimanus*	KF899777
20			*Carcharhinus limbatus*	KF899814
21			*Carcharhinus falciformis*	KF899803
22			*Carcharhinus brevipinna*	KF899802
23			*Carcharhinus amboinensis*	KF899796
24			*Carcharhinus altimus*	KF899786
25			*Carcharhinus sorrah*	KF899821
26			*Carcharhinus amblyrhynchos*	KF899792
27			*Carcharhinus albimarginatus*	KF899782
28			*Carcharhinus leucas*	KF899812
29			*Carcharhinus melanopterus*	KF899824
30			*Lamiopsis temminckii*	KF899563
31			*Prionace glauca*	KF899653
32			*Rhizoprionodon acutus*	KF899687
33			*Galeocerdo cuvier*	KF899436
34			*Triaenodon obesus*	KF899768
35			*Scoliodon laticaudus*	KF899696
36		Sphyrnidae	*Sphyrna lewini*	KF899746
37			*Sphyrna zygaena*	KF899755
38		Scyliorhinidae	*Halaelurus quagga*	KF899715

(continued)

Table 2 (continued)

Sl No.	Order	Family	*Species*	Acc. No.
39			*Cephaloscyllium silasi*	HM467791
40			*Bythaelurus hispidus*	KF899706
41			*Apristurus* sp. A	KF899717
42		Triakidae	*Iago omanensis*	KF899745
43			*Mustelus mosis*	KC175449
44		Proscylliidae	*Eridacnis radcliffei*	KF899425
45		Hemigaleidae	*Hemipristis elongata*	KF899453
46			*Paragaleus randalli*	KF913245
47	Squaliformes	Squalidae	*Squalus* sp. A	KR149162
48		Centrophoridae	*Centrophorus squamosus*	KF899385
49			*Centrophorus atromarginatus*	KF899387
50			*Centrophorus granulosus*	KF899391
51			*Centrophorus zeehaani*	KF899394
52			*Deania profundorum*	KF899382
53		Etmopteridae	*Etmopterus pusillus*	KF899426
54		Somniosidae	*Centroselachus crepidater*	KF899400
55			*Zameus squamulosus*	KF899769
56	Torpediniformes	Torpedinidae	*Torpedo sinuspersici*	KF899724
57			*Torpedo* sp. A	KF899725
58		Narcinidae	*Benthobatis moresbyi*	KJ768662
59			*Narcine oculifera*	KF899605
60			*Narcine maculata*	KF899600
61			*Narcine* sp. A	KF899601
62	Rajiformes	Rhinidae	*Rhina ancylostoma*	KF899663
63		Rhynchobatidae	*Rhynchobatus* cf. *laevis*	KF899693
64			*Rhynchobatus australiae*	JN108019
65		Rhinobatidae	*Glaucostegus obtusus*	KF899439
66			*Glaucostegus thouin*	KF899441
67			*Rhinobatos lionotus*	KF899672
68			*Rhinobatos punctifer*	KF899668
69			*Rhinobatos variegatus*	KF899673
70		Rajidae	*Okamejei powelli*	KF899616
71			*Dipturus* sp. A	KF899402
72			*Dipturus* sp. B	KF899416
73			*Dipturus* cf. *johannisdavisi*	KF899412
74			*Dipturus* cf. *gigas*	KR149208

(continued)

Table 2 (continued)

Sl No.	Order	Family	*Species*	Acc. No.
75	Myliobatiformes	Plesiobatidae	*Plesiobatis daviesi*	HM467801
76		Dasyatidae	*Dasyatis microps*	KJ749657
77			*Himantura undulata*	KF899506
78			*Himantura uarnacoides*	KF899499
79			*Himantura granulata*	KF899471
80			*Himantura leoparda*	KF899501
81			*Himantura jenkinsii*	KF913237
82			*Himantura fai*	KF899475
83			*Himantura bleekeri*	KC508511
84			*Himantura gerrardi*	KF899364
85			*Himantura imbricata*	KF899356
86			*Himantura uarnak*	KF899511
87			*Neotrygon kuhlii*	KF899609
88			*Taeniura meyeni*	HM467797
89			*Pastinachus sephen*	KF899368
90			*Pastinachus atrus*	KF899373
91			*Pteroplatytrygon violacea*	HM239671
92			*Urogymnus asperrimus*	KC508509
93		Gymnuridae	*Gymnura poecilura*	KF899445
94		Myliobatidae	*Aetobatus narinari*	JX978339
95			*Aetobatus ocellatus*	KF899589
96			*Aetomylaeus maculatus*	KF899587
97			*Aetomylaeus vespertilio*	KF899586
98			*Manta birostris*	KF899569
99			*Mobula japanica*	JX978334
100			*Mobula kuhlii*	KF899582
101			*Mobula tarapacana*	KF899580
102		Rhinopteridae	*Rhinoptera jayakari*	KF899683

1.1 Carangids

The family Carangidae comprises 30 genera with 146 species distributed throughout Atlantic, Indian and Pacific Oceans (http://www.fishbase.org). In India approximately 58 species are recorded. Carangids are considered as a most delicious and highly valued fish in India. This group of fishes is categorized into five sub groups i.e., black pomfrets, queen fishes, trevallies, scads, and pompanos. The group has emerged as one of the important resources along Indian coast. A total of 51 fish species, covering 18 genera were barcoded in India from Indian water. NBFGR has completed barcoding of 45 carangid species belonging to 16 genera. Analaysis show that longfin trevally, *Carangoides armatus* is a species complex,

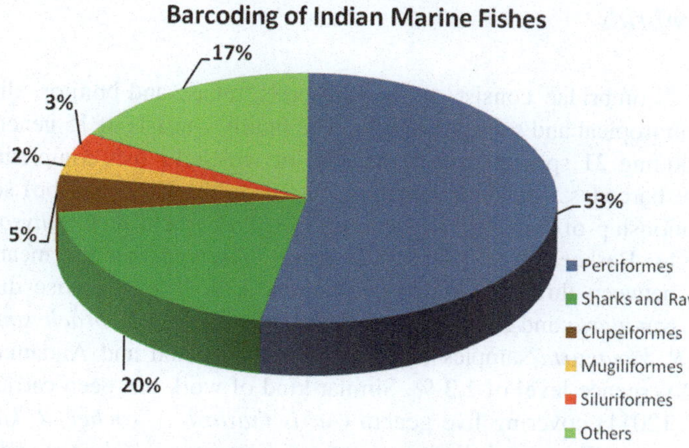

Fig. 1 Different groups of fishes barcoded in India

suggesting cryptic species within the complex. Persis et al. (2009) has carried out the phylogenetic study using COI for 28 carangids from Kakinada coast in India and revealed that all these fishes fall into three distinct groups which are genetically distant from each other and exhibited identical phylogenetic reservation. In addition to that 17 fish species from 13 genera were analysed by Lakra et al. (2011) who observed that the average genetic distance within species was 0.32 % whereas between species it was 16.1 %. The NJ tree also revealed distinct clusters for species of same genera with 94–100 % bootstrap values.

1.2 *Sciaenids*

The family Sciaenidae, known as croakers or drums, is distributed worldwide with approximately 70 genera and 300 species including about 40 species from Indian waters. This group contributes approximately 4.6 % to the total Indian marine fish production. Out of these 14 species were barcoded from Indian waters. Lakra et al. (2009) studied the phylogenetic relationships of six species of Indian sciaenids (*Otolithes cuvieri, O. ruber, Johnius borneensis, J. dussumieri, Dendrophysa russelii* and *Nibea maculata*) based on 16S rRNA and cytochrome c oxidase subunit I which revealed three genetic distinct groups. The average genetic distance within species was 0.28 % whereas the overall mean distance among the species was 18.20 %. The NJ tree clearly distinguished the species having same genus under one cluster with a bootstrap value of 96–100 %.

1.3 Scombrids

The family Scombridae consist of the mackerels, tunas, and bonitos, distributed worldwide in tropical and subtropical seas. The family consists of 15 genera and 54 species including 21 species from India. Out of which 12 fish comprising seven genera were barcodes from Indian waters. Based on the COI (655 bp) sequences genetic relationship of *Rastrelliger kanagurta*, *R. faugni* and *R. brachysoma* were carried out by Basheer et al. (2015) and it was observed that the mean genetic divergence between three mackerel species was 5 %. The pair-wise divergence between *R. kanagurta* and *R. faugni* was 0.08-0.09 and with *R. brachysoma* it was 0.03-0.04. *R. kanagurta* samples between Indian mainland and Andaman waters showed a divergence level of 1.2 %. Similar kind of work has been carried out by Lakra et al. (2011) covering five genera (*Auxis thazard, A. rochei, R. kanagurta, Thunus albacares, T. tonggol, Euthynnus affinis, Katsuwonus pelamis*) and revealed the 0.3 % genetic distance within species and 9.20 between the species. All the species were clearly separated into different clusters in the NJ tree with a bootstrap value ranging from 96 to 100 %.

1.4 Serranids

The family serranidae consist of 537 species, under 75 genera including 70 known species from India. Serranids are distributed throughout tropical and temperate oceans and it is represents one of the most important resources targeted by coastal fisheries in country. A total of 47 species of Serenade family were barcoded from Indian water by NBFGR and other agencies. Traditionally serranid fishes are identified based on visible morphological, meristic, and anatomical characters (Roy and Gopalakrishnan 2011). Sachithanandam et al. (2011) has shown the utility of COI divergences in identifying all the *Plectropomus maculatus* fishes in Andaman Islands with minimum base pair (133 bp). A molecular phylogeny study using COI sequences has shown that the overall mean distance among the species is 12.60 % and within species 0.24 %. In the NJ tree all the species were under one clad with bootstrap values of 94–98 % (Lakra et al. 2011).

At NBFGR, we barcoded 36 grouper species, including seven species listed in the IUCN red list under threatened category. Barcodes from six genera *Aetheloperca*, *Cephalopholis*, *Epinephelus*, *Hyporthodus*, *Plectropomus* and *Variola* were developed out of which *E. polylepis* and *E. miliaris* were developed for the first time. A comparison of the sequences generated were done with sequences available in NCBI to authenticate the species identification and resolve taxonomic ambiguity. Upon the sequence analysis, taxonomic ambiguity exists in *E. polylepis* and *E. cholorostigma*. *E. polylepis* has a distribution in northwest Indian Ocean (Craig et al. 2011) and there is no further report of the species after the description by Randall and Heemstra (1991) from India. It may be misidentification by earlier workers and reported as *E. chlorostigma*. Phylogenetic analysis using neighbour joining tree is given in Fig. 2.

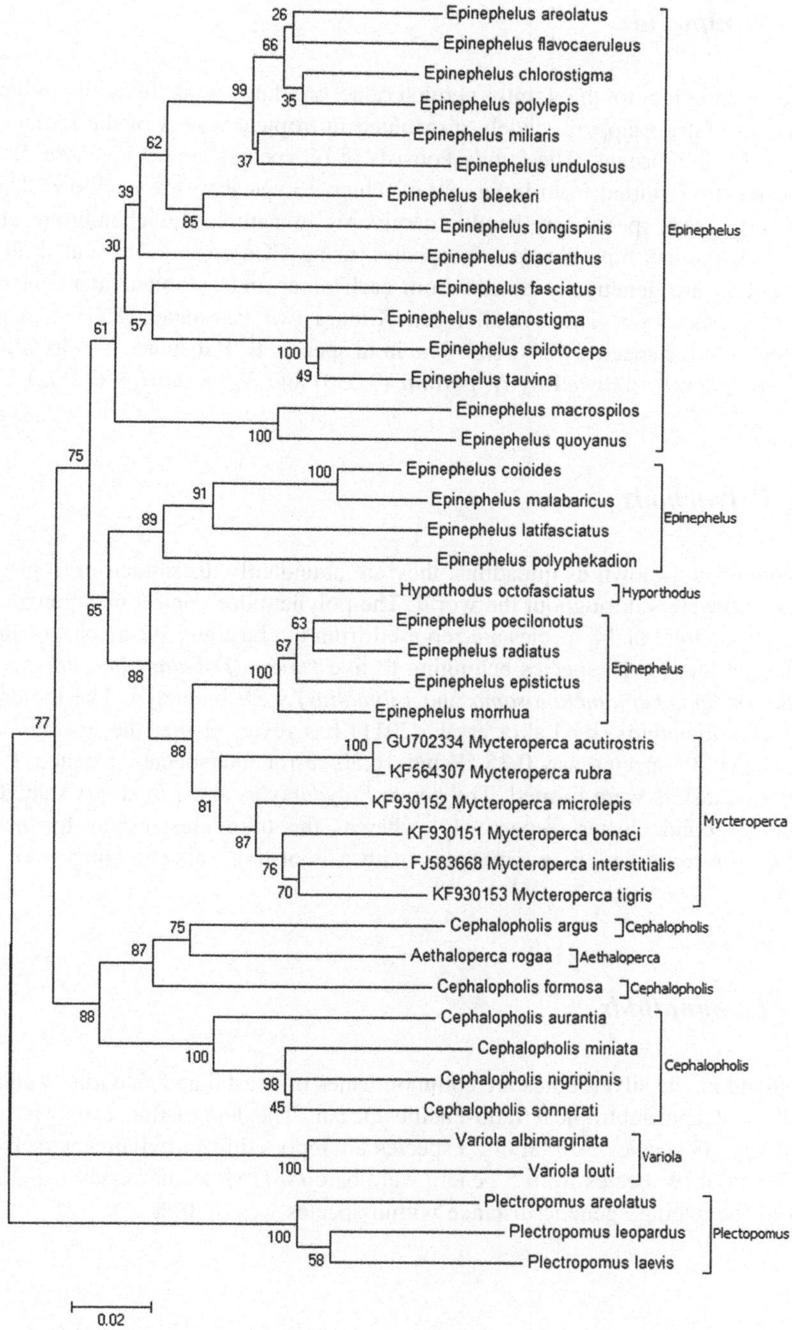

Fig. 2 NJ tree of COI gene sequences computed by using K2P distance of the groupers barcoded

1.5 Nemipterids

Species belonging to the family nemipteridae are known as threadfin, whiptail breams and false snappers, widely distributed in tropical waters of the Indian and Western Pacific Oceans. The family consists of 67 species, under 5 genera. So far 30 species are reported from India, out of which 13 species were barcoded. Under this family most species are benthic carnivores in nature. Ravichandirane et al. (2012), distinguish nine *Nemipterus* species, using COI gene and revealed all the nine species are genetically distant from each other and exhibited identical phylogenetic reservation. The overall mean Kimura two parameter (K2P) distances between the nine species was 0.109. The intra species K2P distance was high in *N. japonicus* (0.069) followed by *N. peronii* (0.050) and *N. mesoprion* (0.002).

1.6 Polynemids

Polynemids are known as threadfins, they are abundantly distributed in tropical to subtropical waters throughout the world. The polynemidae consist of 8 genera and 43 species, a total of 14 species are reported from the East and West coast of India, of which a total of 10 species belonging to five genera (*Polydactylus, Polynemus, Eleutheronema, Leptomelanosoma* and *Filimanus*) were barcoded. The molecular study of polynemids, by Lakra et al. (2011) has revealed that the average K2P distance within species was 0.35 % and 16.30 % for interspecies distance. In NJ tree three clusters were formed. The genus *Polydactylus* and *Filimanus* shared the first and second cluster respectively, whereas the third cluster was formed by *Leptomelanosoma* and *Eleutheronema*, with a bootstrap value ranging from 92–100 %.

1.7 Leiognathids

Leiognathids, the silverbellies are common fishes of coastal and estuarine waters in the tropical and subtropical Indo-Pacific Ocean. The leiognathidae consist of 9 genera and 48 species. A total of 23 species are reported from Indian waters, out of which total of 9 species from 5 genera were barcoded. Molecular study using COI showed the average genetic distance within species was 0.20 %.

1.8 Mullids

The family Mullidae are commonly called goatfish. The family consist of 85 species, under 6 genera. So far 23 species reported from India, out of which 8 species comprising three genera have been barcoded. Molecular phylogeny study by Lakra et al. (2011) of six fish species showed the average genetic distance within species was 0.38 % whereas the overall mean distance among the species was 13.90 %. The NJ tree revealed that the genera *Parupeneus, Mulloidichthys* and *Upeneus* formed three separate clades with a boostrap value of 95–99 %.

1.9 Lutjanids

Snappers are of mainly marine but some are found in estuaries and fresh water. The family includes about 110 species in 17 genera, some of which are considered as important food fish. A total of 47 fishes are reported from Indian waters of which barcoding has been completed for only 10 species. By using COI Victor et al. (2009) has successfully identified the larvae and newly-settled juveniles of the Cubera Snapper (*Lutjanus cyanopterus*) and observed that the nearest neighbor species, *L. analis*, was more than 11 % divergent. Recently *Lutjanus johni, L.lutjanus, L. russelli*, and *L. malabaricus* has been identified using COI by Krishna et al. (2012) and Lakra et al. (2011).

1.10 Pomacentrids

The family pomacentridae comprises damselfishes and clownfishes. They are primarily marine, while a few species inhabit freshwater and brackish environments. About 385 species are classified in this family with 29 genera, and a total of 52 species have been reported from Indian waters. The barcodes for 18 species under 4 genera have been generated in India by Lakra et al. (2011), Dhaneesh et al. (2015) and Bamaniya et al. (2015). The cytochrome oxidase 1 (COI) sequence of 13 clownfish species was used for construction of phylogenetic relationship, and scrutinized species boundaries between four closely related species of the sub-genera *Phalerebus, Amphiprion and Paramphiprion* (Dhaneesh et al. 2015).

1.11 Gobiids

Gobiids constitute a major proportion of fish population in both tropical and temperate freshwater as well as marine ecosystem and is one of the largest families of marine fishes containing 1718 species in 251 genera. Out of these a total of 100 species are reported from India. 14 species have been barcoded by Lakra et al. (2011), and Viswambharan et al. (2015). Due to their small size, cryptic ecology and ambiguous morphological characters, gobiid diversity was not documented completely. Viswambharan et al. (2015) generated COI barcode for 11 species of gobies for accurate species identification. The COI barcodes clearly distinguished all these species with higher interspecific genetic distance values than intraspecific values based on K2P (Kimura 2 Parameter) model. The average genetic distance (K2P model) within species, genus and family was 1.2 %, 22.2 % and 25.3 %, respectively.

1.12 Clupeids

Clupeidae include many of the most important food fishes in the world that includes herrings, shads, sardines, hilsa, and menhadens. The family comprises 198 species under 54 genera. A total of 27 species have been reported from India, of which 10 species have been barcoded by several workers (Lakra et al. 2011). The phylogenetic study of clupeids has been carried out by Lakra et al. (2011) using COI gene of seven species and observed that overall mean distance among the species was very high (20.30 %). The average genetic distance within species was 0.41 %.

1.13 Engraulids

Engraulids are mainly known as a Anchovies, they are schooling fishes, mostly of shallow coastal waters and estuaries in tropical and temperate regions. Some are estuarine in nature. They are widely distributed across Atlantic, Indian and Pacific Oceans. The family consists of 198 species in 54 genera of which 34 species are recorded from Indian waters. Anchovies are considered as a delicious and preferred fish in India. Despite their importance as one of the important fishery resources, only 8 species under this group have been barcoded from India covering three genera. Two species, *Thryssa purava* and *Thryssa setirostris* were barcoded from estuaries of River Krishna by Krishna et al. (2012). The molecular phylogenetic study has been carried out in selected species using COI sequence, and revealed the average genetic distance between *Stolephorus indicus* and *Stolephorus commersonnii* was 9.11 % (Khan et al. 2010) and the average genetic distance within species was 0.41 % (Lakra et al. 2011). In the neighbor-joining tree both the species fall into same clade with a bootstrap value of 98 % (Khan et al. 2010; Lakra et al. 2011).

1.14 Mugilids

Mugilidae, the mullets or grey mullets are distributed worldwide in coastal temperate and tropical waters, with some species in fresh water. Mullets serve as an important source of food in coastal areas. The family includes about 76 species in 20 genera, although half of the species are in just two genera (*Liza* and *Mugil*). A total of 17 species have been reported from Indian waters, of which 14 species were barcoded. Rahman et al. (2013) reported that Canonical Analysis of Principal Coordinates (CAP) failed to separate the 10 species collected from Southeast Coast of India, which were clearly identified and differentiated using COI gene.

1.15 Ariids

Ariidae is a family of catfish, which are marine and esturine in nature, widely distributed in tropical to warm temperate zones. The family comprise about 143 species which includes 24 species from India, of which barcoding of 10 species was completed. Lakra et al. (2011) carried out the COI based molecular study of the catfishes of three genera namely *Osteogeneiosus*, *Netuma* and *Arius* under the family and observed the 0.23 % average K2P distance within species and 8.10 % within family. The NJ tree revealed two clusters, where first cluster is of *Arius subrostratus* and *A. arius*. The second cluster was shared by *Netuma thalassinus* and *Osteogeneiosus militaris* with a bootstrap value of 90 to 99 %.

1.16 Sphyraenids

The barracuda are marine ray-finned pelagic predatory fish of the genus *Sphyraena*, the only genus in the family Sphyraenidae. Globally there are more than 22 species of barracuda, So far seven species have been reported from Indian waters viz., *S. barracuda, S. jello, S. putnamiae, S. qenie, S. forsteri, S. obtusata and S. novahollandae.* All the species from Arabian Sea were characterized by NBFGR using DNA barcodes. Among the species examined, one was confirmed as new species and named *Sphyraena arabiansis* sp. nov. (Abdussamad et al. 2015). The intraspecies genetic distance ranged from 0.000 to 0.007, while interspecies varied from 0.111 to 0.273. COI sequences of *Sphyraena barracuda* and *Sphyraena arabiansis* showed a clear barcode split (11.4 % divergence) congruent with morphological differences. The NJ tree revealed very distinct species clusters. The average interspecies distance among the seven species in the family sphyraenidae was 15.2 % (Jena et al. 2014).

1.17 Sharks and Rays

India is one of the leading Chondrichthyan fishing nations with an estimated landing of 46,471 tonnes (sharks 45.5 %, rays 49.5 % and skates 5 %) in 2013 that accounted for 1.23 % of the total marine fish landings in the country (CMFRI, 2013). Despite rich elasmobranch diversity, only a few detailed studies have been undertaken that makes their management and conservation very difficult. Ward et al. (2005) validated the efficacy of COI barcodes for identifying chondrichthyans by sequencing 61 species of sharks and rays from Australian waters. Recent taxonomic studies that use both morphological and molecular markers resulted in taxonomic resolution of many species complexes and discovery of many new species. A total barcode record of 109 species are available of which 104 species were barcoded by NBFGR, India, representing 10 orders and 33 families of chondrichthyans from Indian waters. This is the largest study using genetic barcodes approach to identify sharks and rays found in the Indian waters. The average Kimura 2 parameter (K2P) distance separating individuals within species observed was 0.32 %, and the average distance separating species within genera 6.73 %. There are 37 and 29 species sequences generated in the present study representing new sequences for GenBank and BOLD respectively. During this study seven species were suggested as putative new species requiring formal descriptions and eleven elasmobranch species were confirmed first records for Indian waters. The present study uncovered the presence of eight undescribed or unrecognized batoid species including three in the genus *Himantura*. Barcode analysis result shows the presence of unrecognized species and highlights the need for further detailed taxonomic examinations of several families of elasmobranchs from Indian waters. Six species of *Carcharinus* were barcoded and phylogentic tree is given in Fig. 3. Barcode analysis shows very clear cut differentiation between these 6 species.

1.18 Other Important Groups

Lizard fishes (Synodontidae), Puffers (Tetradontidae), Half beaks (Hemirahmphidae), Big eyes (Priacanthidae), Tongue fishes (Cynoglossidae), Butterfly fishes (Chaetodontidae), Mojarras (Gerreidae), Cutlass fishes (Trichuridae), Wrasses (Labridae), snake mackerels or escolars (Gempylidae), and Moray eels (Muraenidae) are some of the important groups of marine fishes, distributed along the coast of India. These groups of fishes contribute a substantial amount to total marine landing. The barcoding and species identification of this entire group using COI has already been taken up by several researchers, but in limited scale. Hence there is pressing need of DNA barcoding of all these groups along with other groups available in the country for effective conservation and management purpose.

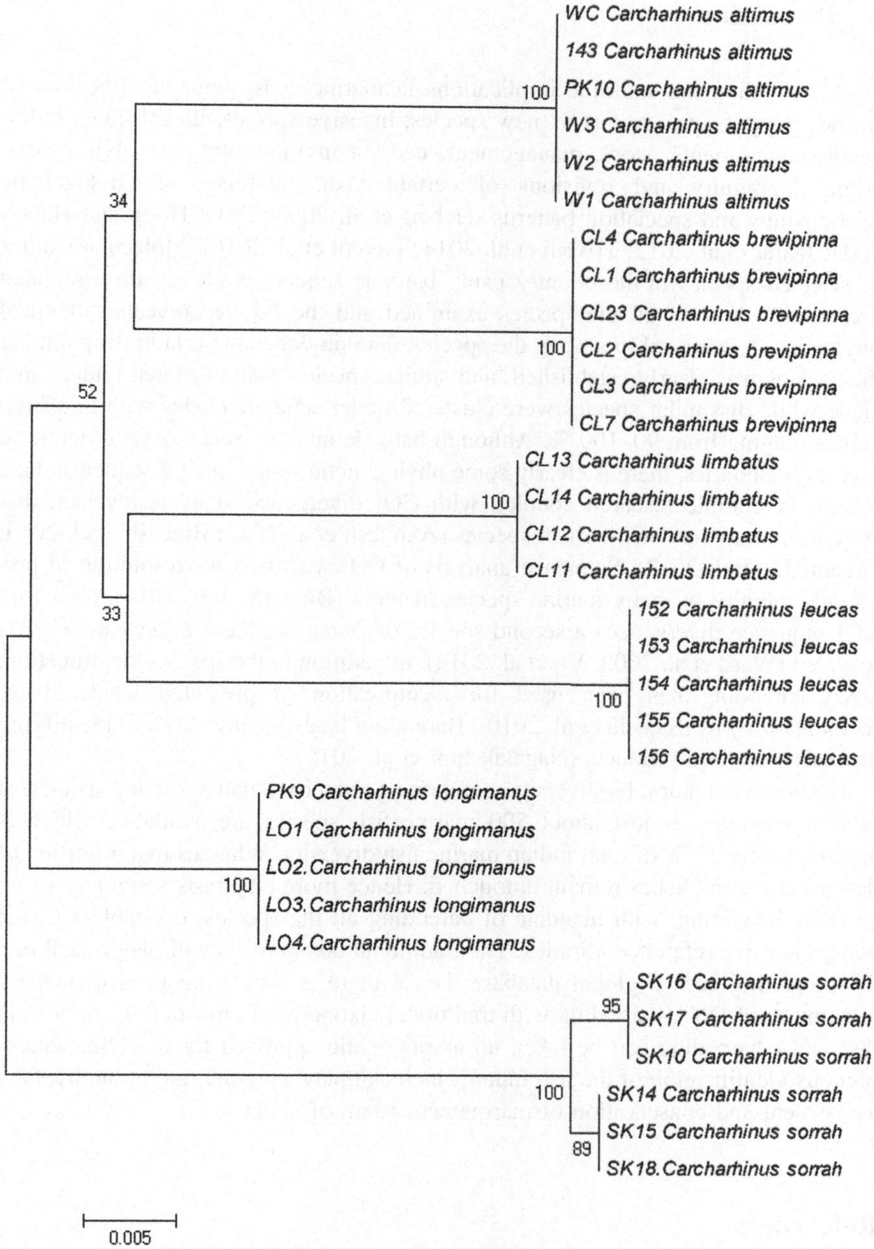

Fig. 3 K2P distance neighbor-joining tree of COI sequence from six species of *Carcharhinus*

2 Discussion

DNA barcoding has multiple implications in marine ecosystem; identification of species, cryptic species, larvae, new species, invasive species, illegal trade, biodiversity assessments, stock management, ecosystem monitoring, resolving taxonomic ambiguity and revisions of certain taxa, inference of phylogenetic relationships and speciation patterns (Hebert et al. 2003, 2004; Hogg and Hebert 2004; Jaafar et al. 2012; Trivedi et al. 2014; Trivedi et al. 2016). Molecular studies of selected species in the country using barcode sequences clearly discriminated taxonomic status of all the species examined and the NJ tree revealed identical phylogenetic relationship among the species. The phylogenetic relationship among the species was clearly established, and similar species were clustered under same clade while dissimilar species were clustered under separate clades with bootstrap values ranging from 90–100 %. Although barcode analysis seeks only to delineate species boundaries, there is clearly some phylogenetic signal in COI sequence data. Recent taxonomic research coupled with COI divergence analysis revealed discovery of many new marine fish species (Akhilesh et al. 2012; Bineesh et al. 2013; Greenfield et al. 2012). Sequence analysis of COI was used for resolution of taxonomic identity of many marine species in India (Bineesh et al. 2014). Based on COI sequence divergences a second species of Asian sea bass, *Lates calcarifer* is revealed (Ward et al. 2008; Vij et al. 2014). In addition to the species identification, DNA barcoding has been used for identification of protected whale shark, *Rhyncodon typus* (Sajeela et al. 2010). Barcoding is also being used for identifying processed seafood products (Nagalakshmi et al. 2016).

Despite rich natural biodiversity, comprising over 1518 native marine species, at present barcodes of just about 500 marine fish species are available, which is approximately 33 % of total Indian marine fish diversity. Whereas major portion of described marine fishes remain untouched. Hence more emphasis should be given to DNA barcoding, with mandate of barcoding all the species to establish global comprehensive reference libraries. The traditional taxonomists will play a vital role in completing such a global database; hence there is a pressing need to make a integration of DNA barcoding with traditional taxonomy. In nutshell it can be said that DNA barcoding can be taken up as pragmatic approach for resolving unambiguous identification of the fish fauna which can play a crucial role in biodiversity assessment and conservation of marine ecosystem of country.

References

Abdussamad EM, Retheesh TB, Thangaraja R, Bineesh KK, Prakasan D (2015) *Sphyraena arabiansis* a new species of barracuda (Family: Sphyraenidae) from the south-west coast of India. Indian J Fish 62(2):1–6

Akhilesh KV, Bineesh KK, White WT (2012) Liopropoma randalli, a new serranid (Teleostei: Perciformes) fish from the Indian Ocean. Zootaxa 3439:43–50

Bamaniya DC, Pavan-Kumar A, Gireesh-Babu P, Sharma N, Reang D, Krishna G, Lakra WS (2015) DNA barcoding of marine ornamental fishes from India. Mitochondrial DNA 27:1–5

Basheer VS, Mohitha C, Vineesh N, Divya PR, Gopalakrishnan A, Jena JK (2015) Molecular phylogenetics of three species of the genus *Rastrelliger* using mitochondrial DNA markers. Mol Biol Rep 42(4):873–879

Bineesh KK, Akhilesh KV, Abdussamad EM, Pillai NGK (2013) *Chelidoperca maculicauda*, a new species of perchlet (Teleostei: Serranidae) from the Arabian Sea. Aqua Int J Ichthyol 19(2):71–78

Bineesh KK, Akhilesh KV, Sajeela KA, Abdussamad EM, Gopalakrishnan A, Basheer VS, Jena JK (2014) DNA Barcoding Confirms the Occurrence Rare Elasmobranchs in the Arabian Sea of Indian EEZ. Middle East J Sci Res 19(9):1266–1271

Craig MT, de Mitcheson YJS, Heemstra PC (2011) Groupers of the world. Published by NISC Grahamstown South Africa 402p

Dhaneesh KV, Kumar TTA, Kumar AB (2015) Barcoding, phylogeography and species boundaries in clownfishes of the Indian Ocean. DNA Barcodes 3(1):5–16

Eschmeyer WN, Fricke R, Fong JD, Polack DA (2010) Marine fish diversity: history of knowledge and discovery (Pisces). Zootaxa 2525:19–50

Greenfield DW, Bineesh KK, Akhilesh KV (2012) *Colletteichthys flavipinnis*, a new toadfish species from Sri Lanka and India (Teleostei: Batrachoididae). Zootaxa 3437:24–31

Hebert PD, Cywinska A, Ball SL (2003) Biological identifications through DNA barcodes. Proc R Soc Lon B Biol Sci 270(1512):313–321

Hebert PD, Penton EH, Burns JM, Janzen DH, Hallwachs W (2004) Ten species in one: DNA barcoding reveals cryptic species in the neotropical skipper butterfly *Astraptes fulgerator*. Proc Natl Acad Sci USA 101(41):14812–14817

Hogg ID, Hebert PD (2004) Biological identification of springtails (Hexapoda: Collembola) from the Canadian Arctic, using mitochondrial DNA barcodes. Can J Zool 82(5):749–754

Jaafar TNAM, Taylor MI, Nor SAM, de Bruyn M, Carvalho GR (2012) DNA barcoding reveals cryptic diversity within commercially exploited Indo-Malay Carangidae (Teleosteii: Perciformes). PLoS ONE 7: e49623

Jena JK, Basheer VS, Bineesh KK, Abdussammad EM, Ratheesh TB, Mohitha C, Akhilesh KV, Gopalakrishnan A (2014) Molecular identification of barracuda (Perciformes, Sphyraenidae) from Indian waters. Abstract book of "MECOS2- Marine Ecosystems- Challenges and Opportunities" conducted by MBAI & CMFRI at Kochi during, 2–5 Dec 2014

Khan SA, Lyla PS, John BA, Kumar CP, Murugan S, Jalal KCA (2010) DNA barcoding of *Stolephorus indicus*, *Stolephorus commersonnii* and *Terapon jarbua* of Parangipettai coastal waters

Krishna PV, Madhusudhana Rao K, Srinivasa Rao D (2012) Identification of selected estuarine fishes by DNA barcoding from river Krishna region, Andhra Pradesh. India. Int JR Pharm Bio Sci 3(3):1044–1049

Lakra WS, Goswami M, Gopalakrishnan A (2009) Molecular identification and phylogenetic relationships of seven Indian Sciaenids (Pisces: Perciformes, Sciaenidae) based on 16S rRNA and cytochrome c oxidase subunit I mitochondrial genes. Mol Biol Rep 36(5):831–839

Lakra WS, Verma MS, Goswami M, Lal KK, Mohindra V, Punia P, Hebert P (2011) DNA barcoding Indian marine fishes. Mol Ecol Resour 11(1):60–71

Nagalakshmi K, Annam PK, Venkateshwarlu G, Pathakota GB, Lakra WS (2016) Mislabeling in Indian seafood: An investigation using DNA barcoding. Food Control 59:196–200

Nagpure NS, Rashid I, Pathak AK, Singh M, Singh SP, Sarkar UK (2012) FBIS: A regional DNA barcode archival & analysis system for Indian fishes. Bioinformation 8(10):483–488

Persis M, Reddy ACS, Rao LM, Khedkar GD, Ravinder K, Nasruddin K (2009) COI (cytochrome oxidase-I) sequence based studies of Carangid fishes from Kakinada coast, India. Mol Biol Rep 36(7):1733–1740

Rahman MAU, Khan SA, Lyla PS, Kumar CP (2013) DNA barcoding resolves taxonomic ambiguity in mugilidae of Parangipettai waters (Southeast Coast of India). Turk J Fish Aquat Sci 13(2):321

Randall JE, Heemstra PC (1991) Revision of Indo-Pacific groupers (Perciformes: Serranidae: Epinephelinae) with descriptions of five new species. Indo Pac Fish 20:1–332

Ravitchandirane V, Geetha V, Ramya V, Janifer B, Thangaraj M, Subburaj J, Ganesan T (2012) Molecular identification and phylogenetic relationships of Threadfin Breams (Family: Nemipteridae) using mtDNA marker. Notulae Sci Biol 4(2):13–18

Roy TSC, Gopalakrishnan A (2011) Identification of groupers based on pyloric caeca differentiation. J Fish Biol 79(5):1334–1339

Sachithanandam V, Mohan PM, Dhivya P, Muruganandam N, Chaaithanya IK, Vijayachari P (2011) DNA Barcoding of Spotted Coral Grouper *Plectropomus maculatus* (Serranidae) From Andaman Islands. India. J Res Anal Eval 2:1–4

Sajeela KA, Rakhi C, Rekha JN, Gopalakrishna A, Basheer VS, Shoba JK, Joe K, Lakra WS (2010) Mitochondrial DNA sequences for forensic identification of the enedangered whale shark, *Rhyncodon typus* (Smith 1828). A case study. In: Nemis PL, Vignes LR (Eds) Tools for identifying biodiversity: progress and problems, pp 353–354

Trivedi S, Affan R, Alessa AHA et al (2014) DNA barcoding of Red Sea fishes from Saudi Arabia—the first approach. DNA Barcodes 2:17–20. doi:10.2478/dna-2014-0003

Trivedi S, Aloufi AA, Ansari AA, Ghosh SK (2016) Role of DNA barcoding in marine biodiversity assessment and conservation: an update. Saudi J Biolog Sci. http://dx.doi.org/10.1016/j.sjbs.2015.01

Victor BC, Hanner R, Shivji M, Hyde J, Caldow C (2009) Identification of the larval and juvenile stages of the Cubera Snapper, *Lutjanus cyanopterus*, using DNA barcoding. Zootaxa 2215 (2):24

Vij S, Purushothaman K, Gopikrishna G, Lau D, Saju JM, Shamsudheen KV, Vinaya Kumar K, Basheer VS, Gopalakrishnan A, Hossain MS, Sivasubbu S, Scaria V, Jena JK, Ponniah AG, Orbán L (2014) Barcoding of Asian seabass across its geographic range provides evidence for its bifurcation into two distinct species. Front Mar Sci. doi:10.3389/fmars.2014.00030

Viswambharan D, Pavan-Kumar A, Singh DP, Jaiswar AK, Chakraborty SK, Nair JR, Lakra WS (2015) DNA barcoding of gobiid fishes (Perciformes, Gobioidei). Mitochondrial DNA 26 (1):15–19

Ward RD, Holmes BH, Yearsley GK (2008) DNA barcoding reveals a likely second species of Asian sea bass (barramundi)(Lates calcarifer). J Fish Biol 72(2):458–463

Ward RD, Zemlak TS, Innes BH, Last PR, Hebert PD (2005) DNA barcoding Australia's fish species. Philos Trans R Soc B Biol Sci 360(1462):1847–1857

DNA Barcoding Identifies Brackish Water Fishes from Nallavadu Lagoon, Puducherry, India

V. Sachithanandam, P.M. Mohan, N. Muruganandam, R. Sivasankar, P. Arunkumar, T. Mageswaran and R. Sridhar

Abstract Animal DNA barcoding has been advanced as a promising tool to aid species identification and discovery through the use of short, standardized mitochondrial gene COI targets in spite of extensive morphological taxonomic studies, for a variety of reasons the identification of fishes can be problematic, time consuming, even for experts. To overcome in this content though DNA barcoding is proving to be a useful tool. The present study use DNA barcoding to genetically identify specimen were collected from Nallavadu Lagoon ecosystem in Puducherry and morphologically identified. The COI was amplified and bi-directionally sequenced from 45 specimens (5 specimens/species) belonging to 9 species, five families. All studied specimens sequence were interrogated using analytical tools developed by Barcode of Life Data System (BOLD). All specimen exhibited discrete clusters of closely related haplogroups in NJ tree, which permitted the discrimination of 96–100 % genetic similarity of all the studied species. These studies represent a significant contribution to the worldwide barcode database from Indian continental regional and demonstrated the utility of COI gene sequences for specific regional species identification.

Keywords Brackish water · DNA barcoding · Nallavadu lagoon · Taxonomy

V. Sachithanandam (✉) · P.M. Mohan
Department of Ocean Studies and Marine Biology, Pondicherry University, Andaman Campus, Port Blair 744112, India
e-mail: pondiunisachin@gmail.com; sachinchem_84@yahoo.com

N. Muruganandam
Regional Medical Research Centre (ICMR), Dollygunj, Port Blair (Andaman Island) 744101, India

R. Sivasankar · P. Arunkumar
Department of Ecology and Environmental Sciences, Pondicherry University, Port Blair 605154, India

Present Address:
V. Sachithanandam · T. Mageswaran · R. Sridhar
National Centre for Sustainable Coastal Management, Ministry of Environment, Forest and Climate Change, Chennai 600025, India

S. Trivedi et al. (eds.), *DNA Barcoding in Marine Perspectives*,
DOI 10.1007/978-3-319-41840-7_16

1 Introduction

Coastal lagoon is a shallow coastal water body separated from the sea by a barrier, connected at least sporadically to the ocean by one or more inlets and usually oriented parallel to shore. Coastal lagoons constitute about 13 % of global coastline (Farmworth and Golley 1974; Poyyamoli et al. 2011) and considered as a different coastal environment from estuaries (Kjerfve 1994). In India, there are 17 noteworthy lagoons are present on the east and west coast of India. Coastal lagoons notably act on the control of freshwater inflow and connections with the ocean (Crivelli and Ximenes 1992; Peja et al. 1996; Quinn et al. 1999), they are found to be highly dynamic, productive and extremely unpredictable systems (Barnes 1980). The fauna that occur in lagoons are of 3 groups, marine species, freshwater species and lagoon species. The presence of this specialist fauna make the conservation of lagoons important in maintaining the biodiversity. These ecosystem areas are important for the marine fisheries, serving as they do as nurseries for many species of fishes. Bouchet (2006) reported that 22, 9602 marine species had been described but the total species living on Earth may exceed more than 10 million as estimated by Grassle and Maciolek (1992). Traditionally taxonomy sciences are described based on the morphological characters, which develop the taxonomy keys. The conventional taxonomic sciences are very tedious to delineate species, in the last 250 years only 25 % of species were identified (Coyne and Orr 2004; Packer et al. 2009). The taxonomic scenario identification keys were initiated by Aristotle and later organized by Linnaeus. The fin fishes have been catalogued to about thirty-two thousand species from the aquatic environments, and it is divided into six classes, 62 orders, and 540 families (Eschmeyer 2010), which occupy >50 % of the all vertebrate species in world wide. Indian waters fishes were listed and identified by Day (1878) and Whitehead and Talwar (1976). Later, Talwar (1990) has reported 2546 fish species belongings to the 969 genera, 254 families and 40 order, among them 57 % of marine fish genera are common to the Indo-Pacific and Atlantic and Mediterranean regions. In Indian marine biodiversity studies are still untouched in minor phyla, non-commercially important fishes, and in remote parts of the islands and minor estuaries (Venkataraman and Wafar 2005). In current taxonomy scenario, the shortage of trained personnel limits development of taxonomic knowledge. The above factors have created a taxonomic impediment to biodiversity studies (Hebert et al. 2003). Despite ongoing scientific debate concerning the role of molecular methods in taxonomy DNA barcoding has emerged as a widely accepted tool for species identification because of its enhanced focus on standardization and data validation.

DNA barcoding is a method, based on short standardized gene sequences of DNA (Hebert et al. 2003), used for species identification. It has been used inter alia for identification of fish species (Hebert et al. 2004; Ward et al. 2005; Hubert et al. 2008; Persis et al. 2009; Steinke et al. 2009; Lakra et al. 2010; Sachithanandam et al. 2011, 2012). The effectiveness of barcoding has been demonstrated in diverse taxa, including butterflies (Hebert et al. 2003), birds (Hebert et al. 2004), fishes

(Ward et al. 2005; Persis et al. 2009; Steinke et al. 2009; Lakra et al. 2010; Sachithanandam et al. 2011, 2012), sponges, bivalves (Jarnegren et al. 2007) and mammals (Clare et al. 2007). DNA barcoding is being applied in the field of fish conservation (Holmes et al. 2009), fisheries managements aspects (Rasmussen et al. 2009) and food safety analysis, with mislabeling being revealed through COI sequences (Wong and Hanner 2008).

In 2005 FISH-BOL (Fish Barcode of Life Initiative) was set up to establish a library of COI gene sequences of all fish species, to enable global taxonomic identification (Ward et al. 2009; Eschmeyer 2010). FISH-BOL's primary work is led by a research team with responsibility for collection of samples, traditional identification using morphological keys, and COI gene sequencing of species in their geographic regions. Current progress of DNA barcoding in the Arctic and Antarctic regions is 74 and 50 %, respectively. Other regions from tropical to subtropical show good progress, with 20 % coverage from Australia, Mesoamerica continental and Oceania. However, species-rich regions of Asia, South America and Africa show low progress (Becker et al. 2011). In India, 11,023 fish species have been morphologically reported (Nelson 2006; Mecklenburg et al. 2010). Reports suggest that only 1918 of these (17.4 %) have been barcoded (Becker et al. 2011).

The aim of this study is to extend barcode coverage to Nallavadu brackish water fishes from Puducherry coastal area. We examined the patterns of barcode sequences genetic divergences among the 45 specimens identified as belonging to 9 species consists of five families representing highly exploitation for commercially values fishes from in this region.

2 Methods

2.1 Study Area

Puducherry located on the Coramandal coast 11° 52′ N, 79° 45′ E and 11° 59′ N and between 79° 52′ E covers an area of 480 km^2. Nallavadu village located at a distance of about 14 km South from Puducherry main town. Nallavadu, Poornankupam, Andiarpalayam, Pillaiarthittu and Panithittu villages are the boundaries of Nallavadu lagoon (Fig. 1). The aerial length of the lagoon is 3.44 km and maximum depth is 10 m in its broadest part (Padmavathy et al. 2010; Poyyamoli et al. 2011). The climate is tropical dissymmetric with the bulk of the rainfall during northeast monsoon from October to December and some amount of rainfall during southwest monsoon from June–September. The lagoon is connected to the sea in its south part by an artificial channel and receives freshwater primarily from Penniar River in the north and east, respectively.

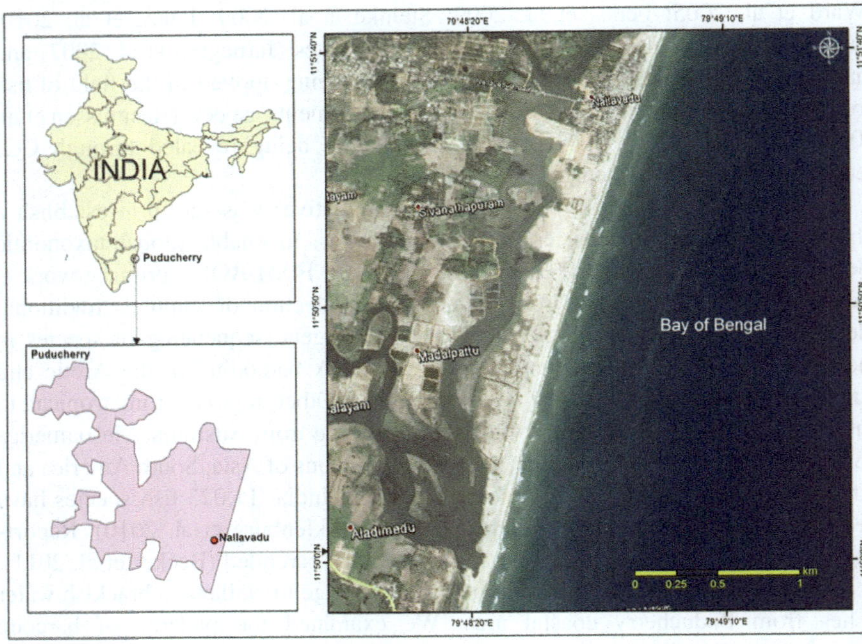

Fig. 1 Collection for specimens examined in this study

2.2 *Sampling and Taxonomic Coverage*

A total of 45 specimens of fishes collected in Nallavadu lagoon near Pooranankuppam village, representing 9 species, six genera, and 5 families were analyzed. Traditional taxonomic analyses were carried out used available literature to identify the specimen (FAO 1974; Rao 2003).

2.3 *DNA Analysis*

Total DNA was extracted from 0.25 g of tissue by using lysis buffer and followed by standard proteinase-K/phenol-chloroform-isoamyl alcohol-ethanol precipitation method (Sambrook et al. 1987) and fish DNA modified isolation protocol followed as described by Sachithanandam et al. (2012). The concentration of DNA was estimated using UV spectrophotometer method at 260/280 nm. Subsequently the DNA was diluted to final concentration of 100 ng/μL for further use. The 650–655 bp section of the mitochondrial (mt) DNA genome from the COI gene was amplified using a published universal degenerated primer set (Ward et al. 2005) synthesized by Sigma Aldrich Chemicals India Pvt. Ltd.

Table 1 Master mix preparation

Reagent	Volume (μL)
10X buffer	5.0
dNTP (5 mM)	1.50
COI gene forward Primer F1 (0.5 mM) 5′TCAACCAACCACAAAGACATTGGCAC3	1.0
COI gene reverse primer R1 (0.5 mM) 5′TAGACTTCTGGGTGGCCAAAGAATCA3′	1.0
Taq polymerase (3 U)	1.0
MgCl2	0.25
RNase free water	10.25
Total volume	20
5 μL DNA was mixed with 20 μL reaction mix	

The polymerase chain reaction (PCR) was carried out in 25 μL consisting of 100 ng/μL of DNA and PCR master mix details described in Table 1. PCR was carried out in an Applied Bio systems AB-2720 Thermal cycler, and PCR thermal condition was followed (Sachithanandam et al. 2014, 2012). PCR products were resolved in 1 % agarose containing 0.5 mg/μL of ethidium bromide and viewed under UV Transilluminator and documented. Nucleotide sequencing was performed using the Sanger et al. (1977) method. Sequencing was performed using a BigDye Terminator Cycle Sequencing kit, following manufacturer's instructions (Applied Biosystems, Foster City, CA, USA). The sequencing was done both in the forward and reverse directions.

2.4 Sequence Analysis

The DNA sequences were aligned both in forward and reverse directions for each individual fish and assembled using the SeqMan II version 5.03 (DNASTAR) and ChromaxSeq Version 3.1. The obtained studied species COI sequences analysis was done along with high similarity genetic identity reference sequences retrieved from the National Centre for Biotechnology Information (NCBI) GenBank. Studied species COI sequences used multiple and pairwise alignment was done using the ClustalW tool, and phylogenetic molecular evolutionary analyses were conducted through MEGA (Molecular Evolutionary Genetics Analysis) version 4.1. Sequence divergences were calculated using the Kimura two parameter (K2P) distance model and unrooted NJ phenograms based on K2P distances were created using MEGA 4.1 and were bootstrapped 1000 times to provide percentage bootstrap values for branch points (Tamura et al. 2007; Saitou and Nei 1987).

3 Results and Discussion

A total of 45 fish specimens belonging to 9 species, 6 genera, and 5 families were barcoded from Pooranankuppam brackish water in Nallavadu lagoon at Puducherry. The present study generated a total of 45 COI sequences for 9 species (No stop codon, insertions or deletions were found in any of the amplified sequences). All the amplified COI sequences were mean length 600 bp. Average nucleotide frequency were C (26.50 %), T (30.25 %), A (25.04 %) and G (18.21 %). The studied species were exhibited unique barcode haplotypes cluster of very closely related reference sequences, which permitted the discrimination of 96–98 % identity of species. All sequences were submitted to the NCBI GenBank.

3.1 COI Sequence Divergence Analysis

Out of the 9 species, 8 species belonging to perciformes and remaining one is belonging to *Anguilla* species were well discriminated though COI barcoding with average intraspecific genetic variability in the range of 0.005–0.020 compared with 0.203–0.354 for species within genera. The K2P genetic distances averaged 0.10 % within species, averaged 6.38 % within genera, 18.25 % within families, 21.55 % within orders and 23.18 % within classes, respectively were also obtained. Hence, overall taxonomic level increasing values, there was a 15 fold more pronounced difference among congeneric species than among conspecific individuals, and this value also produces a high level of resolution between clusters in the NJ tree to group the species to their corresponding genera and families with bootstrap values (Fig. 2).

3.2 Comments on Individual Genera

3.2.1 *Ambassis* spp.

The studied six genera, 9 species were obtained unrooted NJ tree shows that *Ambassis* spp. had separate cluster with three species. Among the three species *Ambassis ambassis* close genetic distance with *Ambassis dussummieri*. These species have clear external morphological features that allow differentiating each other; however *A. commersoni* are synonyms of *A. ambassis* through morphological taxonomy described but COI gene sequences had distinct genetic divergences obtained within *A. commersoni* and *A. ambassis* (K2P = 0.204). Moreover, genetic distance between this genus is 6.38 %, well above typical within genera values. Therefore, the lack of discrimination feature at morphological taxonomy level between these two species is likely misidentification. On the other hand, a reference

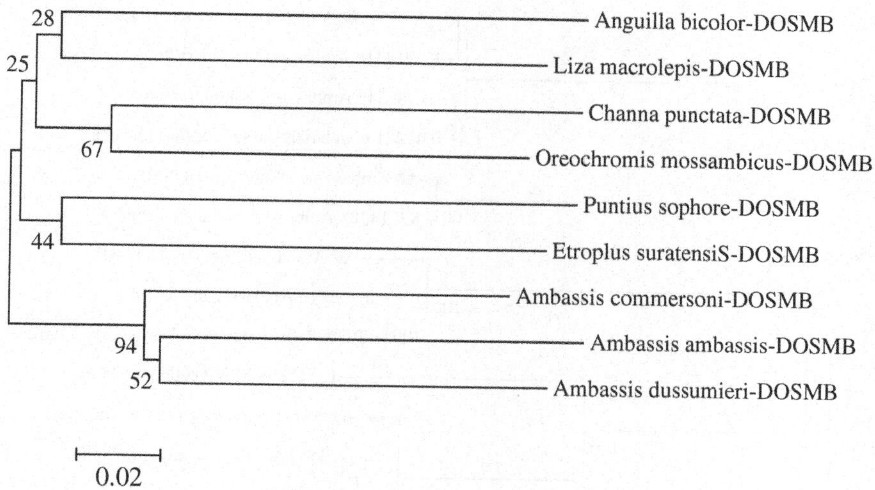

Fig. 2 Neighbour-joining tree for studied species of the mtDNA COI sequences of 9 nominal species of fish in Nallavadu braicksh water. Bootstrap support for NJ probability based on the K2P parameter test is shown above or near the branch. Scale bar represents is 0.02 substitutions per nucleotide position

sequence which was retrieved from NCBI data from Indian water and South Africa region also supports our data and shown in NJ tree (Fig. 3).

3.2.2 *Anguilla* sp.

Tropical eels are widely distributed about four species or subspecies in the Indian Ocean. In India two species of *Anguilla* (*A. bicolor bicolor* and *A. nebulosa*) reported. Based on the morphological characters study suggested that *A. bicolor* historically has been described as two subspecies, *A. bicolor bicolor* in the Indian Ocean and *A. bicolor pacifica* in the western Pacific Ocean and the seas around the northern parts of Indonesia (Ege 1939), but is apparently genetically homogeneous in the Indian Ocean (Minegishi et al. 2011). However, *A. bicolor bicolor* and *A. bicolor pacifica* COI genetic divergences study have been never investigated in details. The present study was to reveal the COI genetic divergency between *A. bicolor bicolor* collected from study site with close genetic identity reference sequences retrieved from GenBank and compared for further confirmation.

Collected specimen was identified as *A. bicolor bicolor* based on the morphological identification keys. Studied species COI gene sequence with nine reference sequence from GenBank used to constructed NJ phylogram formed 4 distinct clades (Fig. 4). The clade 1 in the top of the NJ phylogram contained *Anguilla marmorata* reference sequences JQ665825, HM345929, DQ521000 from different part had a cluster with 100 % identity. Clade 2 consists of (KF182302, KF182302) *Anguilla*

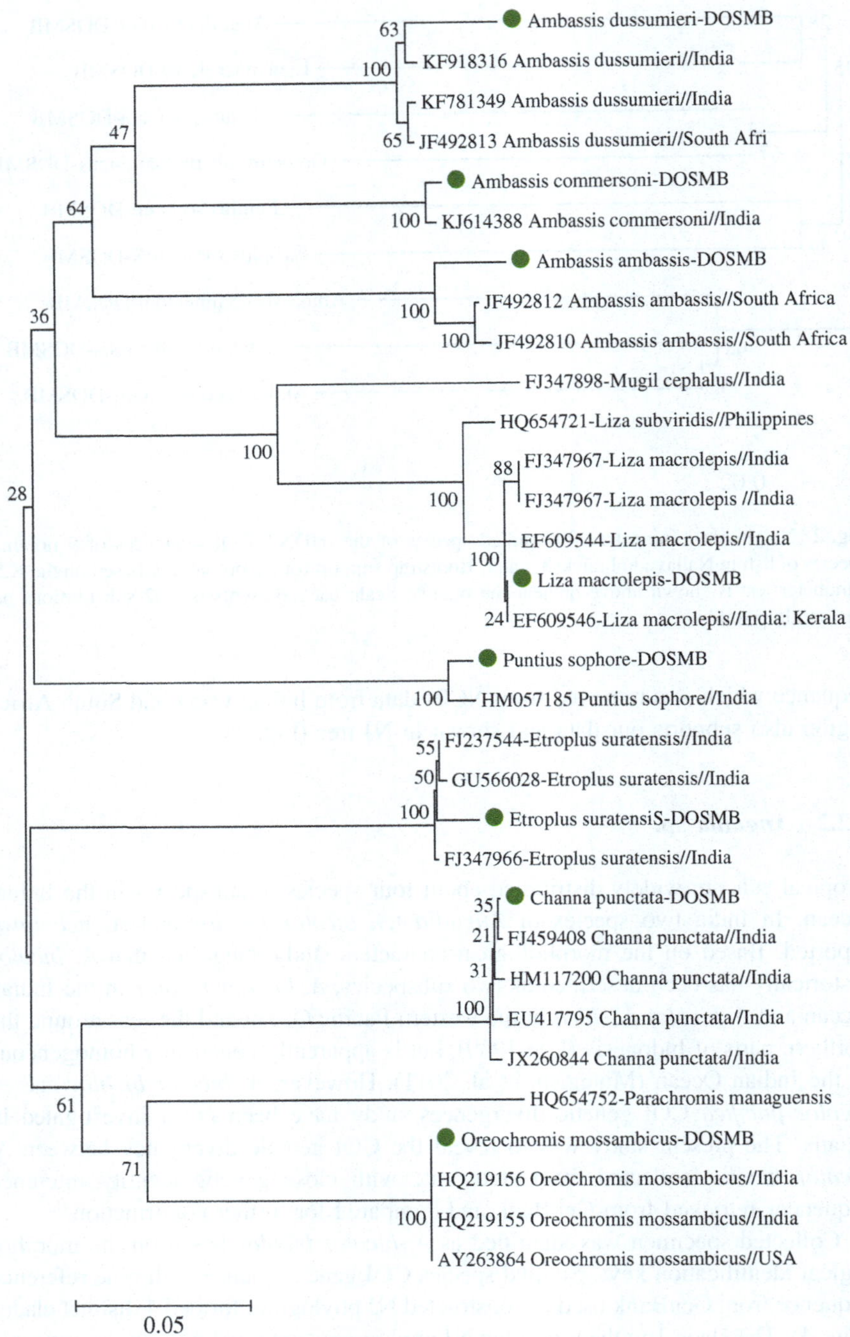

Fig. 3 Neighbour-joining tree of studied species COI gene sequences showing similarities with reference sequences retrieved from GenBank

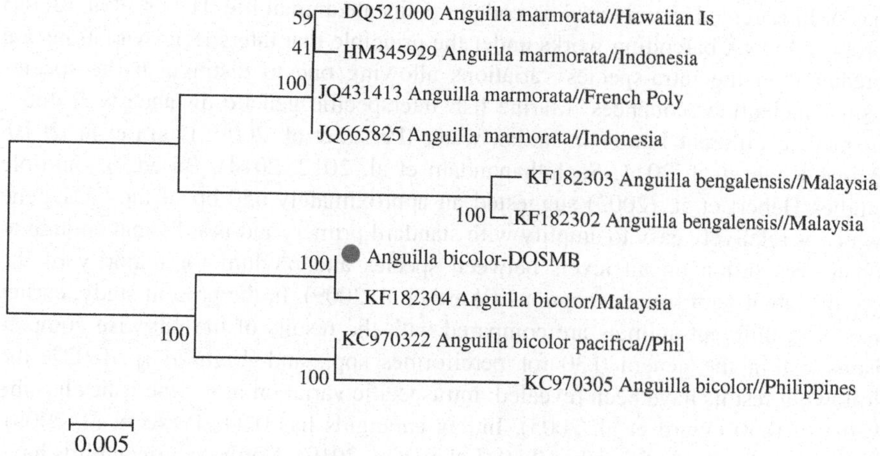

Fig. 4 Neighbour-joining phylogram of COI gene sequences showing similarities between DNA barcodes studied species (*Anguilla bicolor bicolor*) with reference sequences retrieved from GenBank

bengalensis from Indonesian water, which showed 100 % similarity with same species from same region. Clade 3 contains studied species *A. bicolor bicolor* along with reference sequence of same species KF182304 (*A. bicolor bicolor)* from Malaysia with 100 % genetic similarity observed in NJ tree (Fig. 4). The studied species sequences with reference sequences retrieved from GenBank database to form clustered in NJ tree. From this analysis observed that COI gene sequences clearly proving the reliability of in identification of *Anguilla* spp.

4 Genetic Distance

A. bicolor bicolor Four specimen COI gene sequences Intraspecific mean genetic pair-wise distance was calculated and in the range of 0.03 from 0.02 to 0.050 and same species reference sequence (KF182304) from Malaysia also same values. Morphologically subspecies of *A. bicolor pacifica* reference sequence from Philippines water with studied species (*A. bicolor bicolour)* genetic diversity had 0.025. Interspecies genetic diversity analysis between studied species and *A. marmorata* (HM345929; JQ665825; JQ431413) showed 0.050 K2P values. Based on the *A. bicolor bicolour* species COI gene sequences are new reports from India for NCBI GenBank and BOLD system.

DNA barcoding is a well accepted taxonomic method which uses a short genetic marker to facilitate identification of a particular species even by non-specialist. CO1 has been accepted as universal barcode to delineate animal life. Hebert et al. (2004) proposed DNA barcoding works under the principle that interspecies variations are greater than the intra-species variations allowing one to distinguish the species using nucleotide sequences. Marine fish interspecific genetic distance well documented in different family in Indian water (Persis et al. 2009; Lakra et al. 2010; Ajmal Khan et al. 2011; Sachithanandam et al. 2012, 2014). Based on multiple studies Hebert et al. (2003) suggested an approximately 650 bp of the CO1 gene which is relatively easy to amplify with standard primers and is sufficient enough to obtain resolution on all levels between species and phylum for majority of the groups albeit with some exceptions (Ward et al. 2009). In the present study, earlier works in different animals are compared with the results of the pairwise comparisons within the genera 0.30 for perciformes spp., and *Angullia* sp. 0.022, the following results have been revealed: intraspecific variation in marine fishes has the value of 0.30 (Ward et al. 2005), Indian carangids has 0.24 (Persis et al. 2009), Indian Ocean marine fishes has 0.30 (Lakra et al. 2010), North American birds have 0.24, 0.23 and 0.27 (Hebert et al. 2004), Guyanese bats has 0.60 (Clare et al. 2011) and 0.46 for Lepidoptera (Hajibabaei et al. 2006). Atlantic and Brazilian cuzk eels species were discriminated 5.8 % genetic divergence observed (Mabragana et al. 2011). It is evident that congeneric distances are higher than the conspecific distances. In addition to the species identification, DNA barcoding has been used for identification of processed food products (Smith et al. 2008).

5 Conclusion

DNA barcoding is taking great endeavours of biological research genomics, phylogenetics and providing a comprehensive view into the biology, used to screen the large-scale genes, assign unknown individuals and discovery of new species. The present work concluded that studied species DNA barcode tool clearly discriminated in inter and intraspecific level without any ambiguity in genetic divergence of fish; these results support the utility of DNA barcodes for regional species identification of fishes. In future, Indian water need more groups of aquatic species for biodiversity assessment though COI gene sequence approach, will contribute increase biodiversity database. Moreover, COI gene based identification species enabled differentiation of species beyond doubt.

Acknowledgments The Authors express their sincere acknowledgement to, Vice-Chancellor, Pondicherry University and the constant help and encouragement of Dr. P. Vijayachari, Director, Regional Medical Research Centre (ICMR) Port Blair during this study. First Author thanks the UGC-RFSMS fellowship and Ministry of Earth Sciences (MoES) for funding this work. The authors thank the Director, NCSCM, MoEF&CC, Chennai, for providing scientific and administrative support.

References

Ajmal Khan S, Lyla PS, Akbar John B, Prasanna Kumar C, Murugan S (2011) Identifying marine fin fishes using DNA barcodes. Curr Sci 101:9–12

Barnes RSK (1980) Coastal Lagoons. The natural history of a neglected habitat. Camb Stud Mod Biol 1:1–160

Becker S, Hanner R, Steinke D (2011) Five years of FISH-BOL: brief status report. Mitochondrial DNA 22(S.1):3–9

Bouchet P (2006) The magnitude of marine biodiversity. In: Duarte CM (ed) The exploration of marine biodiversity: scientific and technological challenges. Fundación BBVA, Bilbao, pp 31–64

Coyne JA, Orr HA (2004) Speciation. Sinauer Associates, Sunderland, p 545

Clare EL, Lim BK, Engstrom MD, Eger JL, Hebert PDN (2007) DNA barcoding of Neotropical bats: species identification and discovery within Guyana. Mol Ecol Notes 7:184–190

Clare EL, Lim BK, Fenton MB, Hebert PDN (2011) Neotropical Bats: estimating Species Diversity with DNA Barcodes. PLoS ONE 6(7):e22648.http://doi.org/10.1371/journal.pone.0022648

Crivelli AJ, Ximenes MC (1992) Alterations to the functioning of Mediterranean lagoons and their effects on fisheries and aquaculture. In: Finlayson M, Hollis T, Davis T (eds) IWRB. Special Publication, Grado. Managing Mediterranean Wetlands and Their Birds 20:134–140

Day F (1878) The fishes of India; being a natural history of the fishes known to inhabit the seas and fresh waters of India, Burma, and Ceylon. Parts 1–4. William Dawson and Sons Ltd., London, pp 778

Eschmeyer WN (2010) Catalog of fishes. http://research.calacademy.org/research/ichthyology/catalog/speciesByFamily.asp

Ege V (1939) A revision of the genus Anguilla Shaw: a systematic, phylogenetic and geographical study. Dana Rep 16:1–256

FAO (1974) Species identification sheet for fishery purposes. In: Fischer W, Whitehead PJP (eds) Eastern Indian Ocean (fishing area 57) and Western Central Pacific (fishing area 71), vol 1, pp C1–22

Farmworth EG, Golley FB (1974) Fragile ecosytems. Springer-Verlag, New York, pp 258

Grassle JF, Maciolek N (1992) Deep Sea species richness: regional and local diversity estimates from quantitative bottom samples. Am Nat Soc 39:313–334

Hebert PDN, Stoeckle MY, Zemlak TS, Francis CM (2004) Identification of birds through DNA barcodes. PLoS Biol 2:1657–1663

Hajibabaei M, Janzen DH, Burns JM, Hallwachs W, Hebert PDN (2006) DNA barcodes distinguish species of tropical Lepidoptera. Proc Nat Acad Sci USA 103:968–971

Hebert PDN, Cywinska A, Ball SL, DeWaard JR (2003) Biological identifications through DNA barcodes. Proc R Soc Lond B 270:313–321

Hubert N, Hanner R, Holm E, Mandrak NE, Taylor E (2008) Identifying Canadian freshwater fishes through DNA barcodes. PLoS ONE 3(6):e2490

Holmes BH, Steinke D, Ward RD (2009) Identification of shark and ray fins using DNA barcoding. Fish Res 95:280–288

Jarnegren J, Schander C, Sneli JA, Ronningen V, Young CM (2007) Four genes, morphology and ecology: distinguishing a new species of Acesta (Mollusca; Bivalvia) from the Gulf of Mexico. Mar Biol 152:43–55

Kjerfve B (1994) Coastal lagoon. In: Kjerfve B (ed) Coastal lagoon processes. Elsevier Oceanography Series, Amsterdam, pp 1–8

Lakra WS, Verma MS, Goswami M, Lal KK, Mohindra V, Punia P, Gopalakrishnan A, Singh KV, Ward R, Hebert PDN (2010) DNA barcoding Indian marine fishes. Mol Ecol Res 11:60–71

Mabragana E, Astarloa JMD, Hanner R, Zhang J, Castro MG (2011) DNA barcoding identifies Argentine fishes from marine and brackish waters. PLoS ONE 6(12):e28655

Mecklenburg CW, Moller PR, Steinke D (2010) Biodiversity of arctic marine fishes: taxonomy and zoogeography. Mar Biodiv 41(1):109–140

Minegishi Y, Gagnaire PA, Aoyama J, Bosc P, Feunteun E, Tsukamoto K, Berrebi P (2011) Present and past genetic connectivity of the Indo-Pacific tropical eel *Anguilla bicolor*. J Biogeogr. doi:10.1111/j.1365-2699.2011.02603.x

Nelson JS (2006) Fishes of the world, 4th edn. Wiley, New York, pp 601

Packer L, Gibbs J, Sheffield C, Hanner R (2009) DNA barcoding and the mediocrity of morphology. Mol Ecol Res 9:42–50

Padmavathy K, Poyyamoli G, Balachandran N (2010) Coastal dune flora, Nallavadu Village, Puducherry, India. Check List 6(2):198–200

Poyyamoli G, Padmavathy A, Balachandran N (2011) Coastal sand dunes—vegetation structure, diversity and disturbance in Nallavadu Village, Puducherry, India. Asian J Air Water Environ 8:115–122

Persis M, Reddy ACS, Rao LM, Khedkar GD, Ravinder K, Nasruddin K (2009) COI (Cytochrome oxidase—I) sequence based studies of Carangid fishes from Kakinada coast. India Mol Biol Rep 36:1733–1740

Peja N, Vaso A, Miho A, Rakaj N, Crivelli AJ (1996) Characteristics of Albanian lagoons and their fisheries. Fish Res 27:215–225

Quinn NW, Breen CN, Whitfield AK, Hearne JW (1999) An index for the management of South African estuaries for juvenile fish recruitment from marine environment. Fish Manag Ecol 6:421–436

Rao DV (2003) Guide to reef fishes of Andaman and Nicobar Islands. Zoological Survey of India, Kolkata, pp 555

Rasmussen RS, Morrissey MT, Hebert PDN (2009) DNA barcoding of commercially important salmon and trout species in North America. In: 3rd international barcode of life conference, Mexico City, p 84

Sachithanandam V, Mohan PM, Muruganandam N (2014) DNA barcoding of marine venomous and poisonous fish of family Scorpaenidae and Tetraodontidae from Andaman water. In: Venkataraman K, Sivaperuman C (eds) Marine faunal diversity in India taxonomy, ecology and conservation. Academic Press, Elsevier, London

Sachithanandam V, Mohan PM, Dhivya P, Muruganandam N, Baskaran R, Chaaithanya IK, Vijayachari P (2011) DNA barcoding, phylogenetic relationships and speciation of genus: Plectropomus in Andaman coast. J Res Biol 3:179–183

Sachithanandam V, Mohan PM, Muruganandam N, Chaaithanya IK, Dhivya P, Baskaran R (2012) DNA barcoding, phylogenetic study of *Epinephelus* spp. from Andaman coastal region. India Indian J Geo-Mar Sci 41:203–211

Saitou N, Nei M (1987) The Neighbour-joining method—a new method for reconstructing phylogenetic trees. Mol Biol Evol 4:406–425

Sambrook J, Fritsch EF, Maniatus T (1987) Molecular cloning: a laboratory manual. Cold Spring Harbor Laboratory Press, New York, pp 1659

Sanger F, Nicklen S, Coulson AR (1977) DNA sequencing with chain terminating inhibitors. Proc Nat Acad Sci USA 74(12):5463–5467

Smith PJ, Mc Veagh SM, Steinke D (2008) DNA barcoding for the identification of smoked fish products. J Fish Biol 72:464

Steinke D, Zemlak TS, Hebert PDN (2009) Barcoding Nemo: DNA-based identifications for the ornamental fish trade. PLoS ONE 4:e6300

Talwar PK (1990) Fishes of the Andaman and Nicobar Islands: a synoptic analysis. J Andaman Sci Assoc 6(2):71–102

Tamura K, Dudley J, Nei M, Kumar S (2007) MEGA4: Molecular Evolutionary Genetics Analysis (MEGA) software version 4. 0. Mol Biol Evol 24(8):1596–1599

Venkataraman K, Wafar M (2005) Coastal and marine biodiversity of India. Ind J Mar Sci 34 (1):57–75

Ward RD, Zemlak TS, Innes BH, Last PR, Hebert PDN (2005) DNA barcoding Australia's fish species. Philos Trans R Soc Ser Biol 360:847–1857

Ward RD, Hanner R, Hebert PDN (2009) The campaign to DNA barcode all fishes. FISHBOL J Fish Biol 74(2):329–356

Wong EHK, Hanner R (2008) DNA barcoding detects market substitution in North American seafood. Food Res Int 41:828–837

Whitehead D, Talwar PK (1976) Clupeoid fishes of the world (Suborder: clupeoidei): Engraulididae. American museum of natural history, New York, Part 4. Fishes 121:118–119

DNA Barcoding of Marine Fishes: Prospects and Challenges

Annam Pavan-Kumar, P. Gireesh-Babu, A.K. Jaiswar, Aparna Chaudhari, Gopal Krishna and W.S. Lakra

Abstract Marine fishes constitute nearly 50 % of total fish species and exhibit a remarkable diversity of morphological traits and biological adaptations. The inherent limitations of traditional taxonomic tools led to the development of DNA based species identification methods. Among different methods, DNA barcoding approach has got global recognition for fish species identification irrespective of its life stage. This chapter summerizes the DNA barcoding research on marine fishes and discusses the limitations and applications of this tool for effective marine fisheries management.

Keywords DNA barcoding · Cryptic fishes · Cytochrome *c* oxidase subunit I · Conservation

1 Introduction

Oceans act as womb for the life and subsequently the marine ecosystem has turned as a cradle for biodiversity. Among different marine organisms; Fishes, the largest group of vertebrates, originated during the Cambrian period; have radiated/reached their maximum species diversity during the Devonian period. Different ecological and geographical factors of oceans have resulted in radiation/diversification of fish species. Marine fishes constitute nearly 50 % of total fish species (~16,000 species) and exhibit a remarkable diversity of morphological traits and biological adaptations (Eschmeyer et al. 1998; Nelson 2006). New marine fish species are being described at an average rate of 100–150 per year and recent studies estimated that an additional 3000–5000 species are yet to be described (Eschmeyer et al. 2010; Mora et al. 2008). Marine fishes play an important role in maintaining ecological balance, providing animal protein to human at low cost, livelihood for fishermen and up-surging economy of the respective country.

A. Pavan-Kumar (✉) · P. Gireesh-Babu · A.K. Jaiswar · A. Chaudhari · G. Krishna · W.S. Lakra
ICAR-Central Institute of Fisheries Education, Mumbai 400061, India
e-mail: pavanannam@gmail.com

S. Trivedi et al. (eds.), *DNA Barcoding in Marine Perspectives*,
DOI 10.1007/978-3-319-41840-7_17

Species is a distinct biological entity that has a separate evolutionary lineage (Wiley 1978). Each species in an ecosystem has its own role in maintaining and stabilizing ecosystem. Removal/elimination of species from an ecosystem has adverse effects on the function of ecosystem. Accurate and unambiguous identification of species is a prerequisite to address any biological question relating to that species/ecosystem. However, due to anthropogenic activities and other reasons, the biodiversity is decimating before their identification and description (Myers 1996). This "biodiversity crisis" urge the scientific community to develop a methodology for rapid identification/assessment of species. Traditionally, fish species identification is based on external morphological features comprising body shape, pattern of colours, scale size and count, number and relative position of fins, type of fin rays, gill rakers and otoliths (Strauss and Bond 1990; Granadeiro and Silva 2000). However, these tools may not be able to identify/differentiate species during larval stages and if an organism is changing its phenotype in response to environment (Phenotype plasticity; for instance most of the reef associated fishes display phenotype plasticity in life history associated traits). Also, the traditional techniques sometime ignore the genetic variability of the characters used for species recognition. These limitations may lead to misidentification of species causing erroneous grouping of different taxa and faulty synonymous taxon names. This often results in giving less priority to species which are really needed to be conserved and also overlooks morphologically cryptic taxa that are common in many marine fishes (Zemlak et al. 2009; Hubert et al. 2012).

These difficulties in traditional taxonomy for identifying fishes have led to the development of DNA based methods. Initially PCR (Polymerase Chain Reaction)/ hybridization methods such as RAPD (Random Amplification of Polymorphic DNA), Probe hybridization, gene-specific primers and PCR-RFLP (Restriction Fragment Length Polymorphism) were used for species identification (Teletchea 2009). However, these methods have limitations like lack of reference database and universal primers.

Mitochondrial DNA has been used successfully to discriminate species because of its maternal inheritance, rapid evolutionary rate (1–2 % per million years), lack of introns and recombination (Meyer 1993). Different mitochondrial genes namely cytochrome *b*, 12S and 16S rRNAgenes have been used for fish species identification. However, availability of universal primers and presence of phylogenetic signal make the cytochrome *c* subunit I gene more appropriate for fish species identification (Hebert et al. 2003).

2 DNA Barcoding of Marine Fishes

DNA barcoding is an approach where a short fragment of (650 bp) mitochondrial cytochrome *c* oxidase subunit I (COI) is used to identify the species that has been previously described (Hebert et al. 2003). In this methodology, specimen is assigned to a species/taxon on the basis of COI sequence similarity as conspecific individuals have higher sequence similarity and lower genetic distance values.

Conversely, species of different genera, family and order have relatively low sequence similarity and high genetic distance values. Accordingly, a threshold value was proposed on the basis of average intra and inter-specific genetic distances calculated from barcodes of predetermined species to define the species boundaries (Hebert et al. 2003). Initially, a genetic distance value of > 3 % was recommended as a threshold value to flag a specimen as a putative new species/cryptic species. However, several empirical barcoding papers showed that the proposed threshold value is not pertinent to all animal groups and after considering the potential individual/population variation, the genetic distance value of 10 times of the mean intraspecific divergence value was appropriated as a threshold value for flagging genetically divergent specimen as a provisional species (Hebert et al. 2004).

Ward et al. (2005) have successfully identified marine fishes of Australia using DNA barcoding and this early barcoding success with fishes driven to the formation of the Fish Barcode of Life (FISH-BOL) consortium with an objective of generating DNA barcodes for all the fishes of world. FISH-BOL is a sub-database of Barcode of Life Data System (BOLD, Ratnasingham and Hebert 2007) and contains DNA barcodes, images and geospatial co-ordinates for the analysed specimens. The remarkable success of DNA barcoding in species identification led to the formation of several consortia such as Shark Barcode of Life (Shark-Bol), Polar Barcode of Life (PolBOL) and Marine Barcode of Life (MarBOL) to identify/catalogue the fish species from different ecosystem.

3 Taxonomic Coverage

The catalogue of fishes inventoried more than 15,000 marine fish species distributed in 51 orders and 389 families (Nelson 2006). Globally, as of July 2016 more than 6,000 (42 %) fish species have been barcoded and the resulted barcodes with morphological and geographical data have been deposited in BOLD database (Table 1). In most of the marine fishes, DNA barcoding gap was evident between different genera with less value of average intra-specific K2P distance (0.18–0.39 %) than inter-specific distance value (3.75–11 %) (Ward et al. 2005; Steinke et al. 2009; Lakra et al. 2011; Mabragaña et al. 2011; Zhang and Hanner 2012). However, in few orders namely Pristiformes and Torpediniformes (Elasmobranchii); Albuliformes, Batrachoidiformes, and Gobiesociformes (Actinopterygii); Petromyzontiformes (Cephalaspidomorphi) some species have shown absence of barcoding gap (Becker et al. 2011).

4 Cryptic Species Identification

Cryptic species are a group of biological entities that are classified as single species based on morphological similarities, but in real they possess distinct genetic lineages (Bickford et al. 2007). Crypticity of fishes could be due to extreme environmental conditions that impose stabilizing selection on morphology thus reducing

Table 1 List of barcoded marine fish species

S. No	Order	Number of species	Barcoded species	Progress (%)
1.	Carcharhiniformes	230	134	58
2.	Heterodontiformes	9	5	56
3.	Hexanchiformes	5	5	100
4.	Lamniformes	16	15	94
5.	Orectolobiformes	33	18	55
6.	Pristiformes	7	6	86
7.	Pristiophoriformes	5	3	60
8.	Rajiformes	456	197	43
9.	Squaliformes	103	48	47
10.	Squatiniformes	17	9	53
11.	Torpediniformes	60	17	28
12.	Acipenseriformes	18	14	78
13.	Albuliformes	30	15	50
14.	Anguilliformes	833	170	20
15.	Ateleopodiformes	12	2	16
16.	Atheriniformes	97	30	31
17.	Aulopiformes	240	99	41
18.	Batrachoidiformes	73	13	18
19.	Beloniformes	175	110	63
20.	Beryciformes	157	65	41
21.	Cetomimiformes	32	5	16
22.	Characiformes	8	3	38
23.	Clupeiformes	283	87	31
24.	Cypriniformes	28	18	64
25.	Cyprinodontiformes	25	12	48
26.	Elopiformes	8	6	75
27.	Esociformes	1	1	100
28.	Gadiformes	596	154	26
29.	Gasterosteiformes	18	10	56
30.	Gobiesociformes	140	40	29
31.	Gonorynchiformes	7	5	71
32.	Lampriformes	23	16	70
33.	Lophiiformes	316	172	54
34.	Myctophiformes	253	200	79
35.	Notacanthiformes	28	17	61
36.	Ophidiiformes	391	141	36
37.	Osmeriformes	250	87	35
38.	Osteoglossiformes	7	0	0
39.	Perciformes	7421	2894	39
40.	Pleuronectiformes	695	403	58

(continued)

Table 1 (continued)

S. No	Order	Number of species	Barcoded species	Progress (%)
41.	Polymixiiformes	10	9	90
42.	Saccopharyngiformes	29	8	28
43.	Salmoniformes	36	13	36
44.	Scorpaeniformes	1400	691	49
45.	Siluriformes	123	55	44
46.	Stephanoberyciformes	75	23	33
47.	Stomiiformes	403	170	42
48.	Synbranchiformes	3	0	0
49.	Syngnathiformes	278	82	30
50.	Tetraodontiformes	393	340	87
51.	Zeiformes	44	28	63
	Total	15,900	6,665	42

Source BOLD, MarBOL and Fish-BOL accessed on 25th July 2016

or eliminating morphological change that can accompany speciation (Schonrogge et al. 2002). A strong positive relationship was observed between substratum complexity and the density and diversity of cryptic fish assemblages (Willis and Anderson 2003). Marine environment with diverse ecological habitats and niches harbors several cryptic fish species, for instance, coral reefs were reported to represent more number of cryptic fish species (Depczynski 2006). The documentation and description of cryptic species have significant implications for natural resource management and conservation. Further, cryptic species require special attention in conservation planning especially for endangered species complex. Because the species that are considered as endangered might be composed of multiple species that are even fewer than previously supposed and different species might require different conservation strategies (Bickford et al. 2007).

Morphological tools are unable to identify cryptic species due to high similarity in the morphological traits. DNA barcoding approach has flagged the occurrence of cryptic species in several fish species (Table 2). All these cryptic species were flagged on the basis of their high intraspecific COI K2P distance values than the proposed threshold value. Generally, the COI gene divergence values within species cannot increase beyond threshold value even if the samples from different geographical regions are included. For instance, tiger shark *Galeocerdo cuvier* (Peron and Lesueur 1822), with more than 200 specimens from the Atlantic, Pacific and Indian Oceans showed a mean intra specific divergence of 0.09 % and a maximum divergence of only 0.47 %. This pattern is typical of most marine fish species, even those with very wide geographical range (Ward et al. 2008, 2009). However, some species collected from complex ecosystems (coral reefs) or different geographical locations showed higher intraspecific genetic distance values. These values could be due to sympatric or allopatric speciation process. Sympatric speciation, the formation of species in the absence of geographical barriers nurtures morphologically

Table 2 List of species complexes reported to contain cryptic species

Order	Family	Fish species	Reference
Carcharhiniformes	Sphyrnidae	*Sphyrna lewini* (Griffith and Smith 1834)	Zemlak et al. (2009)
	Triakidae	*Mustelus mosis* (Hemprich and Ehrenberg 1899)	Pavan-Kumar et al. (2013)
Lamniformes	Alopiidae	*Alopias pelagicus* (Nakamura 1935)	Pavan-Kumar et al. (2013)
Myliobatiformes	Myliobatidae	*Aetobatus narinari* (Euphrasen 1790)	Pavan-Kumar et al. (2013)
Perciformes	Carangidae	*Scomberoides tol* (Cuvier 1832)	Zemlak et al. (2009)
		Atule mate (Cuvier 1833)	Mat Jaafar et al. (2012)
		Selar crumenophthalmus (Bloch 1793)	Mat Jaafar et al. (2012)
		Seriolina nigrofasciata (Rüppell 1829)	Mat Jaafar et al. (2012)
	Scombridae	*Scomberomorus commerson* (Lacepède 1800)	Zemlak et al. (2009)
		Scomber japonicus (Houttuyn 1782)	Zhang (2011)
	Terapontidae	*Terapon jarbua* (Forsskål 1775)	Zhang (2011)
	Sparidae	*Argyrops spinifer* (Forsskål 1775)	Zemlak et al. (2009)
		Rhabdosargus sarba (Forsskål 1775)	Zemlak et al. (2009)
	Lethrinidae	*Lethrinus nebulosus* (Forsskål 1775)	Zemlak et al. (2009)
	Priacanthidae	*Priacanthus hamrur* (Forsskål 1775)	Zemlak et al. (2009)
	Labridae	*Bodianus perditio* (Quoy and Gaimard 1834)	Zemlak et al. (2009)
		Halichoeres hortulanus (Lacepède 1801)	Hubert et al. (2012)
		Labroides dimidiatus (Valenciennes 1839)	Hubert et al. (2012)
	Acanthuridae	*Zebrasoma scopas* (Cuvier 1829)	Hubert et al. (2012)
	Mullidae	*Parupeneus heptacanthus* (Lacepède 1802)	Hubert et al. (2012)
		Upeneus sulphureus (Cuvier 1829)	Zhang (2011)
	Sciaenidae	*Otolithes ruber* (Bloch and Schneider 1801)	Zemlak et al. (2009)

(continued)

Table 2 (continued)

Order	Family	Fish species	Reference
		Larimus breviceps (Cuvier 1830)	Ribeiro et al. (2012)
		Dendrophysa russelii (Cuvier 1829)	Zhang (2011)
	Trichiuridae	*Trichiurus lepturus* (Linnaeus 1758)	Ribeiro et al. (2012)
Perciformes	Serranidae	*Cephalopholis formosa* (Shaw 1812)	Bamaniya et al. (2014)
		Epinephelus coeruleopunctatus (Bloch 1790)	Bamaniya et al. (2014)
		Pseudogramma polyacantha (Bleeker 1856)	Hubert et al. (2012)
	Haemulidae	*Pomadasys corvinaeformis* (Steindachner 1868)	Ribeiro et al. (2012)
	Ammodytidae	*Ammodytes hexapterus* (Pallas 1814)	Turanov et al. (2014)
	Gempylidae	*Thyrsites atun* (Euphrasen 1791)	Cawthorn et al. (2011)
Perciformes	Gobiidae	*Acentrogobius caninus* (Valenciennes 1837)	Zhang (2011)
		Pseudanthias squamipinnis (Peters 1855)	Steinke et al. (2009)
		Valenciennea species	Steinke et al. (2009)
		Bathygobius cocosensis (Bleeker 1854)	Bamaniya et al. (2014)
	Chaetodontidae	*Heniochus acuminatus* (Linnaeus 1758)	Bamaniya et al. (2014)
	Plotosidae	*Plotosus lineatus* (Thunberg 1787)	Bamaniya et al. (2014)
	Pomacentridae	*Plectroglyphidodon lacrymatus* (Quoy and Gaimard 1825)	Hubert et al. (2012)
	Labrisomidae	*Starksia atlantica* (Longley 1934), *S. lepicoelia* (Böhlke and Springer 1961)	Baldwin et al. (2011)
Scorpaeniformes	Platycephalidae	*Platycephalus indicus* (Linnaeus 1758)	Puckridge et al. (2013)
		Cymbacephalus staigeri (Castelnau 1875)	Puckridge et al. (2013)
		Platycephalus endrachtensis (Quoy and Gaimard 1825)	Puckridge et al. (2013)

(continued)

Table 2 (continued)

Order	Family	Fish species	Reference
		Cymbacephalus nematophthalmus (Günther 1860)	Puckridge et al. (2013)
		Inegocia japonica (Cuvier 1829)	Puckridge et al. (2013)
		Thysanophrys chiltonae (Schultz 1966)	Puckridge et al. (2013)
		Thysanophrys celebica (Bleeker 1855)	Puckridge et al. (2013)
		Sunagocia arenicola (Schultz 1966)	Puckridge et al. (2013)
Scorpaeniformes	Sebastidae	*Helicolenus dactylopterus* (Delaroche 1809)	McCusker et al. (2013)
	Scorpaenidae	*Sebastapistes tinkhami* (Fowler 1946)	Hubert et al. (2012)
		Pterois antennata (Bloch 1787)	Hubert et al. (2012)
Ophidiiformes	Bythitidae	*Dinematichthys iluocoeteoides* (Bleeker 1855)	Hubert et al. (2012)
Angiulliformes	Chlopsidae	*Kaupichthys diodontus* (Schultz 1943)	Hubert et al. (2012)
Aulopiformes	Synodontidae	*Synodus dermatogenys* (Fowler 1912)	Hubert et al. (2012)
	Muraenidae	*Gymnothorax pseudothyrsoideus* (Bleeker 1853)	Zhang (2011)
Ophidiiformes	Ophidiidae	*Neobythites unimaculatus* (Smith and Radcliffe 1913)	Zhang (2011)
Gadiformes	Moridae	*Halargyreus johnsonii* (Günther 1862)	McCusker et al. (2013)
Elopiformes	Elopidae	*Elops hawaiensis* (Regan 1909)	Zhang (2011)
		Elops saurus (Linnaeus 1766)	Valdez-Moreno et al. (2010)
Pleuronectiformes	Achiridae	*Achirus lineatus* (Linnaeus 1758)	Valdez-Moreno et al. (2010)
Albuliformes	Albulidae	*Albula Species*	Valdez-Moreno et al. (2010)
Beryciformes	Holocentridae	*Myripristis berndti* (Jordan and Evermann 1903)	Hubert et al. (2012)
		Sargocentron diadema (Lacepède 1802)	Hubert et al. (2012)

diverse lineages through selective divergence (Barluenga et al. 2006). Whereas in allopatric speciation, same species could be developed into different genetic lineages due to geographical barrier that prevent gene flow between populations and develop large genetic differences with little morphological change (Fig. 1; Leray et al. 2010). Further, these cryptic species should be validated by several independent data sources such as ecology, behaviour and other physiological traits.

Even though the higher intraspecific distance values denote the presence of putative cryptic species, several propositions such as female philopatry (sex biased dispersal; Steinke et al. 2009), hybridization and misidentification need to be considered before placing the species in cryptic species complex.

5 Low Interspecific Genetic Distance Values

Low interspecific genetic distance values were also observed in some of the marine fishes viz., Amphiprion Clownfishes (*Amphiprion akallopisos, A. perideraion, A. sandaracinos*), *Chaetodon multicinctus* and *C. punctatofasciatus; Zebrasoma flavescenes* and *Z. scopas, Harengula jaguana* and *H. clupeola* (Steinke et al. 2009; Valdez-Moreno et al. 2010). Incomplete lineage sorting, recent speciation and hybridization are likely explanation for low interspecific genetic distance among these different fish species (Ward et al. 2009).

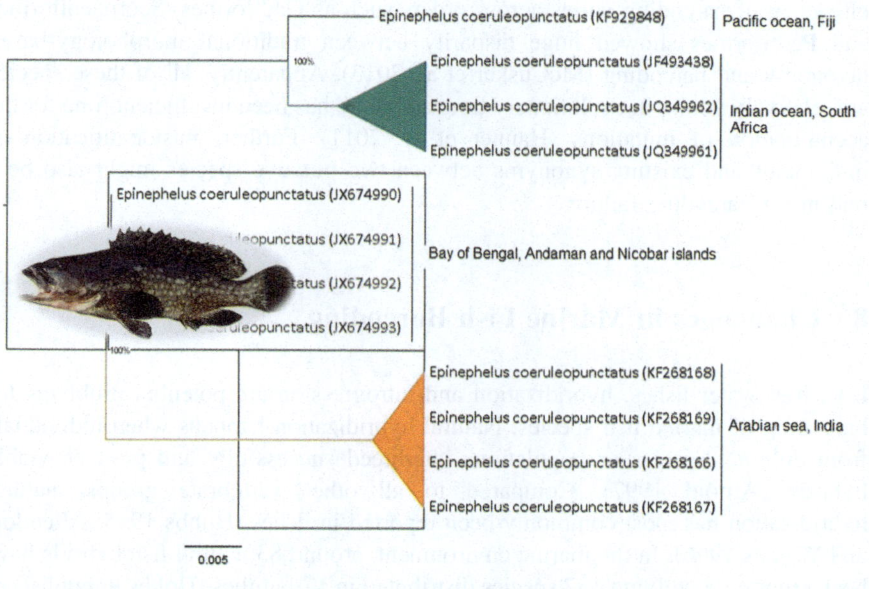

Fig. 1 Neighbour joining tree of COI gene sequences of *Epinephelus coeruleopunctatus* collected from different geographical locations

6 Marine Fish Species Discovery and Validation

Several studies have flagged putative new species based on deep intraspecies divergence of COI gene but these presumptions require detailed analysis by taxonomic experts before they can be confirmed (or rejected) as new species. Once flagged as possible new species, taxonomic validation requires detailed morphological and meristic description and analysis of voucher specimens along with the consideration of behavioural and ecological data (Ward et al. 2009). In these cases, inclusion of sequence data from other mitochondrial or nuclear markers has been advised for species confirmation. Several new species were described in different genera/groups namely goby *Coryphopterus kuna* Victor (2007), sting ray *Urolophus kapalensis* (Yearsley and Last 2006), skate *Dipturus argentinensis* sp. nov. (Mabragaña et al. 2011), spotted bass *Liopropoma olneyi* sp. nov. (Baldwin and Johnson 2014), handfish *Brachionichthys australis* (Last et al. 2007) and five new species of damsel fish (Pyle et al. 2008).

7 DNA Barcoding Failures in Marine Fishes

Apart from success in species identification, there are some failures in the identification of marine fish species. These include some of the bill fishes (*Kajikia audax*, *K. albida*, *Tetrapturus angustirostris*, *T. belone*, and *T. pfluegeri*) and tunas (*Thunnus alalunga*, *T. thynnus*) (Hanner et al. 2011; Cawthorn et al. 2011). The efficiency of barcoding varies across orders such as Gadiformes, Scorpaeniformes and Perciformes showed huge disparity between traditional morphology-based taxonomy and barcoding (McCusker et al. 2013). Apparently, all of these species are relatively young in evolutionary time and there has been insufficient time for the accumulation of mutations (Hanner et al. 2011). Further, misidentification of individuals and existing synonyms between two putative species might also be a reason for barcoding failure.

8 Challenges in Marine Fish Barcoding

Like freshwater fishes, hybridization and introgression are potential problems for barcoding of marine fish species. Natural hybridization happens when individuals from different species or populations interbreed successfully and produce viable hybrids (Arnold 1997). Compared to all other vertebrate groups, natural hybridization has most commonly been reported in fishes (Hubbs 1955; Allendorf and Waples 1996). In the marine environment, around 83 natural fish hybrids have been reported, involving 132 species distributed in 17 families (Hobbs unpublished, cited in Montanari et al. 2012). Some of physiological, ecological, behavioural and geographical factors such as external fertilization (Hubbs 1955), competition for limited spawning grounds (Campton 1987), secondary contact of recently diverged sister taxa (McMillan and Palumbi 1995), spatial or dietary overlap in parental species (van Herwerden et al. 2006; Marie et al. 2007), rarity of one or both parental species (Gosline 1948; Randall et al. 1977; Marie et al. 2007), sneak mating (van Herwerden et al. 2006), and absence of assortative mating (McMillan et al. 1999) have been proposed to explain the hybridization in fishes.

Hybridization is prominent in coral reef fishes and more than 119 fish species of coral reefs were reported to be hybridized (Yaakub et al. 2006). Several studies reported significant marine hybrid zones at Christmas and Cocos Islands (Eastern Indian Ocean) and Western Pacific ocean (Montanari et al. 2012). Even though Introgressive hybridization has been much less documented in the marine environment some recent studies have proved introgressive hybridization between fishes (Roques et al. 2001). Therefore, it is recommended to include other nuclear DNA markers along with COI to avoid misidentification of marine fish species that are suspected to be hybrids.

Apart from these, barcoding is unable to discriminate very recently diverged species because of lack of diagnostic mutations in the COI barcode region. For example, species of the genus Thunnus show little overall divergence, for e.g., Pacific bluefin tuna (*Thunnus orientalis*) (Temminck and Schlegel 1844) and albacore tuna (*Thunnus alalunga*) (Bonnaterre 1788) showed an interspecific genetic distance value of only 0.15 % (Ward et al. 2009). However, such cases of incomplete resolution are rare and perpetually comprise sister taxa that show limited morphological divergence.

9 Applications of DNA Barcoding

DNA barcodes can be used as oligonucleotide probes to prepare DNA chips (DNA array) to identify the commercially important species (Kochzius et al. 2008; Kim et al. 2011) at field level. This could be useful in fine-scale management of fish stocks through the assignment of individual fish to one of the several populations in the same watershed (Hansen et al. 2001; Ruzzante et al. 2004), regional management of fisheries (Swartz et al. 2008) and for ichthyoplankton surveys (Hubert et al. 2015). DNA barcodes can also be useful in identification of invasive species (Briski et al. 2011; Valdez-Moreno et al. 2012), tracing/regulating commercially important fish species (Steinke et al. 2009) and in authenticating species name of processed seafood products (Kannuchamy et al. 2016).

As the DNA sequencing cost is reducing and NGS technologies are taking over the industry, these can be used to barcode the samples from different environmental samples such as marine soil/water and gut content of marine fishes. Especially the gut content analysis of endangered marine fish using metabarcoding approach would give insight about the feeding behaviour and this information would be useful for formulating conservation measures.

10 Role of DNA Barcoding in Conservation

DNA barcoding assist in formulating conservation policies by rapidly assessing the biodiversity at low cost and this information would be helpful to prioritise conservation areas or evaluate the success of conservation actions (Krishnamurthy and Francis 2012). Prioritization of different ecosystems for conservation depends on information of species diversity, its richness and value. Phylogenetic diversity (PD) is an indicator that measures taxonomic divergence between species and an index of phylogenetic diversity can appraise conservation strategies by ignoring tedious species counts and using evolutionary lineages (phylogenies) to boost predictions about biodiversity patterns (Mitchell 2008). Faith and Baker (2006) showed a potential role of DNA barcoding in PD assessments for biodiversity conservation strategies.

11 Conclusion

In this decade of biodiversity (2011–2020), DNA barcodes would assist in characterizing and digitization of marine fish diversity rapidly. These DNA barcodes have applications in marine fisheries management and conservation. Despite some limitations, DNA barcoding has been proved the most successful approach for marine fish species identification. The rapid advancements in informatics and sequencing technology will make mobile DNA barcoding devices a reality in the next decade.

References

Allendorf FW, Waples RS (1996) Conservation and genetics of salmonid fishes. In: Avise JC, Hamrick JL (eds) Conservation genetics: case histories from nature. Chapman and Hall, New York, pp 238–280

Arnold ML (1997) Natural hybridization and evolution. Oxford University Press, New York

Baldwin CC, Castillo CI, Weigt LA, Victor BC (2011) Seven new species within western Atlantic *Starksia atlantica*, *S. lepicoelia*, and *S. sluiteri* (Teleostei, Labrisomidae), with comments on congruence of DNA barcodes and species. ZooKeys 79:21–72

Baldwin CC, Johnson GD (2014) Connectivity across the Caribbean sea: DNA barcoding and morphology unite an enigmatic fish larva from the florida straits with a new species of Sea Bass from deep reefs off Curacao. PLoS ONE. doi:10.1371/journal.pone.0097661

Bamaniya DC, Pavan-Kumar A, Gireesh-Babu P, Sharma N, Reang D, Krishna G, Lakra WS (2014) DNA barcoding of marine ornamental fishes from India. Mitochindrial DNA. doi:10.3109/19401736.2014.1003923

Barluenga M, Stolting KN, Salzburger W, Muschick M, Meyer A (2006) Sympatric speciation in Nicaraguan crater lake cichlid fish. Nature 439:719–723

Becker S, Hanner R, Steinke D (2011) Five years of FISH-BOL: brief status report. Mitochondrial DNA 22:3–9

Bickford D, Lohman DJ, Sodhi NS, Ng PK, Meier R, Winker K, Ingram KK, Das I (2007) Cryptic species as a window on diversity and conservation. Trends Ecol Evol 22:148–155

Briski E, Cristescu ME, Bailey SA, MacIsaac HJ (2011) Use of DNA barcoding to detect invertebrate invasive species from diapausing eggs. Biol Invasion 13:1325–1340

Campton DE (1987) Natural hybridization and introgression in fishes: methods of detection and genetic interpretations. In: Ryman N, Utter F (eds) Population genetics and fishery management. University of Washington Press, Seattle, pp 161–192

Cawthorn DM, Steinman HA, Witthuhn RC (2011) Establishment of a mitochondrial DNA sequence database for the identification of fish species commercially available in South Africa. Mol Ecol Resour 11:979–991

Depczynski M (2006) Small cryptic fishes on coral reefs: a new perspective on reef fish ecology and life histories. Ph.D. thesis, James Cook University

Eschmeyer WN, Ferraris CJ, Hoang D, Long DJ (1998) Species of fishes. In: Eschmeyer WN (ed) Catalog of fishes. California Academy of Sciences, San Francisco, pp 25–1820

Eschmeyer WN, Fricke R, Fong JD, Polack D (2010) Marine fish biodiversity: a history of knowledge and discovery (Pisces). Zootaxa 2525:19–50

Faith DP, Baker A (2006) Phylogenetic diversity (PD) and biodiversity conservation: some bioinformatics challenges. Evol Bioinform 2:121–128

Gosline WA (1948) Speciation in the fishes of the Genus Menidia. Evolution 2:306–313

Granadeiro JP, Silva MA (2000) The use of otoliths and vertebrae in the identification and size-estimation of fish in predator-prey studies. Cybium 24:383–393

Hanner R, Floyd R, Bernard A, Collette BB, Shivji M (2011) DNA barcoding of billfishes. Mitochondrial DNA 22:27–36

Hansen MM, Kenchington E, Nielsen EE (2001) Assigning individual fish to populations using microsatellite DNA markers. Fish Fish 2:93–112

Hebert PDN, Cywinska A, Ball SL, de Waard JR (2003) Biological identifications through DNA barcodes. Proc R Soc Lond Biol 270:313–321

Hebert PDN, Stoeckle MY, Zemlak TS, Francis CM (2004) Identification of birds through DNA barcodes. PloS Biol. doi:10.1371/journal.pbio.0020312

Hubbs CL (1955) Hybridization between fish species in nature. Syst Biol 4:1–20

Hubert N, Espiau B, Meyer C, Planes S (2015) Identifying the ichthyoplankton of a coral reef using DNA barcodes. Mol Ecol Resour 15:57–67

Hubert N, Meyer CP, Bruggemann HJ, Guerin F, Komeno RJL et al (2012) Cryptic diversity in Indo-Pacific coral-reef fishes revealed by DNA-barcoding provides new support to the centre-of-overlap hypothesis. PLoS ONE. doi:10.1371/journal.pone.0028987

Kannuchamy N, Pavan-Kumar A, Venkateshwarlu G, Gireesh-Babu P, Lakra WS (2016) Mislabelling in Indian seafood: an investigation using DNA barcoding. Food Control 59: 196–200

Kim S, Koo H et al (2011) DNA chip for species identification of Korean freshwater fish: a case study. BioChip J 5:72–77

Kochzius M, Nölte M, Weber H et al (2008) DNA Microarrays for identifying fishes. Mar Biotechnol 10:207–217

Krishnamurthy PK, Francis RA (2012) A critical review on the utility of DNA barcoding in biodiversity conservation. Biodivers Conserv 21:1901–1919

Lakra WS, Verma MS, Goswami M, Lal KK, Mohindra V et al (2011) DNA barcoding Indian marine fishes. Mol Ecol Resour 11:60–71

Last PR, Gledhill DC, Holmes BH (2007) A new handfish, *Brachionichthy saustralis* sp. nov., (Lophiiformes: Brachionichthyidae), with a redescription of the critically endangered spotted handfish, B. hirsutus (Lacepede). Zootaxa 1666:53–68

Leray M, Beldade R, Holbrook SJ, Schmitt RJ, Planes S, Bernardi G (2010) Allopatric divergence and speciation in coral reef fish: the three-spot dascyllus, *Dascyllustri maculatus,* species complex. Evolution 64:1218–1230

Mabragaña E, Díaz de Astarloa JM, Hanner R, Zhang J, González Castro M (2011) DNA barcoding identifies argentine fishes from marine and brackish waters. PLoS ONE. doi:10.1371/journal.pone.0028655

Marie A, van Herwerden L, Choat J, Hobbs JPA (2007) Hybridization of reef fishes at the Indo-Pacific biogeographic barrier: a case study. Coral Reefs 26:841–850

Mat Jaafar TNA, Taylor MI, Mohd Nor SA, de Bruyn M, Carvalho GR (2012) DNA barcoding reveals cryptic diversity within commercially exploited Indo-Malay Carangidae (Teleosteii: Perciformes). PLoS ONE. doi:10.1371/journal.pone.0049623

McCusker MR, Denti D, Van Guelpen L, Kenchington E, Bentzen P (2013) Barcoding Atlantic Canada's commonly encountered marine fishes. Mole Ecol Resour 13:177–188

McMillan WO, Palumbi SR (1995) Concordant evolutionary patterns among Indo-West Pacific butterflyfishes. Proc Roy Soc Lond B Biol Sci 260:229–236

McMillan WO, Weigt LA, Palumbi SR (1999) Color pattern evolution, assortative mating, and genetic differentiation in brightly colored butterflyfishes (Chaetodontidae). Evolution 53:247–260

Meyer A (1993) Evolution of mitochondrial DNA in fishes. In: Hochachka PW, Mommsen TP (eds) Biochemistry and molecular biology of fishes. Elsevier, Hague, pp 1–38

Mitchell A (2008) DNA barcoding demystified. Aust J Entomol 47:169–173

Montanari SR, van Herwerden L, Pratchett MS, Hobbs J-PA, Fugedi A (2012) Reef fish hybridization: lessons learnt from butterflyfishes (genus Chaetodon). Ecol Evol 2:310–328

Mora C, Tittensor DP, Myers RA (2008) The completeness of taxonomic inventories for describing the global diversity and distribution of marine fishes. Proc R Soc Lond Biol 275:149–155

Myers N (1996) The biodiversity crisis and the future of evolution. Environmentalist 16:37–47

Nelson JS (2006) Fishes of the World. Wiley, New York

Pavan-Kumar A, Gireesh-Babu PS, Babu PP et al (2013) DNA barcoding of elasmobranchs from Indian coast and its reliability in delineating geographically widespread specimens. Mitochondrial DNA 26:92–100

Puckridge M, Andreakis N, Appleyard AS, Ward RD (2013) Cryptic diversity in flathead fishes (Scorpaeniformes: Platycephalidae) across the Indo-West Pacific uncovered by DNA barcoding. Mol Ecol Resour 13:32–42

Pyle RL, Earle JL, Greene BD (2008) Five new species of the damsel fish genus Chromis (Perciformes: Labroidei: Pomacentridae) from deep coral reefs in the tropical Western Pacific. Zootaxa 1671:3–31

Randall JE, Allen GR, Steene R (1977) Five probable hybrid butterfly fishes of the genus Chaetodon from the central and Western Pacific. Record West Aust Mus 6:3–26

Ratnasingham S, Hebert PDN, Steene R (2007) BOLD: The barcode of life data system (http://www.barcodinglife.org). Mol Ecol Notes 7:355–364

Ribeiro AO, Caires RA, Mariguela TC, Pereira LHG, Hanner R, Oliveira C (2012) DNA barcodes identify marine fishes of São Paulo state, Brazil. Mol Ecol Resour 12:1012–1020

Roques S, Sevigny JM, Bernatchez L (2001) Evidence for broadscale introgressive hybridization between two redfish (genus Sebastes) in the North-west Atlantic: a rare marine example. Mol Ecol 10:149–165

Ruzzante DE, Hansen MM, Meldrup D, Ebert KM (2004) Stocking impact and migration pattern in an anadromous brown trout (*Salmo trutta*) complex: where have all the stocked spawning sea trout gone? Mol Ecol 13:1433–1445

Schonrogge K, Barr B, Wardlaw JC, Napper E et al (2002) When rare species become endangered: cryptic speciation in myrmecophilous hoverflies. Biol J Linn Soc 75:291–300

Steinke D, Zemlak TS, Hebert PDN (2009) Barcoding nemo: DNA-based identifications for the Ornamental fish trade. PLoS ONE. doi:10.1371/journal.pone.0006300

Strauss RE, Bond CE (1990) Taxonomic methods: morphology. In: Schreck CB, Moyle PB (eds) Methods for fish biology. American Fisheries Society, Maryland, pp 109–140

Swartz ER, Mwale M, Hanner R (2008) A role for barcoding in the study of African fish diversity and conservation. S Afr J Sci 104:293–298

Teletchea F (2009) Molecular identification methods of fish species: reassessment and possible applications. Rev Fish Biol Fish 19:265–293

Turanov SV, Kartavtsev YP, Lipinsky VV, Zemnukhov VV, Balanov AA, Lee YH, Jeong D (2014) DNA-barcoding of perch-like fishes (Actinopterygii: Perciformes) from far-eastern seas of Russia with taxonomic remarks for some groups. Mitochondrial DNA. doi:10.3109/19401736.2014.945525

Valdez-Moreno M, Vasquez-Yeomans L, Elias-Gutierrez M, Ivanova NV, Hebert PDN (2010) Using DNA barcodes to connect adults and early life stages of marine fishes from the Yucatan Peninsula, Mexico: potential in fisheries management. Mar Freshw Res 61:655–671

Valdez-Moreno M, Quintal-Lizama C, Gómez-Lozano R, García-Rivas M (2012) Monitoring an Alien invasion: DNA barcoding and the identification of Lionfish and their prey on coral reefs of the Mexican Caribbean. PLoS ONE. doi:10.1371/journal.pone.0036636

van Herwerden L, Choat JH, Dudgeon CL, Carlos G, Newman SJ, Frisch A, van Oppen M (2006) Contrasting patterns of genetic structure in two species of the coral trout Plectropomus (Serranidae) from east and west Australia: introgressive hybridization or ancestral polymorphisms. Mol Phylogenet Evol 41:420–435

Victor BC (2007) *Coryphopterus kuna*, a new goby (Perciformes: Gobiidae: Gobinae) from the western Caribbean, with the identification of the late larval stage and an estimate of the pelagic larval duration. Zootaxa 1526:51–61

Ward RD, Hanner R, Hebert PDN (2009) The campaign to DNA barcode all fishes, FISH-BOL. J Fish Biol 74:329–356

Ward RD, Holmes BH, White WT, Last PR (2008) DNA barcoding Australasian chondrichthyans: results and potential uses in conservation. Mar Freshw Res 59:57–71

Ward RD, Zemlak TS, Innes BH, Last PR, Hebert PDN (2005) DNA barcoding Australia's fish species. Phil Trans R Soc B Biol Sci 360:1847–1857

Wiley EO (1978) The evolutionary species concept reconsidered. Syst Zool 27:17–26

Willis TJ, Anderson M (2003) Structure of cryptic reef fish assemblages: relationships with habitat characteristics and predator density. Mar Ecol Prog Ser 257:209–221

Yaakub SM, Bellwood DR, vanHerwerden L, Walsh FM (2006) Hybridization in coral reef fishes: introgression and bi-directional gene exchange in Thalassoma (family Labridae). Mol Phylogenet Evol 40:84–100

Zemlak TS, Ward RD, Connell AD, Holmes BH, Hebert PDN (2009) DNA barcoding reveals overlooked marine fishes. Mol Ecol Resour 9:237–242

Zhang J (2011) Species identification of marine fishes in China with DNA barcoding. J Evid Based Complement Altern Med. doi:10.1155/2011/978253

Zhang J, Hanner R (2012) Molecular approach to the identification of fish in the South China sea. PLoS ONE. doi:10.1371/journal.pone.0030621

Part IV
DNA Barcoding of Marine Plants

DNA Barcoding in Phytoplankton and Other Algae in Marine Ecosystem: An Effective Tool for Biodiversity Assessment

Farhina Pasha

Abstract Marine biodiversity is a valuable gift of nature, as the marine environment is exceptionally complicated, diverse and of utmost economic value. There is a desperate need to evaluate and protect this treasure because of its multi facet uses and its richness in species composition. Biodiversity protection particularly emphasizes on the ability of quantifying and tracking changes in the marine ecosystem. Phytoplankton and Algal biodiversity too comes under this category because of their diverse species population, vast habitat and most of them having microscopic structure. Thereby both marine phytoplankton and algae play a significant role in marine biodiversity and their taxonomic identification remain a big challenge. With all the merits and limitations of DNA barcoding and amongst the furious debate of researchers in context of its comparison with alpha taxonomic identification, DNA barcoding certainly is a promising tool of future in the field of marine species identification, biodiversity assessment and conservation.

Keywords DNA barcoding · Marine phytoplankton · Biodiversity · Marine algae · Species identification · Cox1 · rRNA

1 Introduction

DNA barcoding certainly is an innovative taxonomic revolution in the field of taxonomic classification system for speedy, precise and automatized identification for a vast number of species, by utilizing a short gene sequence as an inner species identity tag. DNA barcoding fastens the speed not only of species identification but also allow the taxonomists to speedily sort the specimens and make possibility of discovering many new taxa's. Although DNA barcoding is predominantly emerging as a new option it is continuously being criticized by both patrons of taxonomy and taxonomists (Wheeler 2004; Will and Rubinoff 2004; Ebach and Holdrege

F. Pasha (✉)
Faculty of Sciences, Department of Biology, University of Tabuk, Tabuk, Saudi Arabia
e-mail: fbasha@ut.edu.sa; farhinabpl@gmail.com

S. Trivedi et al. (eds.), *DNA Barcoding in Marine Perspectives*,
DOI 10.1007/978-3-319-41840-7_18

2005). DNA barcode is a meticulously standardized sequence of smallest span which has been measured from an identified gene and is present in one of the major database sequences and is attached to a voucher specimen and who's origin and present status is known. Therefore only those *cox1* sequences which fulfill the above criteria are termed as DNA barcode by National Center for Biotechnology Information's Genbank (NCBI, Genbank), the European Molecular Biology Laboratory (EMBL), and the DNA Data Bank of Japan (DDBJ). Another mile stone in DNA barcoding is establishment of "Consortium Barcode of Life" (CBOL) to support global DNA barcoding and is assisted by Life Data System (http://www.barcodinglife.org). When talking of species identification and discovering novel taxa's a relevant term needs to be conferred that's "Biodiversity". Commonly biodiversity is defined as the variety of living organisms including aerial, marine and aquatic ecosystem. The term also includes diversity amongst species within as well as between species in a prevailing ecosystem. Biodiversity though is understood in different context by different experts, such as for an ecologist it refers to the numerical and practical physiognomies of a species, its taxonomic connection and its dispersal amid species (Harper and Hawksworth 1994). Therefore it is essential that Biologists and economists develop a close collaboration to have a better understanding of biodiversity. Esteeming marine biodiversity grieves an additional barrier as the marine environment is exceptionally complicated and diverse. Amongst the total flora and fauna on planet, 32 of plant and 33 of animal phyla are in the marine environment (Ray and Grassle 1991). Another problem is that of monitoring and sampling such vast habitat. There is a desperate need to evaluate marine biodiversity because of its multi facet uses and its richness in species composition. To begin, one should remember that marine biodiversity is more in benthic region as compared to Pelagic and on coast than in Open Ocean as majority of species are habitat in these regions (Pimentel et al. 1997).

Biodiversity protection particularly emphasizes on the ability of quantifying and tracking changes in the marine ecosystem. Phytoplankton biodiversity too comes under this category because of their diverse species population, vast habitat and most of them having microscopic structure (Carstensen and Heiskanen 2007). Even their taxonomic identification needs high level expertise due to their microscopic size, requirement of large sample size, rapid turnover rate and clustering which changes very quickly with even a slight variation in the surrounding environment making community organization erratic (Dybern and Hansen 1989). The applicability of phytoplankton as a measure of eutrophication is long established (Spatharis and Tsirtsis 2010; Spatharis et al. 2011). Another prime indicator of marine ecosystem are the marine algae. An alga is a term basically used for eukaryotes which are perhaps not so closely related. Algae include a large range of organisms for microscopic genera like *Chlorella*, *diatom,* to *brown algae* which grow up to 50 m in length, to the largest algae "sea weeds". They are mostly aquatic and autotrophic and are characterized by lack of tissues present in land plant like stomata, xylem, phloem etc. There by both marine phytoplankton and algae play a significant role in marine biodiversity and their taxonomic identification remain a big challenge.

2 Phytoplankton and Marin Ecosystem

Phytoplankton are represented by genetically diverse organisms with global distribution across the oceans. Due to their wide tolerance range for different climatic conditions and habitat they are predominant in most of the ecosystems. Thereby they also serve as one of the most sensible biological water quality indicators. Taxonomic identification of marine phytoplankton is a complicated task due to lack of prominent morphological structure, their microscopic size, absence of morphological markers and almost no information about their breeding methods (Medlin et al. 2000). Until now such structuring has rarely been quantified in marine phytoplankton. According to Medlin et al. (2000) genetic diversity study in pelagic ecosystem is practically fictional, therefore the genetic and demographic study of phytoplankton is virtually negligent. However molecular techniques based on rRNA gene has provided a new horizon for marine phytoplankton (Medlin et al. 2006; Evans et al. 2007). In many studies it has been reported that morphological similarity and agility has led to an erroneous result by pooling the actual and cryptic species together, thereby not reflecting the proper species composition in field studies.

During the past decade molecular analysis has greatly helped in taxonomic identification of micro aquatic flora and fauna (Lefranc et al. 2005). As a modern taxonomic tool DNA barcoding make use of molecular locus for discrimination at species level and a data base for assessment. The technique relies on comparison of similarities or divergence of a molecular sequence of unknown organisms to voucher sequences present in existing database for species identification. The DNA barcoding of phytoplankton require, firstly sample preservation followed by DNA extraction. The extracted DNA is then amplified by PCR (Polymerase Chain Reaction) and lastly sequence identification of this amplified DNA sequence is performed. The sequences identified are compared with the known sequences from various databases such as GenBank, BLAST (Fig. 1).

The smaller subunit (SSU) 18SrRNA gene is most constantly used marker for phylogenetic study, whereas clone libraries are effectively used from biodiversity identification in different habitat. As many DNA sequences based on 18SrDNA are present in Genbank, therefore offer an advantage for selection of DNA target region to access species.

In 2010 study published in *Nature* about marine phytoplankton states that they have significantly decline in ocean over the past decades. It is estimated almost 40 % decline in phytoplankton population in 1950 alone, possibly in response to global warming (Boyce et al. 2010). The above article began a debate leading to several interactive talks and criticisms also published in *Nature* (Mackas 2011; Rykaczewski and Dunne 2011; McQuatters-Gollop et al. 2011; Boyce et al. 2010). In the follow up article published in 2014 scientists used a varying method for analyzing the phytoplankton concentration and a vast database for their quantification was made but the end result remained the same as published in *Nature* study (Boyce et al. 2010). These alarming facts and the future predictions based on a

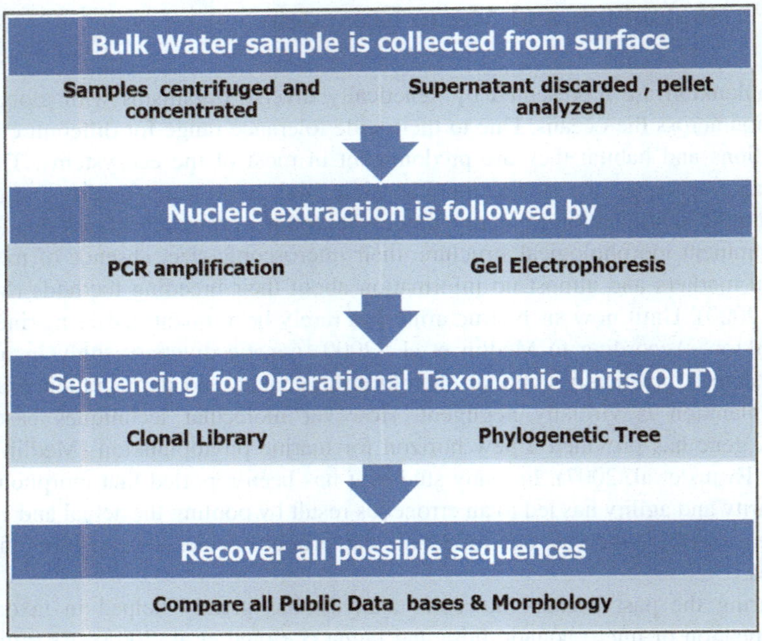

Fig. 1 DNA barcoding modus operand for phytoplankton

current multi model study estimate a 2–20 % primary production decline in phytoplankton population by 2100 A.D.

3 DNA Barcoding: A Predominant Tool for Marine Algae Systematics

Another prime indicator of the marine ecosystem are the marine algae. The online algal database (http://www.algaebase.org) gives an in-depth taxonomic information regarding various algal phyla and classes. Approximately 72,500 algal species are existing among which 44,000 names have been published and 33,248 names have been processed by Algal base till June 2012. Recent Publication on Marine Algae and Phytoplankton from 2000 to 2015 are mentioned in Table 1. It is difficult to identify marine macro algae via molecular techniques because of the morphological variations amongst the group. As a fact the coastal ecosystem is a locale of strong temporal and spatial ascents in salinity, wind patterns, tidal patterns and other ecological influences there by leading to variation and morphological differences in prevailing organisms (Lee 2000; Lee and Frost 2002; Rynearson et al. 2006). The DNA barcode 5′ end of the COI mitochondrial gene has been very successful in identifying red algal species (Saunders 2005, 2008; Robba et al. 2006). As this gene

Table 1 Recent Publication on Marine Algae and Phytoplankton from 2000 to 2015

S. no.	Topic	References	Selection criteria/keyword
1	Marine microbial eukaryote	Massana et al. (2015)	DNA barcoding in marine algae
2	Coralline algae	Pardo et al. (2014)	DNA barcoding in marine algae
3	Brown algae	An et al. (2013)	DNA barcoding in marine algae
4	Photosynthetic sea slug	Krug et al. (2013)	DNA barcoding in marine algae
5	Brown Alga *Saccharine Japonica*	Balakirev et al. (2012)	DNA barcoding in marine algae
6	Bioinformatics method in Brown Algae	Zhang et al. (2012)	DNA barcoding in marine algae
7	Macro algae and Diatoms	Saunders and McDevit (2012)	DNA barcoding in marine algae
8	*Chlorarachniophyte and Lotharella globosa*	Hirakawa et al. (2011)	DNA barcoding in marine algae
9	DNA barcoding in Red alga	Saunders (2005)	DNA barcoding in marine algae
10	Phytoplanktons on the Pesian Gulf	Alemzadeh et al. 2014	DNA barcoding in Phytoplankton
11	Oedogonium species	Lawton et al. (2014)	DNA barcoding in Phytoplankton

Search Source http://www.ncbi.nlm.nih.gov/pubmed

has nonoverlapping intraspecific and interspecific divergence often Called as "barcode gap", it allows to assign explicitly identification to the specie (Samadi and Barberousse 2006). Mitochondrial COI gene and Universal Plastid Amplicon (UPA) of domain V of 23S rRNA gene, are magnificent molecular markers which have been successfully used for identification of red algae of family *Kallymeniaceae.* The marker COI is categorically more sensitive and has managed the identification of a novel specie *Euthoratimburtonii* (Clarkston and Saunders 2010). In another study conducted, taking the advantage of DNA barcoding two of the five classes of *Florideophyceae* and two genera (*Bangia* and *Porphyra*) of the *Bangiales* were identified and therefore DNA barcode was established as a standardized method for red algal species identification and also revealed cryptic divergence within the community (Robba et al. 2006). There are many species of algae which are commercially very important as food alternative of pharmaceutical value therefore making their taxonomic identification more necessary, one such example is *Gracilariaceae* a red algal family important for microbial and biotechnological research as phycocolloidagar. These species are difficult to be identified morphologically and thereby DNA barcode holds a promising aspect in their species identification (Kim et al. 2010). Next in the line of useful algae, *Ahmed* et al. isolated a new microalga from Indian Ocean which can be potentially used as a biofuel. They used 16S rRNA and 23S rRNA gene as DNA barcodes (Ahmad et al. 2013). Not only the beneficial algae identification has been possible by DNA barcoding but also the identification of algae species which cause menace, can also be identified to help monitoring marine and coastal projects such as Harmful Algal Blooms (HAB). There are enormous number of markers available for the three genome, but almost all the researchers have settled down with one of the three or four prototypes, but the absence of an established standard amongst them obstructs the assessment of a novel sequences. Marine microalgae can be extremely difficult to identify due to its simple morphological and anatomical structure, phenotypic flexibility, alternation of generation. Therefore workers are more relaying on genetic tool like DNA barcoding rather than molecular assisted alpha taxonomic methods.

In a study undertaken, for red algae new primers were established for 5′ *cox1* 'barcode' region, which were used for (i) *Mazzaella* species in the Northeast Pacific; (ii) species of the genera *Dilsea* and *Neodilsea* in the Northeast Pacific; and (iii) *Asteromenia peltata* from three oceans (Saunders 2005). These were selected due to their ambiguity in connection to number and distribution of the specie, as well as specie identification because all the identification has been done by molecular markers but DNA barcode study revealed rapid identification and a number of new annotations were revealed. The generalized method for DNA barcoding starts with DNA extraction and purification. The cox1 sequence are acquired from Genbank. The sequences are used in combination with earlier published cox1 barcoding primers (Hebert et al. 2003) to make precise primers so as to amplify this gene region for red algae. Amplified product is then purified and forward and reverse sequences are edited and aligned. Multi sequence alignment is done and nucleotide position analyzed. There is definitely an advantage in implementing a standard marker for specie identification and by selecting *cox1*-5′ and system it gives an added advantage of consistency throughout the domains, which help in giving comparison and offer algal systematic a direct advantage. The prevalence of genetic barcoding does not mean the end of taxonomy to phycologists instead start a revolution in molecular based alpha taxonomy, which will help in recognition of inherited lineage. It should be used in conjunction with anatomical observations and in close integration for accessing mitochondrial data with nuclear markers. The importance of *cox1*-5′ lies in identification and documentation of red algae though, has been adequate but still require a meticulous investigation.

4 Advantages of DNA Barcoding in Marine Perspective

The vast marine environment supports billions of species of flora and fauna, micro and macro organisms all across its realms. There are almost 35 animal phyla, marine representatives from 34 phyla and 14 solely marine animals (Briggs 1994; Gray 1997). DNA barcoding is acts as an efficient tool to identify every specie individually and also isolate them from cryptic species, which represents a specie having similar morphological features but genetically very different. Because of this genetic distinctness the DNA barcode technology stands distinct advantage over molecular taxonomic techniques for identification in marine, estuarine and other aquatic habitat. It also establishes link between the adult and larval stages, as most of the marine organisms have many larval stages and the morphological feature of these larvae are entirely different from adults, thereby unveiling their life cycle. The menace of intrusive species can be accessed via DNA barcoding globally (Molnar et al. 2008). These intrusive species definitely pose a danger to the native species thereby destroying the natural and economic intrinsic system.

Not only this, DNA barcoding is very useful in accessing marine pollution. It is an important tool to identify, verify and access safety vast variety of sea food. As DNA barcoding is used for novel species identification it can also trace the source of many marine products (Galimberti et al. 2013). A study undertaken on Japanese tuna sushi exposed the occurrence of an endangered species served in different restaurants in USA (Lowenstein et al. 2009). DNA barcoding has a prominent role in species taxonomic identification (Ali et al. 2014). Another very important aspect of DNA barcoding is the maintenance of voucher specimens, their grouping and conservation, as they are the perpetual documents for marine biodiversity exploration. As per recent researches in deep sea it has been revealed that the organisms of the deep sea are of tremendous pharmaceutical importance thereby gaining global attention. Census of Diversity of Abyssal Marine Life (CeDAMar) is working thoroughly for deep sea organisms DNA barcoding. Therefore it can be concluded that with all the merits and limitations of DNA barcoding and amongst the furious debate of researchers in context of its comparison with alpha taxonomic identification DNA barcoding certainly is a promising tool of future in the field of marine species identification, biodiversity assessment and conservation.

Acknowledgments The authors would like to acknowledge, University of Tabuk, Tabuk, Saudi Arabia. The author would also like to thanks Department of Biology, Faculty of Sciences, Saudi Digital Library and University Library providing the facility for literature survey and collection.

References

Ahmad I, Fatma Z, Yazdani S, Kumar S (2013) DNA barcode and lipid analysis of new marine algae potential for biofuel. Algal Res 2(1):10–15

Alemzadeh E, Haddad R, Ahmadi AR (2014) Phytoplanktons and DNA barcoding: characterization and molecular analysis of phytoplanktons on the Persian Gulf. Iran J Microbiol 6 (4):296–302

Ali MA, Gyulai G, Hidvegi N, Kerti B, Al Hemaid F, Pandey A et al (2014) The changing epitome of species identification—DNA barcoding. Saudi J Biol Sci 21(3):204–231

An C, Kuda T, Yazaki T, Takahashi H, Kimura B (2013) FLX pyrosequencing analysis of the effects of the brown-algal fermentable polysaccharides alginate and laminaran on rat cecal microbiotas. Appl Environ Microbiol 79(3):860–866. doi:10.1128/AEM.02354-12

Balakirev ES, Krupnova TN, Ayala FJ (2012) DNA variation in the phenotypically-diverse brown alga Saccharina japonica. BMC Plant Biol 12:108. doi:10.1186/1471-2229-12-108

Boyce D, Lewis M, Worm B (2010) Global phytoplankton decline over the past century. Nature 466(7306):591–596

Briggs J (1994) Species diversity: land and sea compared. Syst Biol 43(1):130

Carstensen J, Heiskanen A (2007) Phytoplankton responses to nutrient status: application of a screening method to the northern Baltic sea. Mar Ecol Prog Ser 336:29–42

Clarkston B, Saunders G (2010) A comparison of two DNA barcode markers for species discrimination in the red algal family Kallymeniaceae (*Gigartinales, Florideophyceae*), with a description of *Euthoratimburtonii* specie. Botany 88(2):119–131

Dybern BI, Hansen HP (1989) Baltic Sea Patchiness Experiment (BSPE)—PEX 1986—Part 1: General report. ICES cooperative research report 163:1–10

Ebach M, Holdrege C (2005) DNA barcoding is no substitute for taxonomy. Nature 434 (7034):697

Evans K, Wortley A, Mann D (2007) An assessment of potential diatom "barcode" genes (cox1, rbcL, 18S and ITS rDNA) and their effectiveness in determining relationships in Sellaphora (Bacillariophyta). Protist 158(3):349–364

Galimberti A, De Mattia F, Losa A, Bruni I, Federici S, Casiraghi M et al (2013) DNA barcoding as a new tool for food traceability. Food Res Int 50(1):55–63

Gray J (1997) Marine biodiversity. International Maritime Organization, London

Harper JL, Hawksworth DL (1994) Biodiversity: measurement and estimation. Philos Trans R Soc Lond Ser B Biol Sci 345(1311):5–12

Hebert P, Cywinska A, Ball S, de Waard J (2003) Biological identifications through DNA barcodes. Proc R Soc B Biol Sci 270(1512):313–321

Hirakawa Y, Howe A, James ER, Keeling PJ (2011) Morphological diversity between culture strains of a chlorarachniophyte, Lotharella globosa. PLoS One 6(8):e23193. doi:10.1371/journal.pone.0023193

Kim MS, Kim MY, Yang Cho GY (2010) Applying DNA barcoding to Korean Gracilariaceae (Rhodophyta) *Cryptogamie*. Algologie 31(4):387–401

Krug PJ, Vendetti JE, Rodriguez AK, Retana JN, Hirano YM, Trowbridge CD (2013) Integrative species delimitation in photosynthetic sea slugs reveals twenty candidate species in three nominal taxa studied for drug discovery, plastid symbiosis or biological control. Mol Phylogenet Evol 69(3):1101–1119. doi:10.1016/j.ympev.2013.07.009

Lawton RJ, de Nys R, Skinner S, Paul NA (2014) Isolation and identification of oedogonium species and strains for biomass applications. PLoS One 6:9(3):e90223. doi:10.1371/journal.pone.0090223.eCollection

Lee CE (2000) Global phylogeography of a cryptic copepod species complex and reproductive isolation between genetically proximate "Populations". Evolution 54(6):2014

Lee C, Frost B (2002) Morphological stasis in the Eurytemora affinis species complex (Copepoda: Temoridae). Hydrobiologia 480(1):111–128

Lefranc M, Thenot A, Lepere C, Debroas D (2005) Genetic diversity of small eukaryotes in lakes differing by their trophic status. Appl Environ Microbiol 71(10):5935–5942

Lowenstein J, Amato G, Kolokotronis S (2009) The real maccoyii: identifying tuna sushi with DNA barcodes- contrasting characteristic attributes and genetic distances. PLoS ONE 4(11): e7866

Mackas D (2011) Does blending of chlorophyll data bias temporal trend? Nature 472(7342):E4–E5

Massana R (2015) Getting specific: making taxonomic and ecological sense of large sequencing data sets. Mol Ecol 24(12):2904–2906. doi:10.1111/mec.13210

McQuatters-Gollop A, Reid P, Edwards M, Burkill P, Castellani C, Batten S (2011) Is there a decline in marine phytoplankton? Nature 472(7342):E6–E7

Medlin L, Lange M, Nothig E (2000) Genetic diversity in the marine phytoplankton: a review and a consideration of Antarctic phytoplankton. Antarct Sci 12(03):325–333

Medlin L, Metfies K, Mehl H, Wiltshire K, Valentin K (2006) Picoeukaryotic plankton diversity at the helgoland time series site as assessed by three molecular methods. Microb Ecol 52(1): 53–71

Molnar J, Gamboa R, Revenga C, Spalding M (2008) Assessing the global threat of invasive species to marine biodiversity. Front Ecol Environ 6(9):485–492

Pardo C, Lopez L, Peña V, Hernández-Kantún J, Le Gall L, Bárbara I, Barreiro R (2014) A multilocus species delimitation reveals a striking number of species of coralline algae forming Maerl in the OSPAR maritime area. PLoS One 11:9(8):e104073. doi:10.1371/journal.pone.0104073.eCollection

Pimentel D, Wilson C, McCullum C, Huang R, Dwen P, Flack J et al (1997) Economic and environmental benefits of biodiversity. Bioscience 47(11):747–757

Ray G, Grassle J (1991) Marine biological diversity program. Bioscience 41(7):453–457

Revilla M, Franco J, Bald J, Borja A, Laza A, Seoane S et al (2009) Assessment of the phytoplankton ecological status in the Basque coast (northern Spain) according to the European water framework directive. J Sea Res 61(1–2):60–67

Robba L, Russell S, Barker G, Brodie J (2006) Assessing the use of the mitochondrial cox1 marker for use in DNA barcoding of red algae (Rhodophyta). Am J Bot 93(8):1101–1108

Rykaczewski R, Dunne J (2011) A measured look at ocean chlorophyll trends. Nature 472(7342): E5–E6

Rynearson T, Newton J, Armbrust E (2006) Spring bloom development, genetic variation, and population succession in the planktonic diatom *Ditylum brightwellii*. LimnolOceangr 51 (3):1249–1261

Samadi S, Barberousse A (2006) The tree, the network, and the species. Biol J Linn Soc 89 (3):509–521

Saunders G (2005) Applying DNA barcoding to red macroalgae: a preliminary appraisal holds promise for future applications. Philos Trans R Soc B Biol Sci 360(1462):1879–1888

Saunders G (2008) A DNA barcode examination of the red algal family Dumontiaceae in Canadian waters reveals substantial cryptic species diversity. 1. The foliose Dilsea-Neodilsea complex and Weeksia. This paper is one of a selection of papers published in the Special Issue on Systematics Research. Botany 86(7):773–789

Saunders GW, McDevit DC (2012) Methods for DNA barcoding photosynthetic protists emphasizing the macroalgae and diatoms. Methods in Mol Biol 858:207–222 doi:10.1007/978-1-61779-591-6_10

Spatharis S, Tsirtsis G (2010) Ecological quality scales based on phytoplankton for the implementation of Water Framework Directive in the Eastern Mediterranean. Ecol Indic 10 (4):840–847

Spatharis S, Roelke D, Dimitrakopoulos P, Kokkoris G (2011) Analyzing the (mis)behavior of Shannon index in eutrophication studies using field and simulated phytoplankton assemblages. Ecol Indic 11(2):697–703

Wheeler Q (2004) Taxonomic triage and the poverty of phylogeny. Philos Trans R S B Biol Sci 359(1444):571–583

Will K, Rubinoff D (2004) Myth of the molecule: DNA barcodes for species cannot replace morphology for identification and classification. Cladistics 20(1):47–55

Zhang AB, Feng J, Ward RD, Wan P, Gao Q, Wu J, Zhao WZ (2012) A new method for species identification via protein-coding and non-coding DNA barcodes by combining machine learning with bioinformatic methods. PLoS One 7(2):e30986. doi:10.1371/journal.pone.0030986

A Search for a Single DNA Barcode for Seagrasses of the World

Barnabas H. Daru and Kowiyou Yessoufou

Abstract It has recently been predicted that 91 % of marine species diversity is still unknown. Given that the future of marine habitats is threatened by anthropogenic activities and climate change, there is a pressing need to accelerate the documentation of marine biodiversity. The traditional morphological biodiversity screening could be aided by molecular approach such as DNA barcoding. In this study, we search for single DNA marker that could be used as DNA barcode for all seagrasses, irrespective of the lineages and the geographical locations. We found that the nuclear *phyB* followed by the plastid *matK* emerged as the best candidates. Although both markers have their own strengths and limitations, we suggest they could be prioritised in seagrass biodiversity assessment pending future improvements.

Keywords DNA barcoding · *phyB* · *matK* · Marine biodiversity · Cymodoceaceae · Ruppiaceae

1 Introduction

How many species are there and how do we recognize them? A recent prediction of species richness estimated that global ecosystems harbor 8.7 million species, including 2.2 million marine species (Mora et al. 2011). Of this impressive diversity, 86 % of terrestrial and 91 % of marine species are currently unknown (Mora et al. 2011), raising an urgent need for accelerating the process and increasing our commitment for biodiversity assessment. This need for biodiversity

---

B.H. Daru (✉)
Department of Organismic and Evolutionary Biology,
Harvard University, 22 Divinity Avenue, Cambridge, MA 02138, USA
e-mail: barnabas_daru@fas.harvard.edu; darunabas@gmail.com

K. Yessoufou
Department of Environmental Science, University of South Africa,
Florida Campus, Florida 1710, South Africa

S. Trivedi et al. (eds.), *DNA Barcoding in Marine Perspectives*,
DOI 10.1007/978-3-319-41840-7_19

assessment is even more pressing given the current extinction crisis driven by an unprecedented rate of species loss estimated to be 1,000–10,000 times greater than that recorded in the past (Millennium Ecosystem Assessment 2005; Barnosky et al. 2011). Traditionally, we assess species diversity using morphological features, a long-standing approach that can be very tedious and questionable owing to the potential subjectivity attached to it since the relevance of morphological features to be used is to the discretion of the taxonomist. Although, this approach is irreplaceable, it has its own limitations (e.g. see Packer et al. 2009). For example it took >250 years to describe less than a 1/4 of the world species (see review in Radulovici et al. 2010). Given the unprecedented rate of species extinction, we cannot afford to wait again for more than two centuries to know a tiny proportion of earth's diversity; a quicker and integrative approach (that combines perhaps morphology and molecular data) that can help accelerate biodiversity assessment, the discovery of new species including cryptic species become a matter of urgency.

DNA barcoding has been proposed as an important molecular tool that provides complementary information overlooked in morphology-based biodiversity assessment (Hebert et al. 2003). It is a technique that uses short sequences of DNA to either confirm species identity or assign unknown biological materials (e.g. plants, animals and fungi at any stage of life cycle) to corresponding species or higher taxonomic groups or reveal cryptic species (morphologically similar but genetically distinct species). The technique has witnessed a great application for assessing biodiversity (Smith et al. 2005; Papadopoulou et al. 2015; van der Bank and Greenfield 2015; see also Trivedi et al. 2016 for a comprehensive review). However, more attention has been given to terrestrial ecosystems (see Fig. 1 in Radulovici et al. 2010), although oceans cover more than 70 % of our planet and are potentially as species-rich as terrestrial ecosystems. For example, of the currently known 35 animal phyla, 14 are marine endemics (Briggs 1994; Gray 1997). In general, marine ecosystems provide unique ecosystem services to humanity: foods (e.g. fish, prawns, etc.), biotechnological and non-living resources, as well as indicator of environmental health and ecosystem functioning (food webs), erosion control and carbon sinks (e.g. mangroves) etc. Given major anthropogenic factors that threaten marine ecosystems (e.g. habitat loss, overharvesting, global warming, pollution, invasive species, etc.), there is a need to know the ecosystem engineers that ensure the provision of goods and services for humanity in oceans.

The application of DNA barcoding to assess marine biodiversity is increasingly generating renewed interest (see reviews in Radulovici et al. 2010; Trivedi et al. 2016). However, of the few studies that show interests into marine biodiversity (compared to terrestrial biodiversity), most have focused on marine animals (Radulovici et al. 2010; Trivedi et al. 2016, see also Marine Barcode of Life MarBOL, www.marinebarcoding.org; accessed March 20, 2015), resulting potentially in comparatively poorer knowledge of marine plant diversity. In the present study, we focus on seagrasses, an ecologically important plant taxonomic group in marine ecosystems.

Seagrasses belong to the monocot order Alismatales comprising 72 species represented in 13 genera and five families (Les et al. 1997; den Hartog and Kuo

2006). They have a wide range of vegetative and floral diversity (Fig. 1), and are widely distributed along all marine coastlines worldwide from intertidal to subtidal depths (den Hartog 1970; Green and Short 2003), providing key ecosystem services such as primary productivity, nutrient cycling etc. (Hemminga and Duarte 2000; Duarte 2002; Les et al. 2002; Orth et al. 2006; McGlathery et al. 2007). They are also well-known in traditional medicine for the valuable secondary compounds (e.g. phenolic acids, rosmarinic acid and zosteric acid) widely used as an antioxidant and effective antifouling agent (Trivedi et al. 2016). Nonetheless, these marine plants are undergoing a rapid decline in both species richness and geographic cover: we are losing seagrasses at a rate of 110 km^2 per year (Waycott et al. 2009), prompting the need for documenting the diversity of seagrasses before we lose what we do not yet know about them. However, species boundaries among the lineages are still not resolved (Tomlinson and Posluszny 2001; den Hartogand Kuo 2006). The real challenge lies with the fact that they are submerged plants with high prevalence of cryptic species (Briggs 2003; Trivedi et al. 2016). Because they occur submerged in marine water, they may have acquired adaptations such as reduced morphology in both vegetative and floral structures, making morphological identification difficult. In fact, seagrass species in the field or archived in herbaria are often devoid of

Fig. 1 Representatives of seagrass species showing variation and diversity in the group. **a** *Cymodocea nodosa*; **b** *Lepilaena australis*; **c** *Posidonia oceanica*; **d** *Halophila beccarii*; **e** *Thalassia hemprichii*; **f** Seagrass meadow—a feeding trail for sea cow (*Dugong dugon*) in ChekJawa, Singapore; **g** *Enhalus acoroides* (male flowers); **h** *Cymodocea rotundata*; **i** *Halophila spinulosa*; **j** *Syringodium isoetifolium*; **k** *Halophila ovalis*; **l** shoot of *Thalassia hemprichii*.—Photographs: **a, b** courtesy Y. Ito; **c** J.Á. Rodríguez; **d–l** R. Tan

diagnostic flowers (Trivedi et al. 2016; Personal observations). This calls for an urgent need for a fast, reliable, and cost-efficient technique for recognition and identification of seagrasses especially by non-experts (Cocheret de la Morinière et al. 2003).

Lucas et al. (2012) showed the importance of DNA barcoding in delimiting species boundaries for seagrasses in India. For 14 species examined using *matK* and *rbcL*, sequence divergence for discriminating species is higher for *matK* than *rbcL*. Another study showed the success of DNA barcoding in identifying six seagrass species in the gut of rabbit fish *Siganus fuscescens* in Moreton Bay, Australia (Chelsky Budarf et al. 2011). Other studies have focused on single clades using different markers e.g. *trnK* and *rbcL* for *Zostera* (Les et al. 2002), ITS for *Halophila* (Waycott et al. 2002), 5.8S rDNA and ITS2 for *Halophila* (Uchimura et al. 2008), ITS1, *matK*, *rbcL*, *psbA-trnH* for Zosteraceae (Coyer et al. 2013). As indicated in these studies, they focus either on a single geographic location or a single genus of seagrasses, leaving a knowledge gap on whether a single DNA barcode could help screen seagrass diversity irrespective of the geographic locations or genera.

In this study, we explored this possibility by first assessing the potential of nine markers to discriminate seagrass species of the world, and second, assess the efficacy of barcodes across major seagrass clades.

2 Materials and Methods

2.1 *Taxon Sampling*

We retrieved from GenBank/EBI all available sequences of seagrasses for nine molecular markers, *atp1*, *cob*, ITS, *matK*, NAD5, *phyB*, *rbcL*, *rpoB* and *trnH-psbA*. These sequences are from 44 species belonging to all the five seagrass families Cymodoceaceae, Hydrocharitaceae, Posidoniaceae, Ruppiaceae, and Zosteraceae (Appendix A). Our sampling comprised 95 specimens (see Appendix A). The sequences were aligned using SeaView v.4 (Gouy et al. 2010) and manually adjusted using Mesquite v.2.5 (Maddison and Maddison 2008).

2.2 *Barcoding Analyses*

First, we evaluated the performance of the various plant DNA regions in discriminating seagrass species by applying three criteria commonly used in DNA barcoding literature: the barcode gap of Meyer and Paulay (2005), the level of sequence divergence and the discriminatory power. Barcode gap was assessed by comparing intra-specific variation (i.e. the amount of genetic variation within species) to inter-specific variation (between species). A good barcode should exhibit a significant gap, meaning that sequence divergence within species should be significantly lower than between species. Statistical significance between intra- and inter-specific variation was assessed using Wilcoxon test in R (R Core Team 2013). In addition, we calculated the distribution of range, mean and standard deviation of both intra- and inter-specific distances.

Second, we identified the best DNA barcode using two distance-based methods; near neighbour and best close match (Meier et al. 2006) using the functions *near Neighbour* and *best Close Match* respectively, implemented in the R package SPIDER (Brown et al. 2012). This was done by combining all sequences. Prior to the evaluation of discriminatory power of each barcode candidate, we determined the distance threshold i.e. the optimised genetic distance for species delimitation, given that the 1 % threshold suggested by BOLD does not hold for every organism (Meyer and Paulay 2005). This distance cut-off was identified using the function *local Minima* implemented in SPIDER, which evaluates the transition between intra- and inter-specific distances (Brown et al. 2012). The optimised threshold was used especially in best close match method.

Lastly, given the possibility that the performance of marker could vary between taxonomic levels (Gere et al. 2013), we further assessed the performance of the core barcode within two families, Cymodoceaceae and Ruppiaceae; the other seagrass families were not considered here due to lack of sufficient DNA sequences.

3 Results and Discussion

Information on aligned sequence length, number of species, mean number of substitutions per nucleotide for all DNA regions considered singly or in combination, the range and means of intra- and interspecific distances are summarized in Table 1. The mean interspecific distance for the single and combined regions are lower than 1 %, ranging from 0.011 in *cob* to 0.77 in ITS. The mean intraspecific variation for each and combined DNA regions was also low ranging from 0.00008 in *rbcL* + *matK* + *cob* to 0.024 in ITS (Table 1).

We show that the ranges and mean intraspecific distances for all markers when considered singly or in combination with the core barcodes (*matK* + *rbcL*), are significantly lower than interspecific distances (Wilcoxon test, $p < 0.01$; Fig. 2), suggesting the presence of barcode gap. Comparison of the proportion of sequences with barcode gap showed that *trnH-psbA* (88 %) followed by *rpoB* (78 %) had the highest proportion with the lowest proportion found in *cob* (0 %), *rbcL* + *matK* + *atp1* (0 %), and *rbcL* + *matK* + *cob* (0 %) (Table 2).

We calculated the optimised genetic distance (threshold distance) that is appropriate for species delimitation. The thresholds range from 0.0014 for *rpoB* to 0.29 for ITS. Using these cut-offs, for the best close match method, *phyB* exhibited the highest species identification rate of 71 % for single regions, which improved to

Table 1 Summary statistics of all DNA markers used to delimit seagrasses of the world

Gene region	No. of seq (# spp)	Seq length	K	Range (inter)	Mean inter (±SD)	Range (intra)	Mean intra (±SD)	Threshold (%)
atp1	25 (19)	1,062	0.011	0–0.083	0.022 ± 0.029	0–0.00096	0.00019 ± 0.0004	1.88
cob	25 (18)	1,031	0.17	0–0.044	0.011 ± 0.009	0–0.00097	0.00018 ± 0.00039	2.56
ITS	52 (20)	720	0.053	0–0.12	0.77 ± 0.45	0–0.08	0.024 ± 0.026	29.25
matK	186 (42)	550	0.057	0–0.48	0.12 ± 0.073	0–0.33	0.0055 ± 0.01	4.11
NAD5	22 (16)	1,121	0.0076	0–0.055	0.015 ± 0.013	0–0.017	0.0044 ± 0.0062	1.025
phyB	69 (10)	1,050	0.042	0–0.22	0.12 ± 0.078	0–0.039	0.018 ± 0.016	8.17
rbcL	152 (43)	771	0.28	0–0.21	0.048 ± 0.027	0–0.10	0.0036 ± 0.01	2.077
rpoB	99 (10)	517	0.012	0–0.15	0.026 ± 0.044	0–0.004	0.0024 ± 0.0019	0.14
trnH-psbA	88 (25)	485	0.045	0–0.89	0.44 ± 0.24	0–0.28	0.046 ± 0.041	17.44
rbcL + *matK*	113 (46)	1,321	0.15	0–0.44	0.093 ± 0.06	0–0.19	0.0045 ± 0.013	2.88
rbcL + *matK* + *atp1*	24 (18)	2,383	0.033	0–0.13	0.05 ± 0.034	0–0.00043	0.0001 ± 0.00018	5.44
rbcL + *matK* + *cob*	25 (18)	2,352	0.011	0.00044–0.16	0.051 ± 0.042	0–0.00043	0.00008 ± 0.0002	4.78
rbcL + *matK* + ITS	52 (20)	2,041	0.035	0–0.68	0.26 ± 0.13	0–0.11	0.012 ± 0.012	9.97
rbcL + *matK* + NAD5	22 (16)	2,442	0.0098	0.0004–0.087	0.038 ± 0.023	0–0.0083	0.0017 ± 0.003	2.24
rbcL + *matK* + *phyB*	44 (10)	2,371	0.023	0.0008–0.16	0.074 ± 0.055	0–0.023	0.0096 ± 0.0083	5.63
rbcL + *matK* + *rpoB*	39 (10)	1,838	0.020	0.0006–0.15	0.043 ± 0.053	0–0.017	0.0033 ± 0.0026	2.13
rbcL + *matK* + *trnH-psbA*	59 (22)	1,809	0.038	0–0.36	0.16 ± 0.09	0–0.034	0.012 ± 0.0079	6.58
Combined[a]	27 (18)	7,968	0.0094	0.0003–0.33	0.075 ± 0.068	0–0.0092	0.0029 ± 0.0034	17.21

[a]Combined, *atp1* + *cob* + ITS + *matK* + NAD5 + *phyB* + *rbcL* + *rpoB* + *trnH-psbA*

Table 2 Percentage barcode gap using best close match method (Meier et al. 2006)

Gene region	Number of sequences without gap	Proportion of sequences with gap (%)
atp1	15	40
cob	25	0
ITS	23	56
matK	61	67
NAD5	6	73
phyB	18	74
rbcL	89	41
rpoB	22	78
trnH-psbA	11	88
rbcL + *matK*	84	26
rbcL + *matK* + *atp1*	24	0
rbcL + *matK* + *cob*	25	0
rbcL + *matK* + ITS	22	58
rbcL + *matK* + NAD5	20	9
rbcL + *matK* + *phyB*	24	45
rbcL + *matK* + *rpoB*	18	54
rbcL + *matK* + *trnH-psbA*	27	54
Combined	23	15

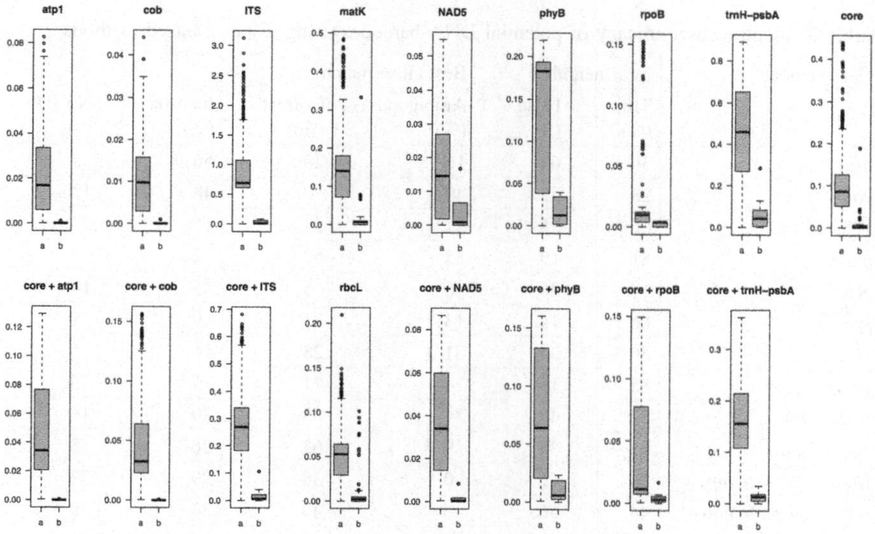

Fig. 2 Comparison of the distribution range of inter- and intra-specific distances using boxplot. The bottom and top of boxes show the first and third quartiles respectively, the median is indicated by the horizontal line, the range of the data by the vertical dashed line and outliers (points outside 1.5 times the interquartile range) by circles. a = interspecific, b = intraspecific

86 % when combined with the core barcodes (i.e. for *phyB* + *rbcL* + *matK*). This was followed by *matK* (52 % for single regions) and 77 % for *rbcL* + *matK* + *rpoB*. The core barcodes alone yielded an identification success of 62 % (Table 3). Similarly, for the near neighbour method, *phyB* followed by *rpoB* yielded the highest identification rates for the single regions (86 % and 85 %, respectively), which improved markedly when combined with the core barcodes (*phyB* + *matK* + *rbcL* = 91 % and *rpoB* + *matK* + *rbcL* = 82 %).

Lastly, at the family level, we found that the combination of *phyB* and the core barcodes (*matK* + *rbcL*) improve species discrimination in Ruppiaceae from 86 to 88 % (*phyB* alone vs *matK* + *rbcL* + *phyB*, respectively), and 78 to 80 % in Cymodoceaceae for *phyB* alone vs *matK* + *rbcL* + *phyB*, respectively (Table 4).

Several criteria have been defined for the identification of the best DNA barcode candidate (Hebert et al. 2004; Kress and Erickson 2007; Lahaye et al. 2008; CBOL Plant Working Group 2009). Firstly, it must provide maximal discrimination between species, and this ability to discriminate depends on the existence of a barcode gap (Meyer and Paulay 2005). All the nine markers tested exhibit significant barcode gap, indicating that they are all good candidates for DNA barcode of seagrasses. To identify the best candidate, we tested their discriminatory power using two distance methods, the near neighbour and best close match methods. In both methods, *phyB* followed by *matK* yielded the best identification rates, thus

Table 3 Identification efficacy of potential DNA barcodes using distance based methods

Gene region	Near neighbor		Best close match			
	True (%)	False (%)	Ambiguous (%)	Correct (%)	Incorrect (%)	No ID (%)
atp1	36	64	16	20	56	8
cob	28	72	40	0	48	12
ITS	65	35	27	35	15	23
matK	81	19	33	52	12	3
NAD5	14	86	36	5	45	14
phyB	86	14	13	71	9	7
rbcL	66	34	51	28	16	5
rpoB	85	15	74	21	2	3
trnH-psbA	52	48	34	33	20	13
rbcL + *matK*	65	35	9	62	26	3
rbcL + *matK* + *atp1*	38	62	0	38	29	33
rbcL + *matK* + *cob*	44	56	0	44	24	32
rbcL + *matK* + ITS	62	38	0	60	19	21
rbcL + *matK* + NAD5	50	50	0	50	50	0
rbcL + *matK* + *phyB*	91	9	0	86	5	9
rbcL + *matK* + *rpoB*	82	18	0	77	10	13
rbcL + *matK* + *trnH-psbA*	66	34	0	63	25	12
Combined	56	44	0	56	18	26

Table 4 Comparisons of the core barcodes (*matK* + *rbcL*) and best barcode within seagrass families, Ruppiaceae and Cymodoceaceae

Family	DNA regions	No of seq	Mean inter (±SD)	Threshold (%)	Best close match Ambiguous (%)	Correct (%)	Incorrect (%)	No ID (%)
Ruppiaceae	core	35	0.0094 ± 0.0037	0.37	0	86	14	0
	core + *phyB*	34	0.025 ± 0.018	1.18	0	88	3	9
Cymodoceaceae	core	27	0.03 ± 0.017	0.84	0	78	22	0
	core + *phyB*	10	0.032 ± 0.031	5.19	0	80	10	10

making them the priority markers for further analyses. We then assessed their performance in combination with the core barcodes. The combination core + *phyB* emerged as the best candidate in both near neighbor and best close match. The core barcodes alone perform poorly and this has already been reported in many cases for different plant taxonomic groups (Hollingsworth et al. 2009; Pettengill and Neel 2010; Roy et al. 2010; Wang et al. 2010; Clement and Donoghue 2012).

Phytochrome B (phyB) is a low copy nuclear DNA marker active in light-grown plants, and plays a key role in regulating circadian rhythm in plants (Somers et al. 1998). Previous phylogenetic studies have shown its utility in resolving relationships in angiosperms (Mathews et al. 2000; Simmons et al. 2001), and in detecting polyploids and hybrids in some seagrass lineages (Ito et al. 2010, 2013). Given that hybridization is very common in aquatic monocots, including seagrasses (Les and Philbrick 1993), our study lends support to the utility of *phyB* as a barcode candidate for identifying a complex taxonomic group like seagrasses. In addition to *phyB, matK* emerged second best in species identification. Although *matK* region has been initially proposed as best plant barcode (Lahaye et al. 2008), some studies have identified potential pitfalls against its suitability (e.g. lack of universal primers; Chase et al. 2007). However more recent studies revealed that such drawback was unjustified for seagrasses. Overall, the nuclear *phyB* and the plastid *matK* are single best candidates that can be used to assess or screen the diversity of seagrasses, but each of both has its own strength and limitations.

Molecular and morphological data do not always concord with regard to species delimitation and this has also been reported for seagrasses (e.g. see Les et al. 2002; Kato et al. 2003; Tanaka et al. 2003 versus den Hartog and Kuo 2006 for the *Zostera capricorni* complex in Australia/New Zealand). Potential reasons for this include mechanisms such as different ecotypes for a single species, ongoing speciation and incomplete lineage sorting or hybridisation through introgression (Coyer et al. 2008). These mechanisms and introgression in particular, obscure taxonomic delimitation caused by deep intraspecific splits in gene trees, resulting in species appearing as paraphyletic or polyphyletic (Pentinsaari et al. 2014). However, introgression concerns less frequently nuclear markers compared to chloroplast (Rieseberg et al. 1991) and mitochondrial markers (Pentinsaari et al. 2014), giving an advantage for nuclear gene (here *phyB*). Also, introgression is more likely to occur between closely related species (Rieseberg et al. 1991; Coyer et al. 2008), suggesting that the differentiation between higher taxa (e.g. genera or families) than species is likely to be more efficient (see Lucas et al. 2012). Our evaluation of the performance of the core barcode at family level for Ruppiaceae and Cymodoceaceae confirms this with a discriminatory power of 86 % and 78 %, respectively. The core barcode performs poorly on all dataset but performs better when its use is limited to diversity within a family. The difference in the barcoding performance between the two could reflect differences in evolutionary history, incomplete lineage sorting, different ecological types, or hybridisation (Coyer et al.

2008, 2013). Ruppiaceae is a monogeneric family widely distributed in brackish waters along tropical and temperate coastlines of the world (Verhoeven 1979), characterised by species with highly similar morphology, and high level of introgression due to polyploidisation and hybridisation (Ito et al. 2010). Similarly, the family Cymodoceaceae is another seagrass lineage with reportedly high level of hybridisation (Ito and Tanaka 2011).

4 Conclusion

Seagrasses are submerged angiosperms that provide important ecosystem services such as nutrient recycling, high primary productivity, and sources of medicinal molecules. However, we are losing them at an alarming rate, in term of diversity and geographical ranges (Waycott et al. 2009; Daru and le Roux 2016), prompting the need for accelerating the screening of seagrass diversity as part of the global campaign for documenting biodiversity. In this need, molecular techniques could complement traditional taxonomic approach, and efforts to identify appropriate marker as DNA barcodes for seagrasses has attracted much attention (Les et al. 2002; Waycott et al. 2002; Uchimura et al. 2008; Lucas et al. 2012). The search for single marker for the entire seagrasses is more convenient as it is cheaper and less time-consuming than the search for multiple markers for each seagrass lineage. Pending future studies with additional sampling and DNA markers, we proposed that the nuclear *phyB* and, secondarily the plastid *matK* as suitable single DNA barcode for genetic identification of seagrass species.

Appendix

See (Table A.1).

Table A.1 Voucher information and GenBank/EBI accession numbers for seagrasses used in this study. APG, Angiosperm phylogeny group

APG III Family	Taxon	GenBank accession numbers								
		atp1	*cob*	*ITS*	*matK*	*NAD5*	*phyB*	*rbcL*	*rpoB*	*trnH-psbA*
Cymodoceaceae	*Amphibolis antarctica*	KF488552	KF488541	–	KF488499	KF488529	–	KF488485	–	–
		–	–	–	–	–	–	U80686	–	–
	Amphibolis griffithii	HQ317985	HQ317978	–	KF488500	HQ267476	–	HQ901574	–	–
	Cymodocea nodosa	DQ859094	DQ859130	–	KF488501	HQ267481	–	KF488486	–	–
	Cymodocea rotundata	–	KF488542	AF102272	KF488502	KF488530	–	KF488487	–	–
		JQ031762	JQ031761	–	JN225358	KF488531	–	U80688	–	–
		KF488553	KF488543	–	JQ031760	KF488532	–	JN225334	–	JN225311
		KF488554	KF488544	–	KF488503	KF488533	–	JQ031763	–	FJ648790
		KF488555	KF488545	–	KF488504	–	–	KF488488	–	–
		–	–	–	KF488505	–	–	KF488489	–	–
	Cymodocea serrulata	AY277801	DQ859131	–	JN225359	KF488534	–	JN225335	–	JN225310
		DQ859095	KF488546	–	KF488506	KF488535	–	KF488491	–	–
		KF488556	KF488547	–	KF488507	–	–	KF488492	–	–
	Halodule pinifolia	KF488557	KF488548	–	JN225368	KF488536	AB571211	AB571198	–	AB571183
		KF488558	KF488549	–	JN225369	KF488537	AB571212	AB571199	–	AB571184
		–	–	–	KF488508	–	AB571213	AB571200	–	AB571185
		–	–	–	KF488509	–	AB571214	AB571201	–	AB571186
	Halodule uninervis	KF488559	KF488550	–	JN225370	KF488538	AB571216	AB571206	–	AB571191
		–	–	–	JN225371	–	AB571219	AB571207	–	AB571192
		–	–	–	KF488510	–	AB571220	AB571208	–	AB571193
	Halodule wrightii	–	–	–	JN225379	–	AB571224	AB571196	–	JN225331
		–	–	–	–	–	AB571225	AB571197	–	–
	Syringodium filiforme	DQ859116	DQ859154	–	KF488511	KF488539	–	KF488496	–	–
		–	–	–	–	–	–	U03727	–	–

(continued)

Table A.1 (continued)

APG III Family	Taxon	GenBank accession numbers								
		atp1	*cob*	*ITS*	*matK*	*NAD5*	*phyB*	*rbcL*	*rpoB*	*trnH-psbA*
	Syringodium isoetifolium	HQ31798		–		HQ26750		AB507901		FJ648793
		–	–	–	JN225372	–	–	JN225342	–	JN225314
		–	–	–	KF488512	–	–	KF488497	–	–
	Thalassodendron ciliatum	KF48856	–	–	KF488513	KF48854	–	KF488498	–	–
	Thalassodendron pachyrhizum	–	–	–	–	–	–	U80692	–	–
Hydrocharitaceae	*Enhalus acoroides*	JF975469	JF975452	AY870347	AB002569	–	–	AB004889	JF975516	–
	Halophila ovalis	JF975473	JF975453	AB243970	AB002570	–	–	AB004890	JF975517	–
	Thalassia hemprichii	JF975480	JF975463	–	AB002577	–	–	AB004897	JF975528	–
Posidoniaceae	*Posidonia australis*	DQ859111	DQ859148	GQ927720	KF488514	–	–	HQ901573	–	–
	Posidonia kirkmanii	–	–	GQ927724	GQ927728	–	–	–	–	–
	Posidonia oceanica	DQ859112	DQ859149	GQ927725	GQ927729	–	–	JQ995767	–	JX028527
		–	–	JQ937106	JQ990941	–	–	U80719	–	JX028527
		–	–	–	KF488515	–	–	–	–	–
Ruppiaceae	*Ruppia cirrhosa*	DQ859114	DQ859151	AB728740	AB728682	HQ267503	AB728706	AB728688	AB728694	AB728724
		–	–	AB728748	KC505607	–	AB728707	DQ859175	KC505613	AB728730
		–	–	AB728749	KF488516	–	–	JN113275	–	JN113267
	Ruppia maritima	–	–	AB728734	AB507905	HQ317975	AB508028	AB507865	AB507945	AB728718
		–	–	AB728735	AB507906	–	AB508031	AB507866	AB507946	AB728720
		–	–	AB728736	AB507907	–	AB508034	AB507867	AB507947	AB728721
		–	–	AB728737	AB507908	–	AB508035	AB507868	AB534790	AB728722
		–	–	AB728738	AB507909	–	AB508037	AB507869	AB534791	AB728723
	Ruppia megacarpa	–	–	JQ034337	AB507929	–	AB508065	AB507889	AB507969	KC505612
		–	–	–	AB507930	–	AB508066	AB507890	AB507970	–
		–	–	–	AB507931	–	AB508067	AB507891	AB507971	–

(continued)

Table A.1 (continued)

APG III Family	Taxon	GenBank accession numbers								
		atp1	*cob*	*ITS*	*matK*	*NAD5*	*phyB*	*rbcL*	*rpoB*	*trnH-psbA*
	Ruppia occidentalis	–	–	–		–		AB507894		–
		–	–	–	AB507935	–	AB508071	AB507895	AB507975	–
	Ruppia polycarpa	–	–	–	AB507938	–	AB508074	AB507898	AB507978	–
	Ruppia tuberosa	–	–	–	AB507939	–	AB508075	AB507899	AB507979	–
		–	–	–	AB507940	–	AB508076	AB507900	AB507980	–
Zosteraceae	*Phyllospadix iwatensis*	–	–	JQ766110	AB096172	–	–	–	–	JX028522
		–	–	–	JQ990933	–	–	–	–	JX028523
	Phyllospadix japonicus	–	–	–	JQ990932	–	–	JQ995760	–	–
	Phyllospadix scouleri	DQ859110	DQ859147	–	–	HQ267497	–	DQ859172	–	–
	Phyllospadix torreyi	–	–	AY077985	EF198333	–	–	JQ995764	–	JX028524
		–	–	EF198346	JQ990934	–	–	U80731	–	–
	Zostera asiatica	–	–	EF198347	AB096161	–	–	AB125352	–	JX028519
		–	–	–	EF198338	–	–	JQ995761	–	–
		–	–	–	JQ990931	–	–	–	–	–
	Zostera caespitosa	–	–	–	AB096162	–	–	AB125351	–	–
		–	–	–	JQ990937	–	–	–	–	–
	Zostera capensis	–	–	–	AB096165	–	–	AM235166	–	JX028515
		–	–	–	JQ990930	–	–	–	–	–
	Zostera capricorni	–	–	AY077995	AB096167	DQ406964	–	AY077963	–	JX028513
	Zostera caulescens	–	–	–	AB096163	–	–	AB125350	–	–
		–	–	–	JQ990936	–	–	–	–	–
	Zostera japonica	–	–	EF198356	AB096166	–	–	AB125353	–	JX028514
		–	–	EF198357	EF198335	–	–	AY077964	–	JX028516
		–	–	EF198358	EF198336	–	–	JQ995758	–	JX028517

(continued)

Table A.1 (continued)

APG III Family	Taxon	GenBank accession numbers								
		atp1	*cob*	*ITS*	*matK*	*NAD5*	*phyB*	*rbcL*	*rpoB*	*trnH-psbA*
		–	–	–	EF198337	–	–	–	–	–
		–	–	–	JQ990922	–	–	–	–	–
		–	–	–	JQ990923	–	–	–	–	–
	Zostera marina	DQ859121	DQ859160	AF102274	AB096164	HQ267511	–	AB125348	–	DQ786516
		–	–	AY077986	EF198339	HQ317970	–	AB125349	–	JN225326
		–	–	EF198349	EF198341	–	–	JN225352	–	JN225327
		–	–	EF198350	EF198342	–	–	JN225353	–	JN225328
	Zostera minima	–	–	–	AJ581456	–	–	–	–	–
	Zostera mucronata	–	–	–	AB096168	–	–	U80732	–	–
		–	–	AY077993	JQ990938	–	–	–	–	–
	Zostera muelleri	–	–	AY077997	AB096169	–	–	JQ995757	–	GU906231
		–	–	AY077998	JQ990921	–	–	–	–	GU906232
	Zostera nigricaulis	–	–	–	JQ990919	–	–	JQ995756	–	JX028510
	Zostera noltii	–	–	JQ677022	JN894021	–	–	JN890769	–	JX028518
		–	–	JQ677023	JN894022	–	–	JN890770	–	JN225329
		–	–	JQ677024	JQ990924	–	–	JN225350	–	JN225330
		–	–	AF102275	AB096170	–	–	JN225351	–	–
		–	–	AY077992	EF198334	–	–	U80733	–	–
	Zostera novazelandica	–	–	–	AB096173	–	–	–	–	–
	Zostera pacifica	–	–	EF198348	EF198340	–	–	–	–	JX028520
		–	–	–	JQ990929	–	–	–	–	–
	Zostera polychlamys	–	–	–	JQ990920	–	–	JQ995759	–	JX028511
	Zostera tasmanica	HQ317987	HQ317980	AY077987	AB096171	HQ267487	–	U80730	–	–

References

Barnosky AD, Matzke N, Tomiya S, Wogan GOU, Swartz B, Quental TB et al (2011) Has the Earth's sixth mass extinction already arrived? Nature 471:51–57
Briggs JC (2003) Marine centres of origin as evolutionary engines. J Biogeogr 30:1–18
Briggs JC (1994) Species-diversity—land and sea compared. Syst Biol 43:130–135
Brown SDJ, Collins RA, Boyer S, Lefort M-C, Malumbres-Olarte J, Vink CJ et al (2012) Spider: An R package for the analysis of species identity and evolution, with particular reference to DNA barcoding. Mol Ecol Resour 12:562–565
CBOL Plant Working Group (2009) A DNA barcode for land plants. Proc Natl Acad Sci USA 106:12794–12797
Chase MW, Cowan RS, Hollingsworth PM, van den Berg C, Madriñán S, Petersen G et al (2007) A proposal for a standardised protocol to barcode all land plants. Taxon 56:295–299
Chelsky Budarf A, Burfeind DD, Loh WKW, Tibbetts IR (2011) Identification of seagrasses in the gut of a marine herbivorous fish using DNA barcoding and visual inspection techniques. J Fish Biol 79:112–121
Clement WL, Donoghue MJ (2012) Barcoding success as a function of phylogenetic relatedness in *Viburnum*, a clade of woody angiosperms. BMC Evol Biol 12:73
Cocheret de la Morinière E, Pollux BJA, Nagelkerken I, Hemminga MA, Huiskes AHL, van der Velde G (2003) Ontogenetic dietary changes of coral reef fishes in the mangrove-seagrass-reef continuum: stable isotopes and gut-content analysis. Mar Ecol Prog Ser 246:279–289
Coyer JA, Hoarau G, Kuo J, Tronholm A, Veldsink J, Olsen JL (2013) Phylogeny and temporal divergence of the seagrass family Zosteraceae using one nuclear and three chloroplast loci. Syst Biodivers 11:271–291
Coyer JA, Miller KA, Engle JM, Veldsink JH, Stam WT, Olsen JL (2008) Eelgrass meadows in the California Channel Islands and adjacent coast reveal a mosaic of two species, evidence for introgression and variable clonality. Ann Bot 101:73–87
Daru BH, le Roux PC (2016) Marine protected areas are insufficient to conserve global marine plant diversity. Glob Ecol Biogeogr 25:324–334
den Hartog C, Kuo J (2006) Taxonomy and biogeography of seagrasses. In: Larkum AWD, Orth RJ, Duarte CM (eds) Seagrasses: biology, ecology and conservation. Springer, Dordrecht, pp 1–23
den Hartog C (1970) Seagrasses of the world. North Holland, Amsterdam
Duarte CM (2002) The future of seagrass meadows. Environ Conserv 29:192–206
Gere J, Yessoufou K, Daru BH, Mankga LT, Maurin O, van der Bank M (2013) Incorporating *trnH-psbA* to the core DNA barcodes improves significantly species discrimination within southern African Combretaceae. ZooKeys 365:127–147
Gouy M, Guindon S, Gascuel O (2010) SeaView Version 4: A multiplatform graphical user interface for sequence alignment and phylogenetic tree building. Mol Biol Evol 27:221–224
Gray JS (1997) Marine biodiversity: patterns, threats and conservation needs. Biodivers Conserv 6:153–175
Green EP, Short FT (eds) (2003) World atlas of seagrasses. University of California Press, Los Angeles
Hebert PD, Cywinska A, Ball SL, de Waard JR (2003) Biological identifications through DNA barcodes. Proc Biol Sci 270:313–321
Hebert PD, Stoeckle MY, Zemlak TS, Franci CM (2004) Identification of birds through DNA barcodes. PLoS Biol 2:e312
Hemminga MA, Duarte CM (2000) Seagrass ecology: an introduction. Cambridge University Press, Cambridge
Hollingsworth ML, Clark A, Forrest LL, Richardson J, Pennington RT, Long DG et al (2009) Selecting barcoding loci for plants: evaluation of seven candidate loci with species-level sampling in three divergent groups of land plants. Mol Ecol Resour 9:439–457

Ito Y, Ohi-Toma T, Murata J, Tanaka N (2013) Comprehensive phylogenetic analyses of the *Ruppiamaritima* complex focusing on taxa from the Mediterranean. J Plant Res 126:753–762
Ito Y, Ohi-Toma T, Murata J, Tanaka N (2010) Hybridization and polyploidy of an aquatic plant, *Ruppia* (Ruppiaceae), inferred from plastid and nuclear DNA phylogenies. Am J Bot 97:1156–1167
Ito Y, Tanaka N (2011) Hybridisation in a tropical seagrass genus, *Halodule* (Cymodoceaceae), inferred from plastid and nuclear DNA phylogenies. Telopea 13:219–231
Kato Y, Aioi K, Omori Y, Takahata N, Satta Y (2003) Phylogenetic analyses of *Zostera* species based on *rbcL* and *matK* nucleotide sequences: implications for the origin and diversification of seagrasses in Japanese waters. Genes Genet Syst 78:329–342
Kress WJ, Erickson DL (2007) A two-locus global DNA barcode for land plants: The coding *rbcL* gene complements the non-coding *trnH-psbA* spacer region. PLoS ONE 2:e508
Lahaye R, van der Bank M, Bogarin D, Warner J, Pupulin F, Gigot G et al (2008) DNA barcoding the floras of biodiversity hotspots. Proc Natl Acad Sci USA 105:2923–2928
Les D, Philbrick CT (1993) Studies of hybridization and chromosome number variation in aquatic angiosperms: evolutionary implications. Aquat Bot 44:181–228
Les DH, Cleland MA, Waycott M (1997) Phylogenetic studies in Alismatidae. II: evolution of marine angiosperms ('seagrasses') and hydrophily. Syst Bot 22:443–463
Les DH, Moody ML, Jacobs SWL, Bayer RJ (2002) Systematics of seagrasses (Zosteraceae) in Australia and New Zealand. Syst Bot 27:468–484
Lucas C, Thangaradjou T, Papenbrock J (2012) Development of a DNA barcoding system for seagrasses: successful but not simple. PLoS ONE 7:e29987
Maddison WP, Maddison DR (2008) Mesquite: a modular system for evolutionary analysis. Version 2.5. http://mesquiteproject.org
Mathews S, Tsai RC, Kellogg EA (2000) Phylogenetic structure in the grass family (Poaceae): evidence from the nuclear gene phytochrome B. Am J Bot 87:96–107
McGlathery KJ, Sundbäck K, Anderson IC (2007) Eutrophication in shallow coastal bays and lagoons: the role of plants in the coastal filter. Mar Ecol Prog Ser 348:1–18
Meier R, Shiyang K, Vaidya G, Ng PKL (2006) DNA barcoding and taxonomy in Diptera: a tale of high intraspecific variability and low identification success. Syst Biol 55:715–728
Meyer CP, Paulay G (2005) DNA barcoding: error rates based on comprehensive sampling. PLoS Biol 3:e422
Millennium Ecosystem Assessment (2005) Ecosystems and human well-being: synthesis. Island Press, Washington, DC
Mora C, Tittensor DP, Adl S, Simpson AGB, Worm B (2011) How many species are there on Earth and in the ocean? PLoS Biol 9:e1001127
Orth RJ, Curruthers TJB, Dennison WC, Duarte CM, Fourqurean JW, Heck KL et al (2006) A global crisis for seagrass ecosystems. Bioscience 56:987–996
Packer L, Gibbs J, Sheffield C, Hanner R (2009) DNA barcoding and the mediocrity of morphology. Mol Ecol Resour 9:42–50
Papadopoulou A, Chesters D, Coronado I, De la Cadena G, Cardoso A, Reyes JC et al (2015) Automated DNA-based plant identification for large-scale biodiversity assessment. Mol Ecol Resour 15:136–152
Pentinsaari M, Mutanen M, Kaila L (2014) Cryptic diversity and signs of mitochondrial introgression in the *Agrilusviridis* species complex (Coleoptera: Buprestidae). Eur J Entomol 111:475–486
Pettengill JB, Neel MC (2010) An evaluation of candidate plant DNA barcodes and assignment methods in diagnosing 29 species in the genus *Agalinis* (Orobanchaceae). Am J Bot 97:1381–1406
R Core Team (2013) R: a language and environment for statistical computing. R Foundation for Statistical Computing, Vienna, Austria. http://www.R-project.org/
Radulovici AE, Archambault P, Dufresne F (2010) DNA barcodes for marine biodiversity: moving fast forward? Diversity. 2:450–472

Rieseberg LH, Choi HC, Ham D (1991) Differential cytoplasmic versus nuclear introgression in *Helianthus*. J Hered 82:489–493

Roy S, Tyagi A, Shulka V, Kumar A, Singh UM, Chaudhary LB et al (2010) Universal plant DNA barcode loci may not work in complex groups: a case study with Indian *Berberis* species. PLoS ONE 5:e13674

Simmons MP, Clevinger CC, Savolainen V, Archer PH, Mathews S, Doyle JJ (2001) Phylogeny of the Celastraceae inferred from phytochrome B gene sequences and morphology. Am J Bot 88:313–325

Smith MA, Fisher BL, Hebert PDN (2005) DNA barcoding for effective biodiversity assessment of a hyperdiverse arthropod group: the ants of Madagascar. Phil Trans R Soc B 360:1825–1834

Somers DE, Devlin PF, Kay SA (1998) Phytochromes and cryptochromes in the entrainment of the *Arabidopsis* circadian clock. Science 282:1488–1490

Tanaka N, Kuo J, Omori Y, Nakaoka M, Aioi K (2003) Phylogenetic relationships in the genera*Zostera* and *Heterozostera* (Zosteraceae) based on *matK* sequence data. J Plant Res 116:273–279

Tomlinson PB, Posluszny U (2001) Generic limits in the seagrass family Zosteraceae. Taxon 50:429–437

Trivedi S, Aloufi AA, Ansari AA, Ghosh SK (2016) Role of DNA barcoding in marine biodiversity assessment and conservation: an update. Saudi J Biol Sci. doi:10.1016/j.sjbs.2015.01.001

Uchimura M, Faye EJ, Shimada S, Inoue T, Nakamura Y (2008) A reassessment of *Halophila* species (Hydrocharitaceae) diversity with special reference to Japanese representatives. Bot Mar 51:258–268

van der Bank H, Greenfield R (2015) A pioneer survey and DNA barcoding of some commonly found gastropod molluscs on Robben Island. ZooKeys 481:15–23

Verhoeven JTA (1979) The ecology of *Ruppia* dominated communities in western Europe. I. Distribution of *Ruppia* representatives in relation to their autecology. Aquat Bot 6:197–268

Wang W, Wu Y, Yan Y, Ermakova M, Kerstetter R, Messing J (2010) DNA barcoding of the Lemnaceae, a family of aquatic monocots. BMC Plant Biol 10:205

Waycott M, Freshwater DW, York RA, Calladine A, Kenworthy WJ (2002) Evolutionary trends in the seagrass genus *Halophila* (thouars): Insights from molecular phylogeny. Bull Mar Sci 71:1299–1308

Waycott M, Duarte CM, Carruthers TJB, Orth RJ, Dennison WC, Olyarnik S et al (2009) Accelerating loss of seagrasses across the globe threatens coastal ecosystems. Proc Natl Acad Sci USA 106:12377–12381